SOIL RECLAMATION PROCESSES

BOOKS IN SOILS AND THE ENVIRONMENT

Soil Biochemistry, Volume 1, edited by A. D. McLaren and G. H. Peterson

Soil Biochemistry, Volume 2, edited by A. D. McLaren and J. Skujiņš

Soil Biochemistry, Volume 3, edited by E. A. Paul and A. D. McLaren

Soil Biochemistry, Volume 4, edited by E. A. Paul and A. D. McLaren

Soil Biochemistry, Volume 5, edited by E. A. Paul and J. N. Ladd

Organic Chemicals in the Soil Environment, Volume 1, edited by C. A. I. Goring and J. W. Hamaker

Organic Chemicals in the Soil Environment, Volume 2, edited by C. A. I. Goring and J. W. Hamaker

Humic Substances in the Environment, by M. Schnitzer and S. U. Khan

Microbial Life in the Soil: An Introduction, by T. Hattori

Principles of Soil Chemistry, by Kim H. Tan

Soil Analysis: Instrumental Techniques and Related Procedures, edited by Keith A. Smith

Soil Reclamation Processes: Microbiological Analyses and Applications, edited by Robert L. Tate III and Donald A. Klein

Additional Volumes in Preparation

SOIL RECLAMATION PROCESSES

Microbiological Analyses and Applications

edited by

ROBERT L. TATE III
Cook College
Rutgers
The State University of New Jersey
New Brunswick, New Jersey

DONALD A. KLEIN
Colorado State University
Fort Collins, Colorado

MARCEL DEKKER, INC. New York and Basel

Library of Congress Cataloging in Publication Data

Main entry under title:

Soil reclamation processes: microbiological analyses
and applications

(Books in soils and the environment)

Includes bibliographies and index.

1. Soil microbiology. 2. Reclamation of land.

I. Tate, Robert L. II. Klein, Donald A. III. Series

QR111.S675 1985 631.4'6 84-26023

ISBN 0-8247-7286-5

MARCEL DEKKER, INC.
270 Madison Avenue, New York, New York 10016

Current printing (last digit):
10 9 8 7 6 5 4 3 2 1

PRINTED IN THE UNITED STATES OF AMERICA

Preface

Land reclamation involves the need to manage a broad spectrum of environments. This can range from strip mine rehabilitation to simple procedures needed to counteract the effects of roadside spills of xenobiotics. Although reclaimed ecosystems are diverse in physical, chemical and biotic traits, a common predictor of successful reclamation relates to the status and potential development of the microbial community. Hence, this treatise was prepared with the overall objective of providing an assessment of the current understanding of soil microbiology and biochemistry as part of reclamation processes. Quite obviously, the entire breadth of the subject cannot be evaluated in a single volume. Thus, a variety of specific examples were selected to provide an indication of the ways in which microorganisms can play important roles in design, management, and monitoring of reclamation.

Impetus for completion of the project has been provided by the current concerns relating to the implications of ecosystem disturbances, especially from the view of impact on aquatic and terrestrial ecosystem development and the resultant effect on human activities. Due to the physically unobtrusive nature of soil microorganisms, reclamation systems can be and have been designed and managed without any intentional consideration of soil microorganisms and their potential contributions (either positive or negative) to the functioning of the particular ecosystems. Procedures for use of topsoil and empirically developed plant systems, when combined with fertilization and irrigation, have made it possible to reclaim a wide range of waste and spoil materials. However, in the last 25 years, it has become increasingly recognized that microorganisms have distinct and unique roles in reclamation, not only in terms of influencing plant system development, but also in regards to the more subtle aspects of reclamation, including the potentiation and inhibition of pollutant releases to the environment. Especially with the more unique wastes which must be reclaimed, such as sulfur-containing materials, oil shales, tar sands, phosphate clays, and radionuclide wastes, possible microbial effects, if considered prior to development of a reclamation management plan, may make reclamation more manageable and cost effective, especially from the view of long-term ecosystem development.

Reclamation systems are described herein where microorganisms have been shown to have particularly unique roles. Common threads from these systems complement the broader literature on soil genesis, ecotoxicological monitoring, toxic waste management, and reconstruction of disturbed ecosystems. From a regulatory viewpoint, reclamation management still appears to emphasize use of chemical analyses of overburden and spoil materials as a basis for planning and management. *Soil Reclamation Processes: Microbiological Analyses and Applications* is an attempt to assemble more specialized literature on reclamation, where application of microbiological concepts has provided an improved understanding of the dynamics of the process.

With the developments in reclamation microbiology occurring in the last decade, this synthesis effort will provide a focus upon which future work can be based. With the increasing concern for trace metal and degradation-resistant organic matter occurrence and effects on native ecosystems, the role of microorganisms, in both positive and negative contexts, must be better understood and utilized as an integral part of reclamation management. It is the hope of the editors and individual chapter authors that this volume will in some way contribute to this goal.

The help of Richard Bartha, Lowell A. Douglas, and Douglas E. Eveleigh in reviewing the manuscripts is gratefully acknowledged. The editors also acknowledge support for a portion of their time devoted to the project from the following sources: New Jersey Agricultural Experiment Station, Publication No. C-15187-1-83, supported by state funds, and Department of Energy under contract DE-AS02-76EV04018, awarded to Department of Range Science, Colorado State University.

Robert L. Tate III
Donald A. Klein

Contents

Contributors

Corale L. Brierley, Ph.D.* Chemical Microbiologist, New Mexico Bureau of Mines and Mineral Resources, New Mexico Institute of Mining and Technology, Socorro, New Mexico

Robert M. Danielson, Ph.D. Research Associate, Department of Biology, The University of Calgary, Calgary, Alberta, Canada

Arokiasmy J. Francis, Ph.D. Microbiologist and Scientist, Department of Applied Science, Brookhaven National Laboratory, Upton, New York

Thomas R. Garland Senior Research Specialist, Environmental Sciences, Battelle Pacific Northwest Laboratories, Richland, Washington

Donald A. Klein, Ph.D. Professor of Microbiology, Department of Microbiology and Environmental Health, Colorado State University, Fort Collins, Colorado

Guy R. Lanza, Ph.D. Associate Professor, Environmental Sciences Program, University of Texas at Dallas, Richardson, Texas

Aaron L. Mills, Ph.D. Associate Professor, Department of Environmental Sciences, University of Virginia, Charlottesville, Virginia

Edward F. Redente, Ph.D. Associate Professor, Department of Range Science, Colorado State University, Fort Collins, Colorado

L. Elliot Shubert, Ph.D. Associate Professor, Department of Biology, University of North Dakota, Grand Forks, North Dakota

---

Current affiliation:

*President, Advanced Mineral Technologies, Inc., Socorro, New Mexico

Darwin L. Sorensen, Ph.D.* Graduate Research Assistant, Department of Microbiology, Colorado State University, Fort Collins, Colorado

Thomas L. Starks, Ph.D.† Post-Doctoral Research Fellow, Department of Biology, University of North Dakota, Grand Forks, North Dakota

Robert L. Tate III, Ph.D.‡ Associate Professor, Department of Microbiology and Cell Science, and The Agricultural Experiment Stations, University of Florida, Belle Glade, Florida

Suzanne Visser Research Associate, Department of Biology, The University of Calgary, Calgary, Alberta, Canada

Raymond E. Wildung, Ph.D. Associate Manager, Environmental Sciences, Battelle Pacific Northwest Laboratories, Richland, Washington

---

Current affiliations:

*Research Assistant Professor, Utah Water Research Laboratory, Utah State University, Logan, Utah
†USDA/ARS, Grand Forks Human Nutrition Research Center, Grand Forks, North Dakota
‡Associate Professor, Department of Soils and Crops, Cook College, Rutgers, The State University of New Jersey, New Brunswick, New Jersey

SOIL RECLAMATION PROCESSES

1

Microorganisms, Ecosystem Disturbance and Soil-Formation Processes

ROBERT L. TATE III* / *University of Florida, Belle Glade, Florida*

I. INTRODUCTION

As a result of their diverse metabolic capabilities, the members of the microbial community play important positive and negative roles in the mining industry, from the initial stages of mineral recovery through reclamation of the mined lands. The soil microbial community, whether it is or is not intentionally managed, is indispensable for the successful reclamation of mine wastes and lands. Essential microbial processes to total ecosystem stability include those activities leading to soil structure development, biogeochemical cycling, and toxin production as well as destruction. These native soil processes may be combined with microbial treatment of waste materials prior to disposal for a complete and effective reclamation process. Specific examples of microbial involvement in reclamation are presented in subsequent chapters. To provide a common background for the discussions without duplicating the more applied presentations, introductory comments will be limited to analysis of some of the basic soil microbial processes that are important in evolution of a stable ecosystem in drastically disturbed areas, such as mine lands. Emphasis will involve the limitations imposed on these processes by the abiotic parameters of mined soils.

This study is not intended to be an inclusive review of the literature, rather, the goal is to provide examples of the types of microbial transformations that are important in soil development. The presentation will be limited to a detailed examination of the microbial reactions

---

**Current affiliation*: Cook College, Rutgers, The State University of New Jersey, New Brunswick, New Jersey

affecting soil organic matter levels. Factors controlling these processes will be contrasted with those affecting the establishment of a functional nitrogen cycle. Both soil organic matter contents and nitrogen cycling are negatively affected by mineral extraction procedures (Cundell, 1977). Although other microbial processes contribute to soil reclamation, soil organic matter and nitrogen transformation will be used herein as models that provide a basic background for analysis of other microbial processes. Discussion of other transformations will be deferred to the appropriate chapters that follow. For example, metal transformations are discussed in Chap. 4 (Wildung and Garland, 1984) whereas sulfur transformations, a highly important process in many minelands, are presented by Mills (1984) and Wildung and Garland (1984). The field applications relating to soil organic matter effects on soil structure and nitrogen fertilization are evaluated by Visser (1984).

II. ROLE OF MICROORGANISMS IN MINELANDS

The primary objectives of mineland reclamation and waste processing, from the view of the casual observer, are to minimize environmental degradation, to facilitate reestablishment of a functional plant-soil system, and to develop an aesthetically pleasing landscape. The final product may or may not resemble the environmental conditions existent prior to site development. Closely allied with the aesthetic goal is the need for long-term stability of the newly established ecosystem. Development of a plant-soil community that exhibits long-term stability under the prevailing environmental conditions would be favorably accepted, whereas loss of viability of the predominant plant (macrophyte) of a particular community would obviously be undesirable. Although the determination of the success of a project may rely in toto on the above-ground aspects of the ecosystem, long-term plant communtity structure, as well as the success in initially establishing a macrophyte community, relies upon the development of a functional soil microbial community. Soil microorganisms contribute to soil structural development, synthesis of various plant nutrients through the mediation of various biogeochemical cycles, and amelioration of adverse chemical and physical limitations, such as soil pH limitations or toxic levels of heavy metals, on macrophyte community development. These basic microbial processes are common to all soil systems supporting a viable macrophyte community. But mine wastes may be devoid of or have limited microbially mediated processes due to the abiotic properties of minelands.

III. SOIL ORGANIC MATTER: SOURCE AND ROLE

Soil organic matter is of unquestionable value for the development of many soil properties conducive to plant growth and development. This has been recognized in the revegetation of minelands (Cundell, 1977; Fresquez and Lindemann, 1982; Stroo and Jencks, 1982). Stroo and Jencks (1982) found that recovery of soil microbial activity in mine soils was directly related to organic matter and nitrogen accumulation. Highest activity was detected in soils receiving high inputs of lime and fertilizers. Activities declined when the exogenously supplied materials were depleted, suggesting the need to reach a steady-state based upon a self-sustaining ecosystem. Cundell (1977) reviewed the need for fertilization and mulching in revegetation of strip-mined land.

Stevenson (1982) recently summarized the general properties of soil humus and the resultant effects on soil properties. Several of these properties are of particular importance in study of highly modified mineland ecosystems: Humus may (1) serve as a plant and microbial nutrient source, (2) provide buffer protection to moderate soil pH changes, (3) chelate metals, (4) combine with clay minerals to augment soil structure, and (5) increase the cation-exchange capacity of the soil.

Although some benefits of soil organic matter are derived merely from its presence in soil, it is the turnover or metabolism of this soil fraction that has the greatest effect on plant growth and nutrition. Indirect effects are derived from the soil structural benefits derived from microbial growth, whereas direct stimulation of plant growth results from mineralization of plant nutrients contained in the organic matter. The rate of nutrient production will depend both on the nature of the organic matter and the abiotic conditions of the soil in that the organic components found in soils are mineralized by the microbial community at rates dependent upon the capability of the microbial community to metabolize the specific carbon source and the in situ limitations imposed by the physicochemical conditions. Compounds such as fulvic and humic acids may have lives of decades or even thousands of years (Stout et al., 1981), whereas readily available nutrients are decomposed in hours to days (Alexander, 1977). Among the latter compounds are proteins, carbohydrates, and lipids common to plant and animal debris. Under the abiotic conditions generally prevalent in well-drained soil systems, the products of complete mineralization are carbon dioxide, mineral nitrogen (as ammonium and nitrate), sulfate, and phsophate. These provide nutrients for both the microbial and plant community (Dalal, 1979). In noncultivated ecosystems, such as grasslands, forests, or deserts, the mineralization or recycling of the nutrients of plant debris provides a vital link in the plant nutrient cycle. Generally, plants are incapable of or have a limited capacity to incorporate organic nitrogen, phosphorus, or sulfur sources. Hence, the plants rely upon the soil microbial community to return the organic nutrients to a more usable mineral state. The importance of this organic reserve of plant nutrients has been demonstrated in forest soils (Jenny, 1980; Gosz, 1981). The quantity of total organic nitrogen per unit of litter far exceeds the supply in standing vegetation.

Buffer capacity, chelation, and cation-exchange properties of a soil are of prime importance in determination of toxicant (such as certain metals) availability in mineland soils as well as in leachates from disposal sites. Although clay components contribute to the buffer and cation-exchange capacities, soil organic matter is of extreme importance in redevelopment of these properties in drastically disturbed soils. The buffering, chelation, and cation-exchange characteristics provided by soil organic matter result from the various acidic and phenolic functional groups found in native soil organic matter. Humification of soil organic matter results in an increase in the oxidation state of the compounds. This results in a total acidity of isolated fractions of humus ranging from 300 to 1400 mEq/100 g (Stevenson, 1982). Stable complexes are found between the resultant anions and Cu^{2+}, Mn^{2+}, and Zn^{2+}, plus other divalent cations (Stevenson, 1982).

All fractions of the soil organic component contribute to the development of soil aggregates. Humic substances bind clay particles (Stevenson, 1982), and polysaccharides cement soil particles together (Jones and Griffiths, 1967; Cheshire et al., 1983), whereas the microbial biomass and plant roots form soil particles into large aggregates (Martin and Anderson, 1942; Gilmour et al., 1948; Tisdall and Oades, 1982). Tisdall and Oades (1982) recently presented a model descriptive of the various types of interactions of these organic components with soil particles. Binding of clay minerals by native soil organic matter is a major contributor to soil structural development. Stable humic acids interact with clays to yield water-stable microaggregates within the soil matrix. These aggregates appear to be a characteristic of the soil and, hence, independent of soil management practices. Other levels of stability described by Tisdall and Oades (1982) result from the microbial metabolism of organic matter. Transient associations result from polysaccharide production primarily, whereas temporary associations are produced by plant roots and fungal mycelia.

As was demonstrated by Schafer et al. (1980), soil structural development in mine soils is slow. They found more rapid structural development in the soil surface (10–50 years) compared with the rate below 10 cm (50–200 years) in mine soils of southeastern Montana. Clayey soils developed structure more rapidly than did sandy soils, suggesting a role of the humus-clay aggregates in accelerating soil development. The slower rate of soil development in sandy soil may also result from the tendency for the abrasive sand particles to disrupt macromolecular and organismal particle associations. The importance of the microbially modified plant polysaccharides in structural development (Martin, 1971), as well as the direct binding of particles together by fungal hyphae (Tisdall and Oades, 1982) and aggregates of sand grains by fungi, bacteria, actinomycetes, and algae (Forster and Nicolson, 1981), highlights the need for an active microbial community in the soil ecosystem.

A. Microbial Transformations of Organic Matter

Two general conclusions are derived from the preceding discussion. First, a portion of the soil organic matter must be metabolized actively. This decomposition of organic compounds provides nutrients for the microbial community development and for the macrophyte community. Second, a reservoir of stable, well-humified organic matter is needed for a variety of functions relating to soil structure and cation-exchange capacity. These divergent needs suggest the existence of a variety of reservoirs (based on biodegradation rates and degree of oxiation) of organic matter within the soil matrix. The quantities of substances in each compartment depend upon the nature and amount of exogenously supplied plant and animal remains, as well as on the activity of the microbial community. The exogenously supplied fixed carbon may result from the primary productivity of the ecosystem itself, or be added in the form of sludge or compost. For this discussion, we shall provide the following four groups of soil organic matter (Jenny, 1980):

1. Recently incorporated plant and animal remains and residues. As these materials are decomposed, it becomes difficult to distinguish them from secondary products synthesized by the microbial community instrumental in their decomposition. A general feature of this class of compounds is that they retain, at least in part, the general cell structure of the parent material.
2. Partially decomposed fixed carbon. The physical structure of the parent material has been lost. Due to their chemical similarity with indigenously synthesized carbonaceous substrates, exogenously supplied and microbially synthesized compounds are frequently indistinguishable.
3. Humified carbon and humic substances. These include the poly aromatic compounds characterized by their biodegradation resistance, as well as the polysaccharides, proteins, etc., attached to the polyaromatics. These materials are extracted from soil as humic and fulvic acids.
4. Soil biological community components. This living portion of the soil ecosystem is also a nutrient reservoir.

Due to the limitations of extraction techniques, it is generally not possible to distinguish between these organic matter groups in a native soil sample. Exceptions are when sufficient plant structure remains that the plant litter or roots can be removed during visual inspection of the sample, or when isotopically labeled materials are added. The quantity of available carbon in soil biomass can be estimated by the chloroform fumigation procedure (Jenkinson, 1966), but determination of the exact source of the carbon requires previous labeling with ^{14}C-labeled materials. Thus, all substances listed above, except those removable by a visual screening, are defined as soil organic matter (Jenny, 1980).

The location of various carbon components in these categories is a dynamic rather than static situation. In an active microbial community, biomass is constantly transformed into carbon dioxide, humic material and microbial biomass (Fig. 1).

The proportion of the carbon substances metabolized for carbon and energy by the microbial community versus the quantity humified appears to depend upon the nature of the substrate as well as the environmental conditions. An example of the latter condition is found in flooded or water-logged soils where partially humified organic matter accumulates (Tate, 1980a). The saturated moisture levels preclude the total mineralization of the accumulating plant biomass. Considering the quantities of organic matter in the world swamp, bog, and organic soil reservoirs, this accumulation of biomass can be quite substantial (Tate, 1980a; Armentano, 1980). Rates of organic matter accumulation in these ecosystems have been estimated to be about 6 cm of vertical depth/100 years (Dawson, 1956; Broadbent, 1960) 10 cm/100 years (Bohn, 1978), or to range between 2 and 20 cm/year (Richardson and Allen, 1979).

The fate of the initial substrate and the metabolic intermediates formed during decomposition relate, at least in part, to their biodecomposition rate. As will be shown in the following discussion, biodegradation-resistance substances such as lignin are more likely to be humified than mineralized whereas easily decomposed substances, such as glucose, are primarily mineralized with a small portion ending up as microbial biomass. The relative decomposition rates in descending order of biodegradation resistance for selected carbon compounds are as follows: humic acids > lignin > simple aromatic compounds > amino acids and proteins. The metabolic pathways for decomposition, the microorganisms involved, and the soil factors controlling the decompo-

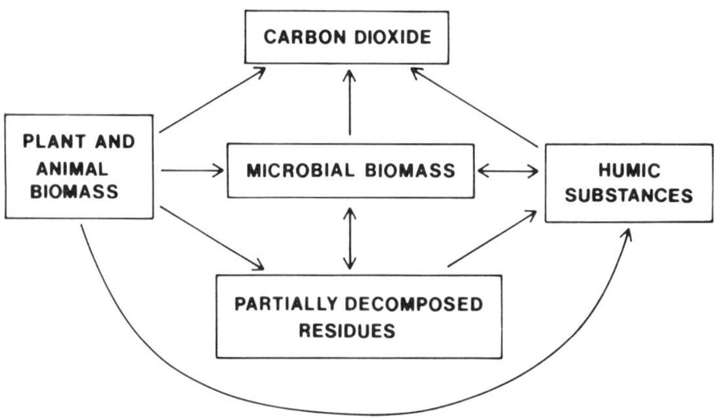

Figure 1 General carbon transformations in soil.

tion of various plant components have been reviewed recently (Alexander, 1977; Crawford, 1981).

In soil environments where biological activity is not precluded by the abiotic conditions, the majority of the carbon contained in readily degradable carbon compounds is oxidized to carbon dioxide by the soil microbial community. For example, Cheshire et al (1969) noted between 60 and 80% of glucose added to soil was converted to carbon dioxide within 14 days after amendment, whereas 84 days of incubation were required for comparable metabolism of starch. Hemicellulose, a more biodegradation-resistant polysaccharide without fine grinding, required 448 days for 70% conversion to carbon dioxide. Fine grinding of the hemicellulose increased the surface area of the substrate particles. Hence, microbial attack of the polymer was accelerated in that 80% of finely ground hemicellulose was converted to carbon dioxide after 448 days of incubation in soil (Cheshire et al., 1974). Soil type can also affect decomposition of polysaccharides (Dormaar, 1975). Complexing of acid-containing polysaccharides with di- and trivalent cations also modified their decomposition rate (Martin et al., 1966). Decomposition of zinc, copper, iron, and aluminum salt complexes of bacterial and plant polysaccharides in Greenfield sandy loam was examined (Martin et al., 1966). The polymers contained between 6 and 83% uronic acid monomers. The exact effect of the various metals on biodecomposition of the polysaccharide varied, but the authors concluded that with the majority of the polysaccharides examined, the decomposition rate and percent decomposition was reduced by some or all of the metals. Mutatker and Wagner (1967) observed 25% of the ^{14}C-labeled glucose amended to soil was incorporated into soil organic matter after 3 months of incubation. The distribution of the ^{14}C-labeled metabolites between microbial biomass and humic substances was not determined. Similar results are found with other easily biodegradable carbon substrates. Sorensen and Paul (1971) noted that 70% of [^{14}C]-acetate-carbon amended to a heavy clay soil was evolved as carbon dioxide after 6 days. Fractionation of the acetate-carbon remaining in the soil demonstrated that 19, 29, and 21% of the carbon was detected in carbohydrates, amino acids, and insoluble residues, respectively. Sorensen (1963) found a variation between the fates of the more readily degradable plant substituents and lignin in soil. Barley straw labeled with ^{14}C and various ^{14}C-labeled components extracted from the barley straw were added to soil. The soil organic matter was extracted and analyzed for ^{14}C after a sufficient incubation time to allow mineralization of the barley residue. The ^{14}C of the straw was distributed rather uniformly between the humic and fulvic acids and humin soil organic matter fractions, whereas the more biodegradation-resistant lignin residues were found predominantly in humic acid and humin fractions. Water-soluble substances of the barley straw (readily decomposable plant components) contributed little to the humin fraction. In each of these examples, at least 60% of the readily decomposable substrates, were mineralized totally to carbon dioxide.

With the development of the chloroform fumigation technique for estimating microbial biomass in soil samples (Jenkinson, 1966), it is now possible to estimate the distribution of carbon metabolic products between living biomass and humified biodegradation-resistant soil organic matter components. Martin et al. (1980) used this procedure, plus soil organic matter extraction techniques to determine the fate of ^{14}C-labeled plant polysaccharides and lignins in soil. The more biodegradation-resistant lignin moieties were metabolized in part to carbon dioxide or incorporated into the biodegradation-resistant portion of soil humic acids, whereas the polysaccharides were metabolized for energy and cell carbon (that is, they were converted primarily to carbon dioxide and cell components), with some polysaccharides stabilized into soil humus. These polysaccharide materials were still acid labile in the humic acid, indicating that the original chemical structure was retained. The loss of free lignin alcohols as carbon dioxide and their distribution in humic acids were intermediate between the values found for the aromatic lignin carbons and polysaccharides. Similar results were found when comparing ferulic, *p*-hydroxycinnamic, anisic, and benzoic acid decomposition with that of glucose in Steinbeck loam and Greenfield sandy loam (Kassim et al., 1982). Relatively higher amounts of the phenolic carbons were incorporated into soil humic materials, whereas the glucose carbon retained in the soil was incorporated into microbial biomass. Greater quantities of the phenolic compounds were not solubilized by 6 N hydrochloric acid than with the glucose. This suggests that the aromatic compounds were incorporated into the polyaromatic structure of the humic acid.

Humification, or attachment of the compounds to humic acids, reduces the bioavailability of readily degradable substrates. This was demonstrated to occur with a wide variety of biologically synthesized carbon compounds (Haider and Martin, 1975; Linhares and Martin, 1978; Martin and Haider, 1976a; Martin et al., 1967; Verma et al., 1975; Verma and Martin, 1976). For example, physical mixing of protein with freeze-dried humic acid reduced the protein decomposition by 35%. A more intimate association resulted from freeze-drying a mixture of protein and humic acid, which reduced the protein decomposition by 60%. Chemical linkage of the protein to model phenolase polymers reduced protein decomposition by 80–90%. These latter polymers apparently mimicked the effect of native humic acids (Verma et al., 1975). Generally, binding of degradable cell components to polyaromatics (be they humic acid, lignin, or melanin) reduces the rate of biodegradation. For example, the naturally occurring linkage of cell components with melanin in certain fungi increases the biodegradation resistance of the fungal cellular material. Martin et al. (1959) found that the decomposition rate of hyaline fungi (nonmelanin containing fungi) > dark fungi (melanin containing) > plant residues > benzene-alcohol extract of dark colored fungi.

Although humic acids are quite stable after formation (Stout et al., 1981), a wide variety of soil microorganisms capable of decomposing humic acid have been isolated. This has been shown for actinomycetes (Fedorov and Il'ina, 1963; Steinbrenner and Mundstock, 1975; Ibrahim and Ibrahim, 1977; Andreyuk and Gordionko, 1978; Sidorenko et al., 1978), fungi (Burges and Latter, 1960; Mathur and Paul, 1967; Mathur, 1969; Biederbeck and Paul, 1971), bacteria (Huntjens, 1972; Taha et al., 1973); Mal'tseva et al., 1975), and associative microbial cultures (Andriiuk et al., 1973). Evidence for humic metabolism generally involves demonstration of growth of the microbe associated, in some cases, with demonstrable changes in the humic acid. Some difficulty in interpretation of the role of the various microbes in humic acid decomposition in native soil samples is caused by the use of crude humic acid preparations for microbial enrichment and growth. In many cases, the microbial isolates may have used the humified substances attached to the aromatic core of humic acid as nutrients rather than the polyaromatic core itself. This polyaromatic core is decomposed in native soil, although the mechanism remains obscure. Evidence for this decomposition is found in studies of the radiocarbon age of humic acids (Stout et al., 1981) and the failure of high levels of humic acids to accumulate in mineral soils. In soils where humic acids are being formed, a steady-state level of soil organic matter is reached. Hence, decomposition and synthesis must be balanced.

B. Abiotic Factors Affecting Carbon Mineralization

Both the chemical nature of the carbon compounds metabolized by the soil microbial community and the species of microorganisms involved are diverse. Thus, carbon mineralization, in general, is less restricted by the environmental conditions than are some other processes in the biogeochemical cycles. (For examples of more environmentally restricted processes, see nitrification and nitrogen fixation discussions below). Thus, although various abiotic conditions may severely limit the overall rate of carbon mineralization in a given soil, carbon mineralization will still occur, albeit at a reduced rate, unless soil conditions are such that life ceases. As is noted in Fig. 2, rates of biologically catalyzed reactions are bounded by maximal and minimal levels of the limiting factor, beyond which the reaction is precluded, and the reaction rate generally increases as the limitation is removed. An optimal level of the limiting parameter at which the reaction proceeds at a maximum rate exists (Dommergues et al., 1978). Examples of abiotic parameters that have been demonstrated to produce rate curves as shown in Fig. 2 are temperature, osmotic pressure, pH, and moisture. Certain divalent cations would also be expected to affect the reaction rate in a similar manner if the metal were both toxic in high concentrations and an essential nutrient. Growth would therefore be limited by bioavailability of the metal at diminished concentrations.

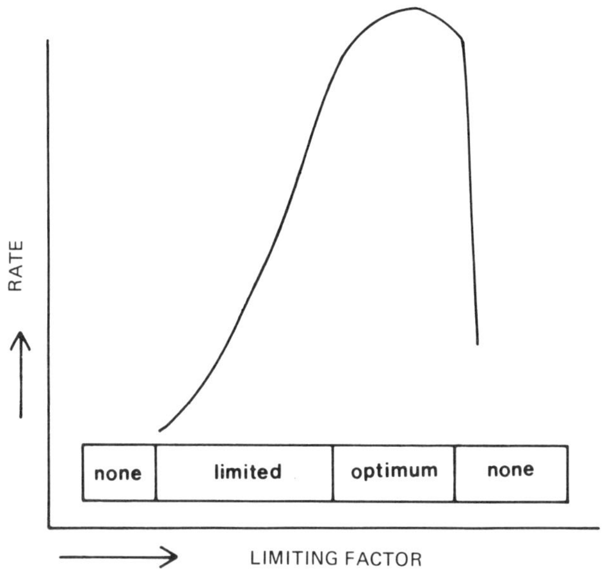

Figure 2 Relationship of rates of biologically catalyzed reactions to limiting factors in soil.

Soil conditions affecting carbon mineralization, of interest to management of mineland reclamation, are pH, clay content, and moisture. Soil temperature is also a significant affector of decomposition (Pal et al., 1975; Bunnell et al., 1977), but since manipulation of this variable is of limited feasibility in the field, it will not be discussed here. Limited variation in soil temperature results from darkening of the soil color (Stevenson, 1982), flooding of the soil, or through the use of ground cover.

Each microbial species involved in carbon mineralization has a characteristic pH range over which optimal growth occurs. Outside of this range, proliferation is decreased and the organism will not be able to grow beyond the tolerance range. Thus, the soil pH selects the members of the microbial community that will be functional. Typically, soil bacteria are more competitive under neutral conditions, whereas fungi predominate in acidic soils (Alexander, 1977). The effect of lime additions in agriculture soils demonstrates that carbon mineralization proceeds more rapidly in near-neutral soils, assuming that soil acidity is the only factor limiting this process. Ivarson (1977) found two- to three-fold stimulation of microbial respiration of Canadian peats limed from pH 3.8 to pHs between 6 and 7. Similar effects of liming would not be unexpected in acidic minelands.

The importance of the amount and type of clay mineral contained within a soil is derived, in part, from (1) the contributory effect of clay-microbe aggregation to soil structure, (2) biological stabilization of organic matter through attachment to clay particles, and (3) the effect of clay minerals on microbial respiration. Recent reviews of the literature pertaining to clay-organic matter interactions are Marshall (1971) (aggregation formation through clay-bacterial interactions) and Stevenson (1982) (organic matter binding to clay materials). The reader is referred to these reviews for a more complete analysis of these subjects.

The role of polyvalent metal ions in linking negatively charged particles, such as soil organic matter, bacterial cells, or clay particles, was discussed by both Marshall (1971) and Stevenson (1982). Although binding of organic matter to clay is augmented by increasing levels of di- and trivalent cations, other types of associations are also operative in the organic matter-clay association. Electrostatic attraction, hydrogen bonding, physical adsorption (Van der Waal's forces), and coordination complexes are all instrumental in the binding of various organic compounds to clay minerals (Stevenson, 1982).

The effects of clay minerals on microbial metabolism and respiratory activity is apparently quite complex. This conclusion is based (at least superficially) on the contradictory data relating specific clay minerals and various microbial activities in a variety of soil ecosystems. Variation in the effect of clay on microbial activity relates to such factors as type of clay, properties of the carbon substrate metabolized, soil pH, etc. Clay complicates the microbial processes by attaching to the microbe itself, adsorbing the metabolic substrates or products, and complexing the extracellular enzymes produced by the microbe. In their study, Galstyan and Avundzhyan (1970) reported that 90% of the soil dehydrogenase was bound to clay particles. Dehydrogenase is an enzymatic activity that provides a measurement of respiratory activity in soil. Zvyagintsev and Velikanov (1968) found between 1.5-and 11-fold inhibition of glucose oxidase and invertase by clay minerals. They concluded that the enzymatic activity was suppressed as a result of the microzone or the interface boundary and a change in the tertiary structure of the proteinaceous enzyme molecules. Both Esterman et al. (1959) and Lynch and Cotnoir (1956) found that bentonite inhibited carbon mineralization. Esterman et al. (1959) examined the proteolytic cleavage of lysozyme, which was attached in monolayers to kaolinite and bentonite. The enzyme was hydrolyzed by chymotrypsin when bound to either clay mineral, but the hydrolysis rate was retarded. In a rather inclusive study, Kunc and Stotzky (1974) studied the effect of montmorillonite, kaolinite, or 5% mica-vermiculite on oxidation of 25 carbon substrates. Montmorillonite and kaolinite concentrations were varied between 0% and 20%. Substrate groups included saccharides, aliphatic and aromatic acids, aldehydes, and amino acids. Except for the aldehyde substrates, montmorillonite stimulated carbon mineralization. Generally, the clay minerals caused all possible types of effect on the rate of carbon oxidation. The initial oxidation rate and/or the extent of oxidation were either increased, not affected or decreased by the amendment of the growth media with various clay minerals. Montmorillonite had the greatest effect on the oxidation rate, whereas kaolinite affected the extent of oxidation to the greatest extent. As a finding related to these observations, montmorillonite inhibits dextran oxidation (Olness and Clapp, 1972). Martin et al. (1976b) found that in culture Calcium-montmorillonite accelerated the rate and increased the amount of growth, glucose consumption, and carbon dioxide evolution by a variety of *Streptomyces*, *Micromonospora*, and *Nocardia* species. The authors attributed the stimulation to adsorption interactions between the substrate and clay mineral. The presence of montmorillonite and kaolinite in cultures of microorganisms using protein as a growth substrate resulted in a diauxic growth pattern (Marshman and Marshall, 1981). With excess protein, clay amendment had no effect on microbial growth, whereas at low protein concentrations, the protein was unavailable to the microorganisms. With intermediate protein levels, the presence of clay resulted in a reduction of growth rate, but not growth yield. The protein-to-clay ratios determining the three growth responses did not correlate with changes in affinity of adsorption of the protein to the clay and differed with microbial species. The authors developed a model explaining these interactions as an association of the availability of the protein to protease as related to the binding strength of the protein substrate to the clay.

Moisture has likely been recognized as a major factor controlling biomass production in the soil ecosystem since the advent of agriculture. Greaves and Carter (1920) found that ammonification, nitrification, and nitrogen fixation had maximal activities at about 60% of the soil water holding capacity. Both high and low soil moisture limits soil microbial respiration. Miller and Johnson (1964) observed increased carbon dioxide evolution from various mineral soils as soil moisture was raised from an air-dried level toward saturation. Maximal carbon dioxide evolution was detected with moisture in the range of -0.5 to -0.15 bar. A decrease in activity occurred between -0.15 bar and saturation. Nitrification peaked in the same moisture range, whereas nitrogen mineralization rates peaked between -0.5 and -0.15 bar, and also at saturation. The dual peaks of nitrogen mineralization demonstrates the versatility of the microorganisms responsible. Both aerobic and anaerobic nitrogen mineralizing microbes commonly occur in soil (Alexander, 1977). Minimal moisture levels allowing for general microbial respiration were investigated by Wilson and Griffin (1975). Bacterial versus mixed natural populations have also been evaluated. With pure cultures of bacteria, activity decreased rapidly below -3 bar, and at -20 bar it was very slight. The greatest decline in the respiratory rate of the bacterial community occurred above −6 bar. In contrast, respiration of a mixed microbial population (bacteria, fungi, protozoa, etc.) was maintained at a reasonably high level between −8 and −30 bar. This rate did decline with decreasing soil moisture, until at −50 bar it became negligible. Amendment of the soil containing the mixed microbial population and equilibrated at −15 bar moisture with glucose caused no increase in respiration. Thus, sufficient water stress existed at this soil moisture level to preclude induction of new metabolic activities. The increased sensitivity of the known bacterial population to reduced soil moisture compared with that of a mixed microbial population was attributed to the restricted microbial mobility as the soil pores drained. Wildung et al. (1975) found that in arid grassland soils, soil microbes responded to changes in soil temperature at water contents as low as −106 to −88 bar. Knight and Skujins (1981), after studying water and microbial activity relationships in arid soils, concluded that the soil microbiota may show diverse physiological responses to water stress in soils from environmentally different areas. These few examples of studies of soil moisture interactions with the microbial community demonstrate the need for soil moisture control during mineland revegetation and the fact that exact moisture needs may vary with geographic region and climate.

Variations in carbon mineralization and microbial respiration in the soil ecosystem are also noted between soils maintained at a constant moisture versus soil experiencing wet-dry cycles. A fluctuating soil moisture level stimulates carbon mineralization. Lund and Goksoyr (1980) documented a rise in bacterial numbers and the length of fungal hyphae in remoistened air-dried soil. Alternate dry-wet cycles also increase organic matter decomposition. Sorensen (1974) reported that air drying and rewetting every 30th day over an incubation period of 260–500 days resulted in a 16–121% increase in carbon dioxide evolution, compared with soils maintained under a continuously moist condition.

C. Accumulation of Organic Matter

The elucidation and control of factors affecting soil organic matter in cultivated soils has long been of concern to the agriculturalist. For examples of studies relating the effects of exogenously supplied organic materials on soil organic matter levels, fertility, and microbial activity, see Conn and Bright (1919), Broadbent and Bartholomew (1948), Pinck et al. (1950), Sinha (1972), Kolenbrander (1974), and Mishra et al. (1974). Soil management practices also affect these variables (Barnette and Hester, 1930; Verstraete and Voets, 1977). Each of these studies reinforces the conclusion that amendment of soil with biodegradable organic materials such as crop debris, sludges, and organic fertilizers, or management of the soil in a manner that allows accumulation of indigenously produced fixed carbon compounds generally results in augmented soil organic matter levels, better soil structure, and, at least in the long run, greater fertility.

Dramatic declines in soil organic matter levels are observed when soil is brought into agricultural production; for instance, organic matter decline from 7.50% to 1.70% over a 60-year period in gleyed alluvial soils following drainage and cultivation (Dent, 1976). Similarly, organic matter decreased from 3.29% to 1.43% over a 25-year period in cultivated Georgia (USA) soils (Giddens, 1957). The latter soils were forested prior to clearing and cultivation. This decline in organic matter has been attributed to the removal of biomass and the fact that plowing and discing accelerates the biological oxidation of humus (Jenny, 1980). The loss of soil humus causes a deterioration in soil structure, fertility, and water-holding capacity, all of which are factors of concern for sustained production in the ecosystem.

Numerous mathematical models have been developed that estimate organic matter transformation in soil (for example, see Hunt, 1977; Smith, 1979a, b). In these models, organic matter is generally divided into groupings such as humus, plant residues, animal remains, etc. Further subdivisions relate to the rate of biodegradation. Plant and animal debris is divided into readily degradable and relatively biodegradation-resistant fractions. These models simulate reasonably well the more rapid, as well as long-term, changes in soil organic matter levels with amendment of the soil with exogenously supplied organic materials.

Of interest to the project manager involved with mineland reclamation, as well as to the general ecologist, is the potential for predicting the time interval needed for development of an equilibrium condition in relation to soil organic matter transformations. Jenny (1980) describes soil organic carbon and nitrogen variation with the equation:

$$\frac{dN}{dt} = A - KN$$

That is, the rate of change (dN/dt) is equivalent to the rate of organic matter influx (A) minus the rate of loss (KN). K is a rate constant specific for the ecosystem. Thus, the level of organic matter in soil is the result of balance between biomass synthesis and mineralization. Development of a steady-state condition can be short (for example, 10 years for an oak and evergreen broad leaf tree forest) or extended, as with a Northern pine site (126 years) (Jenny, 1980). Organic matter addition to cultivated soils may have long-term effects on the level of soil organic matter. Jenkinson and Rayner (1977) record the effect of amendment of cultivated soil with farmyard manure. Manure was added to the soil annually over the first 20 years of the experiment. After 100 years of subsequent cultivation without further manure addition, the plot still contained greater organic carbon and nitrogen than did plots receiving no fertilization.

These data demonstrate the lasting effect of organic amendments on soils. Immediately following amendment, the microbial respiration increases. As the readily degradable carbon sources are exhausted, the respiration rate declines to the low level allowed by the slow min eralization of the more biodegradation-resistant carbon substrates. An extended period of stimulated microbial activity and increased fertility and soil organic matter levels, compared with unamended soil, results from the slow decomposition of the biodegradation- resistant plant components and the slow turnover of humified carbon compounds. The time for these reactions extends over several decades or longer. This explains the slow development of soil structure in reclaimed mine soils (Schafer et al., 1980).

IV. NITROGEN METABOLISM IN MINELANDS

The development and activity of the microbial community of minelands are limited by the more extreme physical and chemical conditions of the soil. For example, in spent shale of Western Colorado and lignite strip mining areas of North Dakota, reclamation is limited by low organic matter content, salinity, fine structure, lack of nitrogen and phosphorus, and a slow rate of soil formation (Cundell, 1977). As was previously discussed, are of these limitation, except one, are ameliorated, at least in part, by amendment of the soil with organic matter. Solutions to the salinity problem relate to the physical and chemical properties of the ecosystem and the capacity of ion transport with water movement. Nutrient problems, nitrogen and phosphorus supplies, are only temporarily solved with organic amendments. Since these nutrients may be removed from the ecosystem with biomass, in ground waters or in volatile forms, more permanent solutions to nutrient deficiencies must include establishment of the appropriate biogeochemical cycles. This would allow introduction of the required nutrient into the ecosystem exclusive of anthropogenic intervention. In many cases, in the absence of a constant nutrient supply, macrophyte development is precluded or at least retarded in the reclaimed areas. This is exemplified by studies of Dancer et al. (1977) on the revegetation of kaolin mine wastes. In their examination of plant succession in these wastes, they found that the principle colonists were woody leguminous plants. In the absence of legumes, the waste soils were slowly colonized by *Calluna vulgaris* and other heathland species, suggesting that the legumes stimulated nitrogen accumulation in the soil and the succession rate.

As has been adequately demonstrated by the monospecific plant ecosystems characteristic of modern agriculture, adequate fixed nitrogen can be supplied from external sources to allow for high plant biomass (crop) yield. Such nutrient amendments are impractical and uneconomical for mineland reclamation. The alternative procedure entails stimulation of the development of a stable nitrogen cycle. Accomplishment of this requires (1) development of a constant carbon supply within the ecosystem and (2) mitigation of many of the extreme chemical and physical conditions typical of many minelands. The degree of moderation of these parameters necessary for sustained nitrogen cycling exceeds that required for carbon mineralization in that several of the nitrogen transformations are of greater sensitivity to factors such as soil pH or aeration than are the microbes responsible for carbon mineralization.

A. Nitrogen Cycle and Carbon Supply

The basic transformations comprising the nitrogen cycle are outlined in Fig. 3. The cycle basically consists of a series of reductive and oxidative reactions allowing stable dinitrogen of the atmosphere to be converted to a metabolizable plant nutrient, then recycled through the soil to the atmosphere. Major processes of the cycle are (1) nitrogen fixation (reduction of the atmospheric nitrogen (dinitrogen) to ammonium, incorporated into the organic compounds comprising the microbe or microbe-plant association responsible for reduction of the dinitrogen); (2) mineralization (conversion of the organic nitrogen to ammonium); (3) nitrification (oxidation of the reduced nitrogen forms to nitrate); (4) denitrification (reduction of nitrate to nitrous oxide

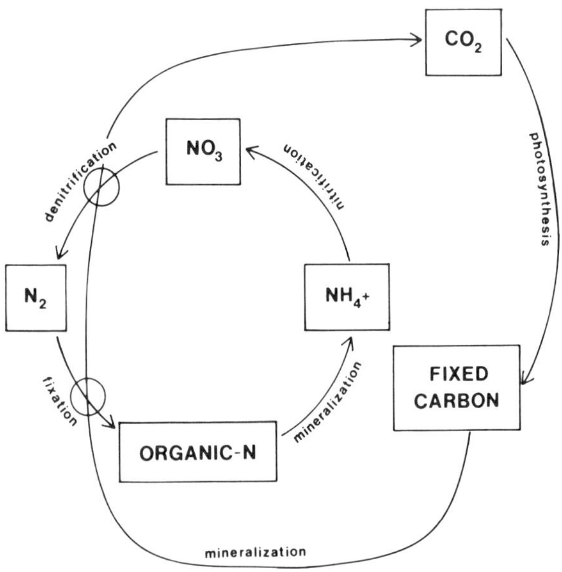

Figure 3 Carbon and nitrogen cycles. Interactions of the two cycles providing energy for nitrogen fixation and dentrification in minelands are emphasized.

and dinitrogen. The proportion of nitrate reduced, which is evolved as nitrous oxide varies with the environmental conditions (Cast, 1976), but dinitrogen is the predominant product of this reaction which returns the fixed nitrogen to the atmospheric pool of dinitrogen). The nitrogen transformtions within the ecosystem and the types of microbes involved are complex, and the reader is referred to the general discussion provided by Alexander (1977), Gibson (1977), or Nutman (1976) for a more detailed description of the processes. This presentation will be limited to examination of the relationship of these nitrogen reactions to the success or failure of mineland reclamation projects.

Neither the carbon cycle discussed above nor the nitrogen cycle operate independently. The microbes functioning in each cycle require the presence and activity of microbes in the other biogeochemical cycles. Two points of interaction of the carbon and nitrogen cycle are shown in Fig. 3. Generally, both the nitrogen-fixing bacteria and the denitrifying bacteria (both processes are catalyzed exclusively by bacteria) require a source of energy supplied by the fixed carbon existent in the ecosystem. Autotrophic nitrogen-fixing bacteria have been shown to be significant in some situations (Child, 1981), but their importance in the harsh soil system of minelands would be questioned. The energy for denitrifying bacteria is supplied by the mineralization of fixed carbon reserves contained within the soil organic matter (Knowles, 1981), whereas the nitrogen-fixing bacteria (diazotrophs) may mineralize soil carbon, or in some species form a loose (*Azospirillum*) or obligate association (*Rhizobium* or actinomycetes) with macrophytes. Energy is provided to the latter organisms through the mineralization of fixed carbon supplied by their host plants. Because of the generally low soil organic matter content of mined soils, nitrogen inputs to the system from free-living diazotrophs are anticipated to be of limited significance to the soil nitrogen budget. Predominant nitrogen inputs would thus result from macrophyte-microbe associations, of which the *Rhizobium*-legume symbiosis is the most studied. Research into the significance of nonlegume plant symbiosis and facultative bacteria-rhizosphere interactions (*Azospirillum*-grass interactions, for example) is required in reclaimed minelands.

The low organic matter content of the mined soils may prevent or limit dentrification in newly reclaimed soils. The stimulation of denitrification potential is arable soils with amendment with organic materials has been adequately demonstrated (Nommik, 1956; Bremner and Shaw, 1958a,b; McGarity, 1961; Ekpete and Cornfield, 1965; Greenland, 1962; Macrae et al., 1968; Nelson and Bremner, 1969; Smid and Beauchamp, 1976; Beauchamp, et al., 1980). Variation of the denitrification rate with levels of soil carbon was demonstrated by Bowman and Focht (1974) when they reported an apparent Michaelis constant (K_m) for glucose of 500 μg/ml with saturating nitrate in a Coachella fine sand. Denitrification potentials from a wide variety of soils correlated with both total organic carbon and available carbon (Stanford et al., 1975b; Burfoed and Bremner, 1975). Highest correlations were detected between the rate of nitrate disappearance and soluble (available) soil organic carbon. Exceptions are observed in mineral soils with sufficiently high organic matter contents to saturate the carbon metabolism enzymes of the denitrifying bacteria (Engler and Patrick, 1974). The situation appears to be different with organic soils. Although Pahokee muck of the Everglades Agricultural Area of south Florida contains 45–51% carbon (approximately 82% organic matter), denitrification is carbon limited (Terry and Tate, 1980).

Associated with these observed effects of exogenous carbon amendment on nitrate reduction is the observation that soil manipulations make occluded organic matter available to denitrifiers. This occluded organic matter may be complexed with other soil components in a manner inhibitory to microbial interaction, or be contained within living cells that are resistant to metabolism by saprophytes until their death. McGarity (1962) found that both freezing and air-drying of soils increased the denitrification potential. Patten et al. (1980) noted a similar effect of air-drying. Significance of these reactions in the field relate to the variation of soil moisture levels. Extreme drying would be anticipated to increase the availability and hence the destruction of native soil organic matter.

Because of the commonly occurring leakage of nutrients from plant roots, plus the decomposition of these root materials upon their death, plant growth also enhances the soil denitrification potential (Bailey, 1976; Lance et al., 1978). Bailey and Beauchamp (1973) found augmented denitrification rates in soils amended with macerated roots, and Smith and Tiedje (1979) noted that the denitrification rates were stimulated in rhizosphere soils containing high nitrate levels. In comparison, diminished rates were found in planted soils with low nitrate concentrations, as compared with those of comparable but unplanted soils. Thus, under low-nitrate conditions, plant growth apparently reduces the denitrification rate.

These data suggest occurrence of a low to nonexistent denitrification rate in newly reclaimed minelands. But, as the soil organic matter level increases, either through decomposition of organic debris or from exudates of macrophyte roots, the denitrification potential will increase. Since nitrate is an obligatory participant in the denitrification process, the development of an active denitrifying bacterial community will require development of both a nitrogen mineralization and nitrifier populations to provide the nitrogenous substrate.

B. Abiotic Limitations of Nitrogen Processes

Of the four major processes in the nitrogen cycle (fixation, mineralization, nitrification, and denitrification), mineralization is least sensitive to environmental perturbations, although the rate may be severely reduced from that anticipated under optimal conditions (Alexander, 1977). The ubiquitous occurrence of mineralization results from the diversity of microbes capable of metabolizing organic nitrogen compounds. Nitrogen-mineralizing microorganisms are aerobes or anaerobes, acid sensitive or acid resistant, and spore forming or non-spore forming (Alexander, 1977). This versatility with respect to aeration was demonstrated by Miller and Johnson (1964) when they observed that mineralization was optimal in soils with moisture levels of −0.50 bar to −0.15 bar and at zero tension (water saturated). Stanford and Epstein (1974) found that with drier soil conditions, greatest mineralization occurred when 80–90% of the soil pores were filled with water. With lower moisture levels, water availability likely became a limiting factor. Thus, mineralization was shown to occur in both aerobic and oxygen-limited conditions.

The low levels and availability of soil organic matter generally found in mined soils and the acidic nature of some soils are inhibitory to nitrogen fixation and denitrification. Because of the high energy requirement to break the triple bond of dinitrogen and produce ammonium, large quantities of fixed carbon must be metabolized by the diazotroph (Child, 1981). For the free-living diazotroph, this carbon substrate must be supplied by the soil environment in which the microbe is functioning. Supplies of metabolizable fixed carbon are limited in mined soils, and free-living diazotrophs thus are anticipated to be of little significance in affecting the nitrogen budget of mined soils. The bulk of the nitrogen fixed in this ecosystem must, therefore, originate from the various microbe-plant associations. Since most of the data relating to this process in minelands involves *Rhizobium*-legume symbiosis, the remaining discussion will be concentrated primarily on this association.

Because of the complexity of the symbiotic nitrogen-fixing association, a threefold effect of environmental perturbations can be anticipated. The overall association (the plant-microbe interaction itself), or the microbe or the macrophyte taken individually, may be affected. Legume growth may be limited or prevented by variation of soil acidity, but the *Rhizobium* or infection process is generally more sensitive. With many *Rhizobium* strains of economic importance, survival of the bacterium in soil outside of the nodule is limited, and infection of the host plant does not occur at pH values much below 5.0. In contrast, the legume has been demonstrated to survive in soil receiving adequate levels of fixed nitrogen at pH values of 4.0 (Alexander, 1977). Munns and Fox (1977) examined the effect soil pH adjustment through liming on 18 tropical and temperate legumes. Growth was stimulated by liming. Yield increases of all 18 species ranged from 30% to 50% up to 2000% with lime amendment ranging from 0.1 ton to 16 tons/hectare. Between 3 and 4 tons per hectare were required to raise the soil pH to 6.0. No significant differences were observed between the lime responses of temperate and tropical legumes.

After the initial plantings of minelands, it is reasonable to assume that the source of bacterial inoculum for establishment of the nitrogen-fixing symbiosis would be the indigenous soil population. Seeds and seedlings could be inoculated prior to planting, but such procedures are impractical in the long-term development of the ecosystem. Hence, the survival of the rhizobia species under the prevailing conditions of minelands needs to be examined. Since few data exist for this situation in minelands, examples will be selected from general laboratory studies and agricultural soils. A significant *Rhizobium* population must be maintained in the soil to allow for infection of young legume roots. Barber (1980) during an examination of the relationship between *Rhizobium meliloti* in alfalfa (*Medicago sativa* L.) stands and soil pH found that the number of *R. meliloti* cells in the soil increased as the soil pH increased. Lowendorf et al. (1981) found differential sensitivity to soil acidity with various strains of *R. meliloti*. Populations of *R. meliloti* strains with greater acid sensitivity declined more rapidly in sterile soil (pH 4.6) than did those that were more acid resistant. The acid-sensitive strain was eliminated from native soil of pH 4.8, whereas some survival was detected with the more acid-tolerant variety. Increased stability at pH 4.6 of the tolerant, but not the sensitive strain, was observed in soil containing growing alfalfa.

Two distinct sources of inhibition of microbial growth and survival of bacteria occur in acidic soils. The first, which was discussed above, results directly from the acidity itself. The second stems from the effect of acidic pH values on the solubility of polyvalent cations which can stimulate or retard microbial growth. Keyser and Munns (1979a) examined the toxic effects of manganese and aluminum (two cations of increased solubility in acid) and of calcium deficiency (an ion of diminished availability under acidic conditions) on 23 cowpea rhizobia capable of growth at pH 4.5 and 10 strains of *Rhizobium japonicum* which grew at pH 4.8. They concluded that in acid soils the aluminum toxicity and the direct effects of the acidity itself had greater inhibitory effects on rhizobial growth and survival in soil than manganese toxicity or calcium deficiency. In a companion study (Keyser and Munns, 1979b), these workers noted that acid tolerance did not necessarily relate to aluminum tolerance. Aluminum at 50 μM increased the lag time for growth or slowed the growth rate of nearly all acid-tolerant strains. Essentially complete inhibition of growth was recorded for 40% of the rhizobia. Franco and Munns (1982) extended these observations in noting that both survival and the infection process are inhibited under acidic soil conditions.

Because of the economic importance of many legumes and the relative ease of isolation and study of the *Rhizobium* endophyte, the legume-*Rhizobium* symbiosis is the more studied and understood of the nitrogen-fixing associations. But the actinomycete nonleguminous angiosperm nitrogen-fixing symbiosis may prove to be as important, if not more important, in mineland reclamation. Actinomycetes, particularly of the genus *Frankia*, form nitrogen-fixing nodules on roots of a variety of trees and shrubs, including members of the genera *Alnus*, *Myricia*, *Elaeagnus*, and *Hippophae* (Alexander, 1977). The efficiency of nitrogen fixation by the two symbiotic systems is apparently comparable (Tjempka and Winship, 1980). *Myricia asplenifolia* (*comptonia peregrina*) and *Alnus glutinosa* colonized anthracite mine wastes in Pennsylvania (Schramm, 1966). Primary value of the nonlegume associations may result from their acid tolerance (Silver, 1971). Nodulation of *Myricia gale* and *Alnus glutinosa* occurs at pH 4.2 (Bond et al., 1954). With the development of procedures to isolate and study the endophyte (Benson, 1982), and the capability to infect susceptible hosts and form nitrogen-fixing nodules (Rogers and Wollum, 1974), the barriers to extensive study of this association are at least partially removed. This may prove to be a particularly rewarding research area for mineland reclamation.

Nitrification, an oxidative portion of the nitrogen cycle, is catalyzed by a limited number of obligate aerobic bacterial species of the family *Nitrobacteriaceae*. These organisms gain their energy through oxidation of reduced nitrogen substrates and their cell structure through reduction of carbon dioxide. For a more complete discussion of the biology and biochemistry, see a recent review of these organisms prepared by Belser (1979). The nitrifying bacteria are extremely tolerant of extremes in temperature and aeration, but are highly sensitive to acidic soil conditions (Dommergues et al., 1978). Nitrification is severely limited, if not precluded, below a pH of 5.5. Some nitrification has been detected in soils of pH 4–5, but it is not known if this nitrate production results from activity of the autotrophic nitrifiers, described above, growing in soil microsites under conditions more conducive to function of the microbe (Dommergues et al., 1978), or if it results from growth of the heterotrophic nitrifier. A variety of heterotrophic microorganisms exist that are capable of oxidizing reduced nitrogen to nitrite and/or nitrate, among other products (Eylar and Schmidt, 1959; Doxtander and Alexander, 1966; Odu and Adeoye, 1970; Verstraete and Alexander, 1972). Although it is reasonably easy to demonstrate heterotrophic nitrification in culture and large populations of these organisms exist in some soil ecosystems (Tate, 1977), heterotrophic nitrification has not been shown to occur in a native soil sample. The basis for questioning the role of the heterotrophic nitrifier in situ is that nitrification is not obligatory for growth of the heterotroph, whereas it is with the autotroph (Marshall and Alexander, 1962; Obaton et al., 1968; Tate, 1980b). Verstraete and Alexander (1973) suggested a role for the heterotroph in nitrification in sewage, lake water, and soil when they amended natural samples with ammonium and a carbon source and observed production of nitrogenous intermediates characteristic of heterotrophic nitrification.

Since the autotrophic nitrifiers are obligate aerobes, variation in soil moisture would be anticipated to affect nitrification rates. Reichman et al. (1966) studied soil moisture relationship and nitrification in samples of chestnut and chernozem soil. Nitrification was essentially directly proportional to soil water content between −15 bar and −0.2 bar. Measurable nitrification occurred at moisture levels as low as −15 bar.

Denitrification, the reduction of nitrate or nitrite to nitrous oxide and dinitrogen, occurs in sites where free oxygen (O_2) availability is limited. Extremely small quantities of free oxygen distributed throughout the soil pores completely inhibit denitrification (Pilot and Patrick, 1972). This raises the important question of how does this anoxic process occur simultaneously in soils with the aerobic nitrogen oxidation process of nitrification. Early reports of "aerobic denitrification" (Broadbent and Stojanovic, 1952) as an explanation of this phenomenon led to numerous studies with the objective of determining the miniumum oxygen tension necessary for the *N*-oxide reductase activities associated with denitrification (Wijler and Delwiche, 1954; Allison et al., 1960; Cady and Bartholomew, 1961; Greenwood, 1962). Greenwood (1962) successfully elucidated the role of oxygenated and anoxic microsites within the soil matrix. This can be pictured as a soil granule that is located in the aerobic soil atmosphere. The activity of the aerobic microbes on the surface prevent free oxygen from diffusing past the granule surface. Hence, the internal portions of the granule are anoxic. Nitrification can occur on the surface concurrently with denitrification in the granule interior. Because of the poor soil structure of most minelands, it may be postulated that denitrification could tend to be limited by the shortage of anoxic microsites.

Soil denitrification potential increases with soil pH within the range of approximately pH 3.8–9.0 (Ekpete and Cornfield, 1965; Dubey and Fox, 1974; Focht, 1974; Muller et al., 1980; George and Antoine, 1982). Denitrification was totally inhibited in some tropical soils by the interaction of low soil organic matter and the high soil acidity (Dubey and Fox, 1974). Valera and Alexander (1961) noted that the population density of soil denitrifiers correlated with soil pH and that this population was more sensitive to low pH conditions than was the total bacterial populations.

This abbreviated discussion of the nitrogen cycle in minelands emphasizes the importance of this cycle to overall ecosystem development and stability, the sensitivity of various processes to the prevalent conditions of many minelands, and the importance of the interactions between the carbon and nitrogen cycle. Of the two cycles, nitrogen metabolism appears to be the most vulnerable to environmental perturbations in these disturbed soils. This results from the limited number of microorganisms capable of catalyzing certain key reactions of the cycle. These limitations have been overcome with common procedures used in mineland reclamation. Practical examples are provided in the ensuing chapters.

V. CONCLUSION

The preceding discussion emphasizes the role of the oft neglected soil microbial community to total ecosystem development and stability in mineland soils and provides an indication of the complexity of the reactions catalyzed by the soil microorganisms. Carbon mineralization and the nitrogen cycle, both of which are important in mineland reclamation and which are indicative of the sensitivity of the microbial community to the prevalent environmental conditions of these disturbed lands, were considered. Major points include the following:

1. Successful reclamation of minelands relies upon the development of an active soil microbial community.
2. This microbial community is responsible for the development of a soil structure conducive to macrophyte growth, production of plant nutrients through contributions to the various biogeochemical cycles, and mitigation of adverse chemical and physical limitations to macrophyte community development.
3. Induction of many of the functions of the microbial community in mined soils can be accomplished through amendment of the soil with organic materials such as sludges or compost. The abiotic conditions of such amended soils must not be so extreme as to preclude microbial growth and development.
4. The amendment of soil with organic materials may temporarily overcome the problems of inactive communities, but long-term stability of the system depends upon the development of an indigenous microbial community that can contribute to biogeochemical cycling processes.
5. Microbial processes in soil are highly sensitive to the environmental parameters characteristic of mined lands. In many cases, pretreatment of the soil is needed for development of the requisite microbial populations.

With the above principles in mind, the reader is now invited to explore the more applied chapters that follow. These chapters reveal how the environmental limitations alluded to above are overcome in the field during a successful reclamation project, and discuss the technological and basic informational blocks to successful reclamation.

ACKNOWLEDGMENTS

This paper is New Jersey Agricultural Experiment Station Publication No. F-15187-1-83, and is supported by state funds and Hatch funds.

REFERENCES

Alexander, M. (1977). *Introduction to Soil Microbiology.* John Wiley & Sons, New York, 472 pp.

Allison, F. E., Carter, J. N., and Sterling, L. D. (1960). The effect of partial pressure of oxygen on denitrification in soil. *Soil Sci. Soc. Am. Proc. 24*: 283–285.

Andreyuk, E. I., and Gordienko, S. A. (1978). Transformation of humic acids by soil actinomycetes. *Mikrobiol. Zh. (Kiev) 40*: 690–697 (Russian) (*Chem. Abstr. 90*: 102586 w.).

Andriiuk, K. I., Hordienko, S. O., Havrysh, I. N., Konotop, H. I., and Martynenko, V. A. (1973). Decomposition of peat humic acids by associative cultures of microorganisms. *Mikrobiol. Zh. 35*: 554–559. (Russian)

Aremntano, T. V. (1980). Drainage of organic soils as a factor in the world of carbon cyle. *Bioscience 30*: 825–830.

Bailey, L. D. (1976). Effects of temperature and root on denitrification in a soil. *Can. J. Soil Sci. 56*: 79–87.

Bailey, L. D., and Beauchamp, E. G. (1973). Effect of moisture, added NO_3^- – N, and macerated roots on NO_3^- transformations and redox potential in surface and subsurface soils. *Can. J. Soil Sci. 53*: 219–230.

Barber, L. E. (1980). Enumeration, effectiveness, and pH resistance of *Rhizobium meliloti* populations in Oregon soils. *Soil Sci. Soc. Am. J. 44*: 537–539.

Barnette, R. M., and Hester, J. B. (1930). Effect of burning upon the accumulation of organic matter in forest soils. *Soil Sci. 29*: 281–284.

Beauchamp, E. G., Gale, C., and Yoemans, J. C. (1980). Organic matter availability for denitrification in soils of different textures and drainage classes. *Commun. Soil Sci. Plant Anal. 11*: 1221–1233.

Belser, L. W. (1979). Population ecology of nitrifying bacteria. *Ann. Rev. Microbiol. 33*: 309–333.

Benson, D. R. (1982). Isolation of *Frankia* strains from alder actinorrhizal root nodules. *Appl. Environ. Microbiol. 44*: 461–465.

Biederbeck, V. O., and Paul, E. A. (1971). Fungal degradation of soil humic nitrogen. *Agron. Abst. 1971*: 80.

Bohn, H. (1978). On Organic Soil carbon and CO_2. *Tellus 30*: 472–475.

Bond, G., Fletcher, W. W., and Ferguson, T. P. (1954). The development and function of the root nodules of *Alnus, Myrica,* and *Hippophaë*. *Plant Soil. 5*: 309–323.

Bowman, R. A., and Focht, D. D. (1974). The influence of glucose and nitrate concentrations upon denitrification rates in sandy soils. *Soil Biol. Biochem. 6*: 297–301.

Bremner, J. M., and Shaw, K. (1958a). Denitrification in soil. I. Methods of investigation. *J. Agric. Sci. 51*: 22–39.

Bremner, J. M., and Shaw, K. (1958b). Denitrification in soil. II. Factors affecting denitrification. *J. Agric. Sci. 51*: 40–52.

Broadbent, F. E. (1960). Factors influencing the decomposition of organic soils in the California delta. *Hilgardia. 29*: 587–612.

Broadbent, F. E., and Bartholomew, W. V. (1948). The effect of quantity of plant material added to soil on its rate of decomposition. *Soil Sci. Soc. Am. Proc. 13*: 271–274.

Broadbent, F. E., and Stojanovic, B. F. (1952). The effect of partial pressure of oxygen on some soil nitrogen transformations. *Soil Sci. Soc. Am. Proc. 16*: 359–363.

Bunnell, F. L., Tait, D. E. N., Flanagan, P. W., and Van Cleve, K. (1977). Microbial respiration and substrate weight loss. I. A general model of the influences of abiotic variables. *Soil Biol. Biochem. 9*: 33–40.

Burford, J. R., and Bremner, J. M. (1975). Relationships between the denitrification capacities of soil and total, water-soluble and readily decomposable soil organic matter. *Soil Biol. Biochem. 7*: 389–394.

Burges, A., and Latter, P. (1960). Decomposition of humic acid by fungi. *Nature 186*: 404–405.

Cady, F. B., and Bartholomew, W. V. (1961). Influence of low pO_2 on denitrification processes and products. *Soil Sci. Soc. Am. Proc. 25*: 362–365.

Cast, (1976). Effect of increased nitrogen fixation on stratospheric ozone. *Council for Agricultural Science and Technology. Report 53.* Iowa State University. Ames, Iowa.

Cheshire, M. V., Mundie, C. M., and Shepherd, H. (1969). Transformation of ^{14}C glucose and starch in soil. *Soil Biol. Biochem. 1*: 117–130.

Cheshire, M. V., Mundie, C. M., and Shepherd, H. (1974). Transformation of sugars when rye hemicellulose labeled with ^{14}C decomposes in soil. *J. Soil Sci. 25*: 90–98.

Cheshire, M. V., Sparling, G. P., and Mundie, C. M. (1983). Effect of periodate treatment of soil on carbohydrate constituents and soil aggregation. *J. Soil Sci. 34*: 105–112.

Child, J. J. (1981). Biological nitrogen fixation. In *Soil Biochemistry*, E. A. Paul and J. N. Ladd (Ed.). Marcel Dekker, Inc., New York, Vol. 5, 297–322.

Conn, H. J., and Bright, J. W. (1919). Ammonification of manure in soil. *J. Agric. Res. 16*: 313–350.

Crawford, R. L. (1981). *Lignin biodegradation and transformation.* John Wiley and Sons, New York, 154 pp.

Cundell, A. M. (1977). The role of microorganisms in the revegetation of strip-mined land in the western United States. *J. Range Management 30*: 299–305.

Dalal, R. C. (1979). Mineralization of carbon and phosphorus from Carbon-14 and Phosphorus-32 labeled plant material added to soil. *Soil Sci. Soc. Am. J. 43*: 913--916.

Dancer, W. S., Handley, J. F., and Bradshaw, A. D. (1977). Nitrogen accumulation in kaolin mining wastes in Cornwall. I. Natural communities. *Plant Soil. 48*: 153--167.

Dawson, J. E. (1956). Organic Soils. *Adv. Argon. 8*: 377–401.

Dent, D. L., Downing, E. J. B., and Rogaar, H. (1976). Changes in the structure of marsh soils following drainage and arable cultivation. *J. Soil Sci. 27*: 250–265.

Dommergues, Y. R., Belser, L. W., and Schmidt, E. L. (1978). Limiting factors for microbial growth and activity in soil. *Adv. Microbiol. Ecol. 2*: 49--104.

Dormaar, J. F. (1975). Susceptibility to organic matter of chernozeimic Ah horizons to biological decomposition. *Can. J. Soil Sci. 55*: 473--480.

Doxtander, K. G., and Alexander, M. (1966). Nitrification by heterotrophic soil microorganisms. *Soil Sci. Soc. Am. Proc. 30*: 351–355.

Dubey, H. D., and Fox, R. H. (1974). Denitrification losses from humid tropical soils of Puerto Rico. *Soil Sci. Soc. Am. Proc. 38*: 917–920.

Ekpete, D. M., and Cornfield, A. H. (1965). Effect of pH and addition of organic materials on denitrification losses from soil. *Nature 208*: 1200.

Engler, R. M., and Patrick, W. H., Jr. (1974). Nitrate removal from floodwater overlying flooded soils and sediments. *J. Environ. Qual. 3*: 409–413.

Estermann, E. F., Peterson, G. H., and McLaren, A. D. (1959). Digestion of clay-protein, lignin-protein, and silica-protein complexes by enzymes and bacteria. *Soil Sci. Soc. Am. Proc. 23*: 31–36.

Eylar, O. R., and Schmidt, E. L. (1959). A survey of heterotrophic microorganisms from soil for ability to form nitrite and nitrate. *J. Gen. Microbiol. 20*: 473–481.

Fedorov, M. V., and Il'ina, T. K. (1963). Utilization of humic acid by soil actinomycetes as a sole source of carbon and nitrogen. *Microbiology 32*: 234–237.

Focht, D. D. (1974). The effect of temperature, pH, and aeration on the production of nitrous oxide and gaseous nitrogen—a zero-order kinetic model. *Soil Sci. 118*: 173–179.

Forster, S. M., and Nicolson, T. H. (1981). Microbial aggregation of sand in a maritime dune succession. *Soil Biol. Biochem. 13*: 205–208.

Franco, A. A., and Munns, D. N. (1982). Acidity and aluminum restraints on nodulation, nitrogen fixation, and growth of *Phaseolus vulgaris* in solution culture. *Soil Sci. Soc. Am. J. 46*: 296–301.

Fresquez, P. R., and Lindemann, W. E. (1982). Soil and rhizosphere microorganisms in amended coal mine spoils. *Soil Sci. Soc. Am. J. 46*: 751–755.

Galstyan, A. Sh., and Avundzhyan, Z. S. (1970). Dehydrogenases in the clay fraction. *Soc. Soil Sci. 1970*: 721–723.

George, U. S. and Antoine, A. D. (1982). Denitrification potential of a salt marsh soil: Effect of temperature, pH, and substrate concentration. *Soil Biol. Biochem. 14*: 117–125.

Gibson, A. H. (Ed.) (1977). *A treatise on dinitrogen fixation.* IV. *Agronomy and Ecology.* John Wiley and Sons, Inc., New York, 527 pp.

Giddens, J. (1957). Rate of loss of carbon from Georgia soils. *Soil Sci. Soc. Am. Proc. 21*: 513–515.

Gilmour, C. M., Allen, O. N., and Truog, E. (1948). Soil aggregation as influenced by the growth of mold species, kind of soil, and organic matter. *Soil Sci. Soc. Am. Proc. 13*: 292–296.

Gosz, J. R. (1981). Nitrogen cycling in coniferous ecosystems. In *Terrestrial Nitrogen Cycles, Ecol. Bull. (Stockholm)*, F. E. Clark and T. Rosswall (Ed.), *33*: 405–426.

Greaves, J. E. and Carter, E. G. (1920). Influence of moisture on the bacterial activities of the soil. *Soil Sci. 10*: 361–387.

Greenland, D. J. (1962). Denitrification in some tropical soils. *J. Agric. Sci. 58*: 227–233.

Greenwood, D. J. (1962). Nitrification and nitrate dissimilation in soil. II. Effect of oxygen concentration. *Plant Soil 17*: 378–391.

Haider, K. and Martin, J. P. (1975). Decomposition of specifically Carbon-14 labeled benzoic and cinnamic acid derivatives in soil. *Soil Sci. Soc. Am. Proc. 39*:657–662.

Hunt, H. W. (1977). A simulation model for decomposition in grasslands. *Ecology 58*: 469–484.

Huntjens, J. L. M. (1972). Availability of microbial and soil organic nitrogen to a *Pseudomonas* strain and the effect of soil organic matter on the availability of casein nitrogen. *Soil Biol. Biochem. 4*: 347–358.

Ibrahim, A. N., and Ibrahim, I. A. (1977). Biodegradation of soil humus by streptomycetes. *Agrokim. Talajtan. 26*: 415–423 (Hungarian).

Ivarson, K. C. (1977). Changes in decomposition rate, microbial population and carbohydrate content of an acid peat bog after liming and reclamation. *Can. J. Soil Sci. 57*: 129–137.

Jenkinson, D. S. (1966). Studies on the decomposition of plant material in soil. II. Partial sterilization of soil and the soil biomass. *J. Soil Sci. 17*: 280–302.

Jenkinson, D. S., and Rayner, J. H. (1977). The turnover of soil organic matter in some of the Rothamsted classical experiments. *Soil Sci. 123*: 298–305.

Jenny, H. (1980). *The soil resource.* Springer-Verlag, New York, 377 pp.

Jones, D., and Griffiths, E. (1967). Microbiological aspects of soil structure. II. Soil aggregation by the extracellular polysaccharide of *Lipomyces starkeyi.* *Plant soil 27*: 187–200.

Kassim, G., Stott; D. E., Martin, J. P., and Haider, K. (1982). Stabilization and incorporation into biomass of phenolic and benzenoid carbons during biodegradation in soil. *Soil Sci. Soc. Am. J. 46*: 305–309.

Keyser, H. H., and Munns, D. N. (1979a). Effects of calcium, manganese, and aluminum on growth of rhizobia in acid media. *Soil Sci. Soc. Am. J. 43*: 500–503.

Keyser, H. H., and Munns, D. N. (1979b). Tolerance of rhizobia to acidity, aluminum, and phosphate. *Soil Sci. Soc. Am. J. 43*: 519–523.

Kolenbrander, G. J. (1974). Efficiency of organic manure in increasing soil organic matter content. *Trans. Int. Congr. Soil Sci. 10th 2*: 129–136.

Knight, W. G., and Skujins, J. (1981). ATP concentration and soil respiration at reduced water potentials in arid soil. *Soil Sci. Soc. Am. J. 45*: 657–660.

Knowles, R. (1981). Denitrification in *Soil Biochemistry*, E. A. Paul and J. N. Ladd (Ed.) Marcel Dekker, Inc. 5: 323–369.

Kunc. F., and Stotzky, G. (1974). Effect of clay minerals on heterotrophic microbial activity in soil. *Soil Sci. 118*: 186–195.

Lance, J. C., Rice, R. C., and Whisler, R. D. (1978). Effects of vegetation on denitrification and phosphate movement during rapid infiltration of soil columns. *J. Water Pollut. Contr. Fed. 50*: 2183–2188.

Linhares, L. F., and Martin, J. P. (1978). Decomposition in soil of the humic acid-type polymers (melanins) of *Eurotium echlinulatum, Aspergillus glaucus* sp. and other fungi. *Soil Sci. Soc. Am. J. 42*: 738–743.

Lowendorf, H. S., Baya, A. M., and Alexander, M. (1981). Survival of *Rhizobium* in acid soils. *Appl. Environ. Microbiol. 42*: 951–957.

Lund, V., and Goksoyr, J. (1980). Effects of water fluctuations on microbial mass and activity in soil. *Microbial Ecol. 6*:115–123.

Lynch. D. L., and Cotnoir, L. J., Jr. (1956). The influence of clay minerals on the breakdown of certain organic substrates. *Soil Sci. Soc. Am. Proc. 20*: 367–370.

Macrae, I. C., Ancajas, R. R., and Sulandanan, S. (1968). The fate of nitrate nitrogen in some tropical soils following submergence. *Soil Sci. 105*: 327–334.

Mal'tseva, N. N., Gordienko, S. A., and Izzheurova, V. V. (1975). Use of humic acids by oligonitrophilous microorganisms. In

Tr. S'ezda Mikrobiol. Ukr. 4th, D. G. Zatula (Ed.), pp. 61–62 (Russian). "Naukova Dumko," Kiev, USSR, (*Chem. Abstr. 85*: 188971.)

Marshall, K. C. (1971). Sorptive interactions between soil particles and microorganisms. In *Soil Biochemistry,* A. D. McLaren and J. Skujins (Ed.). Marcel Dekker, New York, Vol. 2, pp. 409–445.

Marshall, K. C., and Alexander, M. (1962). Nitrification by *Aspergillus flavus. J. Bacteriol. 83*: 572–578.

Marshman, N. A., and Marshall, K. C. (1981). Bacterial growth on proteins in the presence of clay minerals. *Soil Biol. Biochem. 13*: 127–134.

Martin, J. P. (1971). Decomposition and binding action of polysaccharides in soil. *Soil Biol. Biochem. 3*: 33–41.

Martin, J. P., Ervin, J. O., and Shepherd, R. A. (1959). Decomposition and aggregating effect of fungus cell material in soil. *Soil Sci. Soc. Am. Proc. 23*: 217–220.

Martin, J. P., Ervin, J. O., and Shepherd, R. A. (1966). Decomposition of the iron, aluminum, zinc and copper salts or complexes of some microbial and plant polysaccharides in soil. *Soil Sci. Soc. Am. Proc. 30*: 196–200.

Martin, J. P., and Haider, K. (1976a). Decomposition of specifically carbon-14 labeled ferulic acid; free and linked into model humic acid-type polymers. *Soil Sci. Soc. Am. J. 40*:377–380.

Martin, J. P., Filip, Z., and Haider, K. (1976b). Effect of montmorillonite and humate on growth and metabolic activity of some actinomycetes. *Soil Biol. Biochem. 8*: 409–413.

Martin, J. P., Haider, K., and Kassim, G. (1980). Biodegradation and stabilization after 2 years of specific crop, lignin, and polysaccharide carbons in soils. *Soil Sci. Soc. Am. J. 44*: 1250–1255.

Martin, J. P., Richards, S. J., and Haider, K. (1967). Properties and decomposition and binding action in soil of "Humic Acid" synthesized by *Epicoccum nigrum. Soil Sci. Soc. Am. Proc. 31*:657–662.

Martin, T. L., and Anderson, D. A. (1942). Organic matter decomposition, mold flora, and soil aggregation relationships. *Soil Sci. Soc. Am. Proc. 7*: 215–217.

Mathur, S. P. (1969). Microbial use of podzol Bh fulvic acids. *Can. J. Microbiol. 15*: 677–680.

Mathur, S. P., and Paul, E. A. (1967). Microbial utilization of soil humic acids. *Can. J. Microbiol. 13*: 573–580.

McGarity, J. W. (1961). Denitrification studies on some South Australian soils. *Plant Soil. 14*: 1–21.

McGarity, J. W. (1962). Effect of freezing of soil on denitrification. *Nature 196*: 1342–1343.

Miller, R. D., and Johnson, D. D. (1964). The effect of soil moisture tension on carbon dioxide evolution, nitrification, and nitrogen mineralization. *Soil Sci. Soc. Am. Proc. 28*:644–647.

Mills, A. L. (1984). Microbiology and the recovery of environments acidified by mining wastes. In *Soil Reclamation Processes: Microbiological Analyses and Applications*, R. L. Tate III and D. A. Klein (Ed.). Marcel Dekker, New York.

Mishra, M. M., Neelakantan, S., and Khandelwal, K. C. (1974). Buildup of soil organic matter under continuous manuring and cropping practices. *J. Res. Haryana Agric. Univ. 4*: 225–230.

Muller, M. M., Sundman, V., and Skujins, J. (1980). Denitrification in low pH spodosols and peats determined with the acetylene inhibition method. *Appl. Environ. Microbiol. 40*: 235–239.

Munns, D. N., and Fox, R. L. (1977). Comparative lime requirements of tropical and temperate legumes. *Plant Soil. 46*: 533–548.

Mutatker, V. K., and Wagner, G. H. (1967). Humification of carbon-14 labeled glucose in soils of Sanborn field. *Soil Sci. Soc. Am. Proc. 31*: 66–70.

Nelson, D. W., and Bremner, J. M. (1969). Factors affecting chemical transformations of nitrite in soils. *Soil Biol. Biochem. 1*: 229–239.

Nommik, H. (1956). Investigations on denitrification in soil. *Acta. Agr. Scand. 6*: 195–228.

Nutman, P. S. (Ed.) (1976). *Symbiotic nitrogen fixation in plants.* Cambridge University Press, New York, 584 pp.

Obaton, M., Amarger, N., and Alexander, M. (1968). Heterotrophic nitrification in *Pseudomonas aeruginosa*. *Arch. Mikrobiol. 63*: 122–132.

Odu, C. T. I., and Adeoye, K. B. (1970). Heterotrophic nitrification in soils: A preliminary investigation. *Soil Biol. Biochem. 2*: 41–45.

Olness, A., and Clapp, C. E. (1972). Microbial degradation of a montmorillonite-dextran complex. *Soil Sci. Soc. Am. Proc. 36*: 179–181.

Orlov, D. S. (1974). Variability of composition of humic acids and some problems of their structure. *Trans. Int. Congr. Soil Sci. 10th 2*: 97–104.

Pal. D., Broadbent, F. E., and Mikkelsen, D. S. (1975). Influence of temperature on the kinetics of rice straw decomposition in soils. *Soil Sci. 120*:442–449.

Patten, D. K., Bremner, J. M., and Blackmer, A. M. (1980). Effects of drying and air-drying storage of soils on their capacity for denitrification of nitrate. *Soil Sci. Soc. Am. J. 44*:67–70.

Pilot, L., and Patrick, W. H., Jr. (1972). Nitrate reduction in soils: Effect of soil moisture tension. *Soil Sci. 114*: 312–316.

Pinck, L. A., Allison, F. E., and Sherman, M. S. (1950). Maintenance of soil organic matter: II. Losses of carbon and nitrogen from young and mature plant materials during decomposition in soil. *Soil Sci.* *69*: 391–401.

Reichman, G. A., Grunes, D. L., and Viets, F. G., Jr. (1966). Effect of soil moisture on ammonification and nitrification in two northern plains soils. *Soil Sci. Soc. Am. Proc.* *30*: 363–365.

Richardson, C. J., and Allen, L. H. (1979). Estimates of carbon accumulation rates for some natural peatlands. In *The Role of Organic Soils in the World Carbon Cycle—Problem Analysis and Research Needs. Report to the U.S. Dept. of Energy. No. DOE/ET/10040-2.* T. V. Armentano (Ed.). The Institute of Ecology (TIE), Indianapolis, Indiana, p. 26.

Rogers, R. D., and Wollum, A. G. II. (1974). Virulence of the Alnus endophyte after *in vitro* cultivation. *Soil Sci. Soc. Am. Proc.* *38*: 756–759.

Schafer. W. M., Nielsen, G. A., and Nettleton, W. D. (1980). Minesoil genesis and morphology in a spoil chromosequence in Montana. *Soil Sci. Soc. Am. J.* *44*: 802–807.

Schramm, J. R. (1966). Plant colonization studies on black wastes from anthracite mining in Pennsylvania. *Trans. Amer. Phil. Soc.* *56*: 1–64.

Sidorenko, D. D., Aristarkhova, V. I., and Chernikova, V. A. (1978). Changes in composition and properties of humic acids upon Nocardia microorganisms action. *Izv. Akad. Nauk SSSR. Ser. Biol.* *2*: 195–202 (Russian).

Silver, W. S. (1971). Physiological chemistry of non-leguminous symbiosis. In *The Chemistry and Biochemistry of Nitrogen Fixation*, J. R. Postgate (Ed.) Plenum Press, New York, pp. 246–281.

Sinha, M. K. (1972). Organic matter transformation in soils. I. Humification of C^{14} tagged oat roots. *Plant Soil* *36*: 283–293.

Smid, A. E., and Beauchamp, E. G. (1976). Effects of temperature and organic matter on denitrification in soil. *Can. J. Soil. Sci.* *56*: 385–391.

Smith, M. S., and Tiedje, J. M. (1979). The effect of roots on soil denitrification. *Soil Sci. Soc. Am. J.* *43*: 951–955.

Smith, O. L. (1979a). An analytical model of the decomposition of soil organic matter. *Soil Biol. Biochem.* *11*: 585–606.

Smith, O. L. (1979b). Application of a model of the decomposition of soil organic matter. *Soil Biol. Biochem.* *11*: 607–618.

Sorensen, H. (1963). Studies on the decomposition of C^{14}-labeled barley straw in soil. *Soil Sci.* *95*: 45–51.

Sorensen, L. H. (1974). Rate of decomposition of organic matter in soil as influenced by repeated air-drying—rewetting and repeated additions of organic matter. *Soil Biol. Biochem.* *6*: 287–292.

Sorensen, L. H., and Paul, E. A. (1971). Transformation of acetate carbon into carbohydrate and amino acid metabolites during decomposition in soil. *Soil Biol. Biochem.* *3*: 173--180.

Stanford, G., and Epstein, E. (1974). Nitrogen mineralization-water relations in soils. *Soil Sci. Soc. Am. Proc.* *38*: 103--107.

Stanford, G., Vander Pol, R. A., and Dzienia, D. (1975). Denitrification rates in relation to total and extractable soil carbon, *Soil Sci. Soc. Am. Proc.* *39*: 284–289.

Steinbrenner, K., and Mundstock, I. (1975). Untersuchungen zum huminstoffabbau durch Nokardien. *Arch. Acker-pflanzenbau Bodenkd.* *19*: 243–255.

Stevenson, F. J. (1982). *Humus chemistry*. John Wiley and Sons, New York, 443 pp.

Stout, J. D., Goh, K. M., and Rafter, T. A. (1981). Chemistry and turnover of naturally occurring resistant organic compounds in soil. In *Soil Biochemistry*, E. A. Paul and J. N. Ladd (Ed.) Marcel Dekker, New York, Vol. 5, pp. 1–73.

Stroo, H. F., and Jencks, E. M. (1982). Enzyme activity and respiration in mine soils. *Soil Sci. Soc. Am. J.* *46*: 548--553.

Taha, S. M., Zayed, M. N., and Zohdy, L. (1973). Studies in humic acids decomposing bacteria in soil. II. Isolation and identification of microorganisms. *Z. Bakt. Parasitenk. Infekt.* *128*: 168–172.

Tate, R. L. III. (1977). Nitrification in Histosols: A potential role for the heterotrophic nitrifier. *Appl. Environ. Microbiol.* *33*: 911--914.

Tate, R. L. III. (1980a). Microbial oxidation of organic matter of Histosols. *Adv. Microbial Ecol.* *4*: 169–201.

Tate, R. L. III. (1980b). Variation in heterotrophic and autotrophic nitrifier populations in relation to nitrification in organic soils. *Appl. Environ. Microbiol.* *40*: 75–79.

Terry, R. E., and Tate, R. L. III. (1980a). Denitrification as a pathway for nitrate removal from organic soils. *Soil Sci.* *129*: 162–166.

Tisdall, J. M., and Oades, J. M. (1982). Organic matter and water-stable aggregation in soils. *J. Soil Sci.* *33*: 141–163.

Tjepkema, J. D., and Winship, L. J. (1980) Energy requirement for nitrogen fixation in actinorrhizal and legume root nodules. *Science.* *209*: 279--280.

Valera, C. L., and Alexander, M. (1961). Nutrition and physiology of dentifying bacteria. *Plant Soil* *15*: 268–280.

Verma, L., Martin, J. P., and Haider, K. (1975). Decomposition of carbon-14-labeled proteins, peptides, and amino acids; free and complexed with humic polymers. *Soil Sci. Soc. Am. Proc.* *39*: 279–284.

Verma, L., and Martin, J. P. (1976). Decomposition of algal cells and components and their stabilization through complexing with model humic acid-type phenolic polymers. *Soil Biol. Biochem.* *8*: 85--90.

Verstraete, W., and Alexander, M. (1972). Heterotrophic nitrification by *Arthrobacter* sp. *J. Bacteriol.* *110*: 955–961.

Verstraete, W., and Alexander, M. (1973). Heterotrophic nitrification in samples of natural ecosystems. *Environ. Sci. Technol.* *7*: 39--42.

Verstraete, W., and Voets, J. P. (1977). Soil microbial and biochemical characteristics in relation to soil management and fertility. *Soil Biol. Biochem.* *9*: 253--258.

Visser, S. (1984). Microbial aspects of the reclamation of surface-mined land in western Canada. In *Soil Reclamation Processes: Microbiological Analyses and Applications*, R. L. Tate III and D. A. Klein (Ed.). Marcel Dekker, New York.

Wijler, J., and Delwiche, C. C. (1954). Investigations on the denitrifying process in soil. *Plant Soil* *5*: 155–169.

Wildung, R. E., and T. R. Garland. (1984). Microbial development on oil shale wastes: Influence on geochemistry In *Soil Reclamation Processes: Microbial Analyses and Applications*, R. L. Tate III and D. A. Klein (Ed.). Marcel Dekker, New York.

Wildung, R. E., Garland, T. R., and Buschbom, R. L. (1975). The interdependent effects of soil temperature and water contents on soil respiration rate and plant root decomposition in arid grassland soils. *Soil Biol. Biochem.* *7*: 373–378.

Wilson, J. M., and Griffin, D. M. (1975). Water potential and the respiration of microorganisms in the soil. *Soil Biol. Biochem.* *7*: 199--204.

Zvyagintsev, D. G., and Velikanov, L. L. (1968). Influence of soils and clay minerals on the activity of glucose oxidase and invertase. *Soviet Soil Sci.* *1968*: 789–794.

2

Acid Mine Waste Drainage: Microbial Impact on the Recovery of Soil and Water Ecosystems

AARON L. MILLS / *University of Virginia, Charlottesville, Virginia*

I. INTRODUCTION

The mining of coal and metal ores in the United States is associated with a variety of environmental problems including scarred land and particulate air pollution from strip-mine operations, underground mine subsidence, tailings and overburden accumulation, and acid mine drainage (Down and Stocks, 1977). The latter two items are of prime interest to the microbial ecologist. Water pollution by acid mine drainage (AMD) is an extremely severe problem in areas of the country where mining activity is high and the amount of land covered by mine spoils of various types is extensive. In both of these cases, the effect of the pollution on the microbes may be severe. At the same time, the recovery of the polluted ecosystem relies on the activities of the microbial community.

Most of the acidic mine waste problems have been centered in the Ohio River valley and Appalachian regions of the eastern part of the nation, espeically in areas of Pennsylvania, Maryland, Ohio, West Virginia, Virginia, and Kentucky (Fig. 1). Approximately 17,000 km of streams of various sizes are degraded by AMD and over half are under a regime of continual pollution (Council on Environmental Quality, 1981).

Water flowing from abandoned or active mines or seeping from beneath tailings piles frequently displays pH values of 3.0 or less. The high acidity of the water increases the capacity of the stream or lake to carry metals in a soluble form. Mine drainage generally has elevated levels of metals due to the natural abundance of metals in these materials and an enhanced leaching potential due to a low pH induced

Figure 1 Major streams in the United States affected by acid mine drainage. Compare with location of major coal deposits as shown in Fig. 2. (Redrawn from Council on Environmental Quality, 1982.)

by the oxidation of reduced sulfur compounds. The total dissolved solids content of acid mine drainage is normally estimated by measuring the specific conductance of the solution. Values of 2000 $\mu S\ cm^{-1}$ are not infrequent (compared with values of $\leq 100\ \mu S\ cm^{-1}$ for unpolluted lakes). The U.S. Environmental Protection Agency (1973) has divided drainage from coal mines into four major groups ranging from highly acidic to neutral (Table 1).

Over the next 20–30 years, coal production is predicted to increase dramatically to keep pace with the demands for domestically produced energy. The United States still has substantial coal reserves (Fig. 2), but exploitation of those sources means that most of the new production will involve surface mining activities in the western parts of the country, particularly in the northern Great Plains (Griffith and Clark, 1979). Although water pollution problems are also associated with strip mining, an additional burden for cleanup and rehabilitation results from the greater portion of land seriously disturbed by this method as opposed to underground procedures.

Table 1 United States EPA Classification of Mine Drainage

Parameter	1 Acid	2 Partially oxidized and/or neutralized	3 Oxidized and neutralized and/or alkaline	4 Neutralized and not oxidized
pH	2–4.5	3.5–6.6	6.5–8.5	6.5–8.5
Acidity[a]	1000–15000	0–1000	0	0
Fe^{2+}	500–10000	0– 500	0	50–1000
Fe^{3+}	0	0–1000	0	0
Al^{3+}	0–2000	0– 20	0	0
$SO_4^{=}$	1000–20000	500–10000	500–10000	500–10000

[a]Concentrations expressed as mg/liter, acidity as mg $CaCO_3$/liter.
Source: U.S.E.P.A. (1973).

Figure 2 Location of major coal reserves in the United States. Average sulfur content of coal by rank. (Spiro and Stigliani, 1980, State University of New York Press. Reprinted by permission.)

Although coal mining provides the most extensive mine waste pollution, extraction of metal ores creates severe problems in more localized areas, predominantly in the metal-rich western United States. Mining of iron and precious metal sulfides has led to severe pollution of areas in northern California (Nordstrom et al., 1977) and Colorado (Moran and Wentz, 1974; Wentz, 1974), as well as in many other places. In 1971 alone, 14,700 ha of land were disrupted for the mining of metals, 38,500 ha in nonmetal extraction, and 30,300 ha in the mining of fossil fuels, excluding oil and gas (Paone et al., 1974). The total land mined in the United States is extensive in terms of actual area, but in 1971, the land used for mining represented only 0.16% of the total land area. As a point of comparison, the land areas affected by mining operations were only about one-sixth of those used for highway construction (Down and Stocks, 1977). Thus, while mining activities are widespread and even increasing, their associated problems must be considered as localized. Within those areas, however, the impact of the problem may be extremely severe.

The present chapter examines acidic mining waste generation, effects on the microbiology of aquatic and terrestrial ecosystems, and the homeostatic recovery of the microbial communities in contaminated areas. It is the latter point that serves as the focus for this review, that is, what types of naturally occurring processes allow the communities to recover, and finally, how might those processes be included in a reclamation scheme for contaminated environments. It must be pointed out that investigations of microbiological recovery from the effects of mine wastes are limited in number and scope. Nevertheless, those few studies have generated a pattern of results that allows for a generalized understanding of how the recovery takes place in both terrestrial (soil) and aquatic habitats.

II. DESCRIPTION OF MAJOR WASTES FROM MINING ACTIVITIES

A. Tailings and Overburden

Two distinct types of solid mine wastes can be identified—tailing and overburden. The tailings are associated with the extraction of base metal minerals, and are materials removed from underground mines, milled to a fine powder along with the ore, and then removed from the ore material by a flotation separation process. Although large total amounts of metal and sulfides remain in the tailings, further extraction of those materials has frequently not been considered to be an economical process because of the usually low concentration of the metal of interest. A notable exception is in the copper extraction industry where leaching of tailings and low-grade ores accounts for 10% of the total copper recovery (Ehrlich, 1981; Madsen and Wadsworth, 1981; Brierly, 1982).

Another type of solid material frequantly causing problems is the overburden left from strip mining operations. Overburden is the soil and rock, mixed with low-grade ores or coal, that overlay the mineral vein being extracted. Overburdens (frequently referred to as minesoils) vary widely in their character depending on the underlying geology and the depth at which the extracted material lay. In shallow-mined areas, such as those of the Appalachian area, the amount of coal that remains in the overburden relative to the total amount of material is high. Biological and chemical action on the reduced sulfur compounds in that material by mechanisms described later frequently causes the evolving soil to be highly acidic. In the northern Great Plains, on the other hand, the coal seams are frequently deep (10–40 m), thereby reducing the proportion of the overburden that is potentially acid forming (Schafer, et al. 1980). These overburdens also often contain large amounts of carbonate, which tends to maintain the developing soil at a neutral to alkaline pH. The composition of the overburden materials varies as do the soils of the region. In the Appalachians, for example, the soils tend to be moderately acid, whereas in the northern Great Plains, the soils are often very slightly to slightly alkaline. The specific nature of the native soil will help determine the capacity of the material to neutralize the acid products as they are formed. Furthermore, differences in soil moisture regimes due to precipitation differences between the regions may also promote variations in microbial activities in these soils.

Both tailings and overburden present major pollution problems. The piles are unsightly in place, and they provide an opportunity for increased sediment loading in nearby streams due to erosion. Improperly graded slopes occasionally fail, causing loss of property, personal injury, and death. Oxidation of reduced sulfur compounds remaining in the piles frequently results in seeps of extremely concentrated AMD, which can pollute streams, lakes, or groundwater systems, especially in eastern areas.

The most effective means of reducing the impact of such potential problems is rapid pile revegetation. Revegetation accelerates soil formation, leading to the establishment of a stable soil body that contributes minimal pollution to the remaining environment. The same problems that revegetation is intended to combat, however, work against the establishment of a reasonable plant cover on waste heaps. Goodman (1974) lists the main environmental factors that inhibit plant growth in waste material from mineral extraction activities. Presented here are those shared by both coal and metal mining wastes:

Instability (susceptibility to erosion)
Spontaneous combustion
Unworkably steep slopes
Inhibitory water regimes
Compaction and cementation
Inhibitory surface temperature regime
Wind turbulence causing "sand blasting"
Excessive stoniness (absence of fine material for soil formation)
Broken, uneven surface (no seed bed)
High levels of toxic elements
Low nutrient status
Absence of soil microflora and soil fauna

All but the last three of the items specify physical problems related to the piles. Of the last three, the final one may be directly attributed to the previous two, i.e., the absence of soil microbes is a direct effect of the high concentration of toxic elements (including H^+) and the low nutrient status.

The high metal content and high acidity of tailings material generally renders plant growth marginal at best. The lack of cover and roots allows for erosion of the material into the water increasing both the sediment load and, frequently, the metal and acid content. The lack of appropriate water uptake by plants promotes percolation and leaching of the metals and acid from the spoils into groundwater systems, as well as into surface creeks and lakes. Seepage of leachate from beneath tailings piles accounts for some of the most severe acid mine drainage contamination of lakes, rivers, and streams. The most effective means of preventing these various problems is to stabilize piles of tailings and/or overburden material physically and do whatever is necessary to enhance the process of revegetation.

A key factor in the reclamation of mine spoils is organic matter accumulation. As with most newly deposited soils, the minespoils are often largely devoid of organic matter (Schafer et al., 1980; Stroo and Jencks, 1982). Many attempts have been made to add organic matter to the surface layers of the soil by the amendment with topsoil from other areas, or often by the addition of some other waste material such as sewage sludge (Sopper and Kerr, 1980) or dredge material (Harrison and Van Luik, 1979). Such amendments can be extremely important to the successful reclamation of both tailings and overburden (Down and Stocks, 1977; Stroo and Jencks, 1982).

One successful microbiological means of encouraging plant growth on minespoils involves the use of mycorrhizal fungal associations with young plants; a topic that is covered in detail elsewhere in this volume.

B. Acid Mine Drainage: Generation and Acidification Reactions

Many deposits of sulfide-bearing minerals were formed abiotically, and these generally comprise the ores that are mined for their metal content. Some nonferrous and many ferrous sulfide deposits, however, are biogenic in origin. Those associated with coal were formed by bacterial reduction of sulfate in the reducing environment that accompanied decomposition and initiation of coal formation. Similar reduction reactions may be important in recovery of aquatic communities from the effects of AMD and they will be discussed later in that context. Most coal deposits contain appreciable quantities of such minerals, predominantly pyrite (FeS_2), as noted in Table 2.

When exposed to an aerobic environment, sulfide minerals are oxidized by a series of chemical and biological reactions to form a highly acidic effluent. The reactions associated with the generation of the pollution are complex, and the exact mechanisms are not completely understood. The overall reactions are evident, however, and the oxidation of sulfide proceeds as follows (Stumm and Morgan, 1970; Ehrlich, 1981):

Table 2 Average Sulfur Content of American Coal

Type	Percent of total coal reserves	Percent with sulfur content >1%
Anthracite	0.9	2.9
Bituminous	46.0	70.2
Subbutuminous	24.7	0.4
Lignite	28.4	9.3
Total	100	35.0

Source: P. Averitt, U.S. Geological Survey, Bulletin 1136, January 1960.

Table 3 Consituents in Acid Mine Waters from Several Geographical Locations and Mineral Sources

	Balaklala CA metal	Weil CA metal	Mammoth CA metal	Argo CO metal	Contrary VA sulfur
pH	2.75	1.61	2.83	2.60	2.85
Sulfate	7,450	120,000	10,400	23,700	—
Fe^{2+}	68	13,000	32	100	35
Fe^{3+}	1,840	35,000	2,970	3,090	2,010
Cu	255	4,890	330	109	46
Al	626	5,000	447	663	1,073
Cd	—	15.0	2.7	1.0	—
Zn	274	4,800	449	675	172
Pb	—	3.4	—	—	—

Units of concentration are in μ*M*/liter.
[a]Davis et al. (1972).
[b]Wilmoth et al. (1972).
[c]Klein (1962).
[d]Wakao et al. (1978).

$$2FeS_2 + 2H_2O + 7O_2 \rightarrow 2Fe^{2+} + 4SO_4^{=} + 4H^+ \qquad (1)$$

Although Reaction (1) occurs abiotically, catalysis of the oxidation by bacteria of the genus *Thiobacillus* can accelerate the reaction rate by several orders of magnitude (Stumm and Morgan, 1970).

A second reaction associated with the AMD formation process is the oxidation of ferrous iron to the ferric form (Stumm and Morgan, 1970; Ehrlich, 1981):

$$2Fe^{2+} + 1/2O_2 + 2H^+ \rightarrow 2Fe^{3+} + H_2O \qquad (2)$$

The oxidation of iron in AMD is accelerated by the chemoautotrophic bacterium *Thiobacillus ferrooxidans*.

An important abiological reaction is the subsequent oxidation of the sulfur moiety in the pyrite by the biogenic ferric iron (Stumm and Morgan, 1970; Ehrlich, 1981):

Lee VA coal	Davis USA coal	Wilmoth PA coal	Klein UK coal	Wakao et al. Japan sulfur
3.15	3.00	3.40	3.20	1.70
37,000	—	8,000	—	44,500
99	8,000	1,040	2,305	14,900
1,780	4,000	250	4,000	500
—	—	—	—	—
—	—	—	—	6,555
—	—	—	—	—
—	—	—	—	—
—	—	—	—	—

$$8Fe^{3+} + S^{2-} + 4H_2O \rightarrow 8Fe^{2+} + SO_4^{2-} + 8H^+ \quad (3)$$

The resultant ferrous iron is, of course, avilable for bacterial reoxidation. Ferric iron may also be precipitated as $Fe(OH)_3$ which results in the generation of three additional moles of acidity (from the hydrolysis reaction) for each mole of iron precipitated. The oxidized iron compounds are responsible for the reddish-brown color of AMD.

The resulting acidity of the effluent increases the solubility of most of the metals contained in the ores along with the pyrite. Thus, AMD is characterized by high acidity, high SO_4^{2-} concentrations, and high concentrations of iron and other metals. The ionic content of acid mine drainage from a variety of mineral sources and ore deposits is listed in Table 3.

C. Treatment of Acidified Soils and Waters

AMD is frequently released from both active and abandoned underground mines. In general, treatment of this source of the pollutant consists of filling the shaft with the tailings or of simply sealing the portal. The former treatment may slow or stop the flow of AMD by preventing oxygen from reaching the site of the oxidation. The latter process, however, frequently stops only the water flow, and the AMD often reappears through another portal or seeps through a fissure in the rock formation. On occasion, the buildup of water in the shafts will even cause a breakout in another portion of the hillslope.

The most severe AMD problems result from water flow through tailings piles. Oxygenated water percolates through the pile and emerges from the bottom as a highly concentrated effluent. In some cases, the leaching is encouraged as a means of recovering additional metals from low-grade ores or tailings (Madsen and Wadsworth, 1981; Brierly, 1982). Preventing this process to retard pollution is difficult, and is most often accomplished by successful revegetation of the spoil. Revegetation, in turn, is usually a difficult task due to the high concentrations of toxic metals in the spoil, and due to the low pH of the spoil as a result of iron and sulfur oxidation.

The treatment of AMD is a subject of much research, and the most successful methods are those that effectively prevent effluent formation. Polluted streams or lakes are sometimes amended with large amounts of lime and/or limestone to neutralize the acidity and induce metals precipitation. In studies of lakes acidified by acid precipitation, long-term liming elevated the productivity of the lake and allowed the establishment of a stocked population of brown trout (*Salmo trutta*), but many species present prior to acidification could not restablish upon neutralization (Hultberg and Andersson, 1982). Bacterial communities or activities were not mentioned in the report.

It has been suggested that bacterial reduction of the sulfate in the AMD may be used as an active means of reducing the impact of the AMD (McCoy and Dugan, 1968; Tuttle et al., 1968,1969b). Although this process is thought to function in such a capacity in some waters receiving acid precipitation (Schindler et al., 1980; Schindler et al., 1982) and AMD (as described later), no active exploitation of the scheme has been attempted.

Some authors have speculated that soil may provide a means of cleansing AMD and AMD-contaminated water. Natural flooding or supplemental irrigation of soils with water that contains acid mine drainage is known to generate problems with plant growth due to the enhancement of available Al^{3+} (Blevins et al., 1970; Terkeltoub, 1971). In spite of this, soils have been evaluated as to their suitability for use in the reclamation of acid mine waters. Beers et al. (1974) reported a decrease in the activity of AMD by percolation through soil column, but Ciolkoz et al. (1979) pointed out that treated surface soil horizon's pH values dropped from between 6.1 and 6.5 to 4.0 or lower. The percent base saturation of those horizons was also lowered in a consistent pattern from between 50% and 60% to less than 10%. Thus, although water quality might be enhanced somewhat by treatment, the suitability of the soil for any agricultural purposes was destroyed.

III. EFFECTS OF ACID MINE WASTES ON MICROBES AND MICROBIAL PROCESSES

The general effect of AMD on various environments is well documented (Carpenter, 1924; Carpenter, 1925; Carpenter, 1926; Campbell et al., 1965 a,b; Harp and Campbell, 1967; Campbell and Lind, 1969; Brezina, 1970; Smith and Frey, 1971; Tyler and Buckney, 1971, 1973; Wollitz, 1972; Minshall and Andrews, 1973; Mount, 1973; Thorpe and Lake, 1973; Ayers, 1974; Down and Stocks, 1977; Parsons, 1977; Goto et al., 1978; Chadderton and Shane, 1978; Letterman and Mitsch, 1978; Koryak et al., 1979; Yan, 1979; Norris et al., 1981 a,b; Olem, 1981). Most reports deal with the physicochemical changes in the environment (e.g., Koryak et al. 1979; Reed, 1980), or the overall effect on life forms such as plankton, plants, benthic invertebrates, or fish.

To date, most of the published microbiological work dealing with acid mine drainage has focused on the formation of the effluent and the distribution of thiobacilli and other sulfur- and iron-oxidizing organisms. Much research has dealt with general questions of the effect of pH, of metal ions, and other potential toxicants on the organisms. Comparatively little literature is available that directly addresses the microbial response to acid mine wastes and the succession of events that occurs after exposure of the community to such pollutants. Often, the available literature consists of lists of organisms or types or organisms that were recovered from the polluted water or soil by various culture techniques (Weaver and Nash, 1968; Guthrie et al., 1978). In general, it is well recognized that many heterotrophic microbes and their attendant functions are inhibited at pH values below 5 or above 9 (Alexander, 1977). Furthermore, other components of AMD, namely, metals such as Cu, Cd, Pb, and Zn, are known to be highly inhibitory to the function of a variety of microorganisms in both soil and water (Bhuiya and Cornfield, 1974; Mikkelsen, 1974; Babich and Stotzky, 1977; Jensen, 1977; Lawrey, 1977a,b; Mills and Colwell, 1977; Bitton and Freihofer, 1978; Gadd and Griffiths, 1978; Bollag and Barabasz, 1979; Sunda and Gillespie, 1979; Zevenhuisen et al., 1979; Sterritt and Lester, 1980).

In a comprehensive study of the combined effects of iron (the dominant metal in acid mine waters), H^+, and $SO_4^=$ on bacterial growth, acidity was determined to be the most significant factor, with some concern for the amount of $SO_4^=$ present (McCoy and Dugan, 1968). Using test organisms (strains of *Pseudomonas*, *Bacillus*, and *Flavobacterium*) isolated from an unpolluted stream, it was concluded that the bacteria should grow if the pH were above 5.3, the iron concentration between 1 mg/liter and 100 mg/liter, and the $SO_4^=$ concentration between 5 mg/liter and 500 mg/liter. The authors indicated that organism growth increased with increasing iron concentrations up to 100 mg/liter at pH values greater than 5. They inferred that iron would not inhibit heterotrophic growth and activity unless the concentrations were greatly in excess of 100 mg/liter. In acid mine water, the iron concentration may range as high as 10,000 mg/liter (see Tables 1 and 3), so that the parameters applied by McCoy and Dugan would not be appropriate until the pH had increased to above 5 and the iron level had dropped to around 100 mg/liter. In that state, the system might still be considered polluted, but certainly not as severely as the majority of AMD-contaminated waters.

Although heterotrophic bacteria exist that are capable of growth in acid mine streams (Wichlacz and Unz, 1981), the bacteria appear to be specialized organisms able to function in these extreme environments. Almost nothing is known of their activity, of their role in the ecosystem, or of their contribution, if any, to the rehabilitation of the polluted water. Some heterotrophs are capable of recovering and regrowing after exposure to concentrated AMD (Hackney and Bissonnette, 1978). In cases where heterotrophs are cultured from AMD, they are frequently considered to be transients that have entered the system from other aquatic or terrestrial sources (Wassel and Mills, 1983; Tuttle et al., 1968). When heterotrophs are recovered from contaminated waters, their abundance and diversity are observed to decrease with decreasing pH. Guthrie et al. (1978) observed a 44% reduction in total culturable bacteria as the pH was lowered from 6.8 to 5.4 by fly ash contamination of a stream, and a concomitant reduction in diversity (number of colony types) of 30%. Furthermore, a shift in the prevalent organisms inhabiting the system was observed. *Bacillus, Sarcina, Achromobacter, Flavobacterium,* and *Pseudomonas* dominate at the higher pH, and *Pseudomonas, Flavobacterium, Chromobacterium, Bacillus,* and *Brevibacterium,* dominate at the lower pH. A noticeable increase in the number of pigmented forms accompanied the reduction in pH. Similar data were obtained in an impoundment by Wassel and Mills (1983). The planktonic bacterial community was smaller and less diverse in the mouth of an acid mine stream (pH 3.2) than at the mouth of a similar but uncontaminated stream (pH 6.2).

Few measures of microbial activities in acid mine waters are available, but Carpenter et al. (1984) observed a reduction in the exponential decay rate of leaf material from several plant species under acidic conditions as compared with nonacidic controls. The sites used were the same as those of Wassel and Mills (1983). Furthermore, glucose turnover in the acidic waters was over nine times longer than at a station where the acid mine drainage was more dilute, and was about 100 times longer than at the unpolluted station. These observations are quite similar to those obtained in ponds and streams that have undergone natural and artificial acidification.

IV. RECOVERY OF MICROBIAL COMMUNITIES FROM MINE WASTE EFFECTS

A. Definition of Terms

A variety of terms are used to describe the procedure used in the attempt to return polluted areas to some semblance of their preimpaction status. Two terms that are especially applicable for the coal and metal mining industries are restoration and reclamation. Down and Stocks (1977) separate the two terms as follows: restoration is the recreation of the original topography and reestablishment of the original land or water use, whereas reclamation is any treatment that is not restoration. In most cases, the activities discussed herein constitute reclamation.

Recovery, as used in the present context, refers to the return of an ecosystem to a reasonable level of preimpaction function. A recovered location may be quite different from the original state with regard to the inhabitant species, but for full recovery, all of the important ecosystem functions must be present. Those functions are shared largely by plants, which provide an input of energy to the system through photosynthesis, and microorganisms, which carry out the processes of decomposition and nutrient regeneration. Any ecosystem may be considered to be recovered (in the purely ecological sense) when all of the important functions of that system are present and operating with little or not inhibition. A general model for the recovery of any perturbed ecosystem is presented in Fig. 3. The horizontal axis may represent units of time if the environment is static, such as in a soil system, or it may represent distance from the pollution source in a dynamic environment if the inputs are chronic. It is this general pattern that, when observed in a perturbed environment, indicates that the ecosystem is recovering from the effects of the pollution.

B. Recovery of Microbial Communities in Minesoils and Tailings

When the solid mining residues are deposited, they are somewhat similar to any other freshly deposited parent material. The microbiological processes that are to occur are closely related to the general processes of soil formation. As described above, freshly deposited material is generally nutrient poor, and lacks organic matter. A major portion of the reclamation of tailings piles and overburden involves revegetation. Over a period of time, the plant cover not only physically stabilizes the surface of the piles, but also serves as a source of organic matter to the upper horizons of the young soil. The establishment of vegetation on the wastes is not always a simple matter, however, since the low acidity and high metal content of the spoils presents a

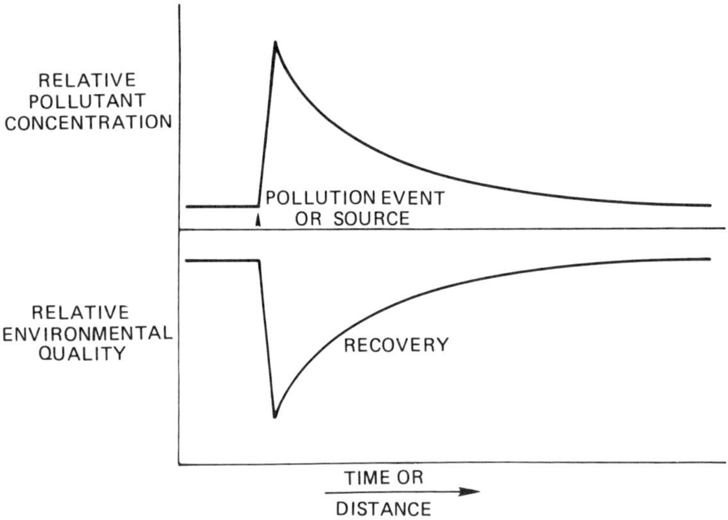

Figure 3 General model of the abatement of pollution and the recovery of an ecosystem along a gradient of either time or space.

situation of high toxicity to many of the plants that might be used as a cover.

The reclamation of mine tailings and overburden is intimately associated with the recovery of a stable microbial community resembling that of a normal soil. There is a mutualistic relationship, such that the community recovers as the soil profile develops, but the reclamation of the spoil material to allow a completely reclaimed site is also dependent on the recovery of that microbial community. The reestablishment of stability and productivity in minesoils relies on such microbial activities as humification of organic matter, aggregation of soil particles, and appropriate nutrient cycling (Cundell, 1977; Lawrey, 1977a,b; Hersman and Temple, 1979; Stroo and Jencks, 1982). The importance of these processes is such that the condition of the microbial community may be considered to be an index of the progress of soil genesis in mine spoils (Stroo and Jencks, 1982).

When removed from the mine and placed in a pile, the tailings or overburden material can be considered to be essentially sterile. Tailings that have passed through a milling process may contain small numbers of microorganisms; however, due to the lack of organic matter and other nutrients in the waste material, that small inoculum cannot be considered to provide even the beginnings of a microbial community.

The initial processes occurring in spoil and overburden may be quite different depending on the source of the material. In highly sulfurous wastes, the processes are often those that are associated with the formation of AMD. However, those processes are also important in terms of the removal of potentially toxic materials, and in removal of the acidity and acid-forming potential from the material itself. Sulfur-oxidizing bacteria such as *Thiobacillus* sp. are the important early colonizers of the material. Although they are responsible for the formation of acid mine drainage, AMD formation also represents the removal of sulfides and heavy metals from the solid material causing a reduction in the toxicity of the developing spoil. The first reactions that occur cause a great decrease in the pH of the spoil material which renders the material even less habitable for plants and microorganisms. Once conversion of the reduced sulfur compounds to H_2SO_4 has occurred, the acid leaches from the pile and the remaining material approaches a regime of neutral pH and reduced concentrations of heavy metals, making the site more habitable for other microbes and vegetation.

The recovery of communities in minesoils may occur naturally over a period of time, although that period may be long, often 150 years or more. Although the initial pH of the spoil may be high (pH 6.5–8.5), oxidation of residual pyrite can cause a general increase in acidity, reducing pHs to as low as 2 within 5 years (Down and Stocks, 1977). Eventually, leaching of the acid from the minesoil increases the pH to a level where many plants and most microbial activities are only minimally affected (Fig. 4). The magnitude, rate, and duration of the acidification varies, depending on the relative sulfide concentration of the waste.

In many cases, acidification is extremely localized. Olson et al. (1981) reported the presence of acidic zones associated with chunks of coal; included in graded overburden in alkaline spoils near Decker, Montana. The affected areas were often small, some less than 10 cm in diameter, but showed a distinct zonation from highly acidic (pH <3) through an intermediate area (pH ca. 5.5) which was rust colored from ferric iron precipitation, to the unaltered original alkaline material (pH 7.8–8.0). These small but numerous zones of sulfur and iron oxidation were considered to be the source of large numbers of *Thiobacillus* sp., both acidophilic and otherwise, found in local streams and wells. The general observation that acidification may occur in localized areas in spoils undergoing reclamation is corroborated by observations of acidification of neutral coal and overburden when inoculated with cells of *Thiobacillus* (Kleinmann and Crerar, 1979). Olson et al. (1981) considered that moisture limited sulfur oxidation in the region more than any other factor measured. They also suggested that the bacterially induced acidification of mine soils is responsible for the high levels of iron, manganese, and zinc found in many drinking water supplies of the area.

In the absence of intense acidification, recovery of the microbial community in a minesoil proceeds as one would expect the community to develop in any newly deposited parent material (Fig. 5) (Shafer

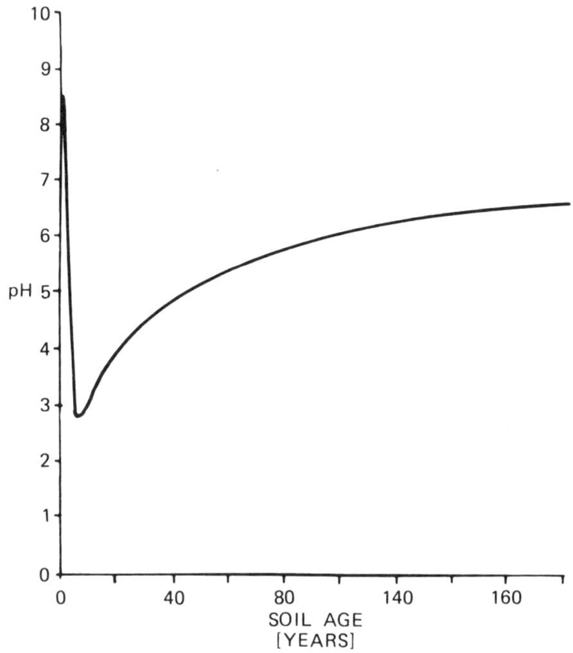

Figure 4 Changes in pH occurring with time in spoil piles containing pyrite of sulfur waste. (From Down and Stocks, 1980.)

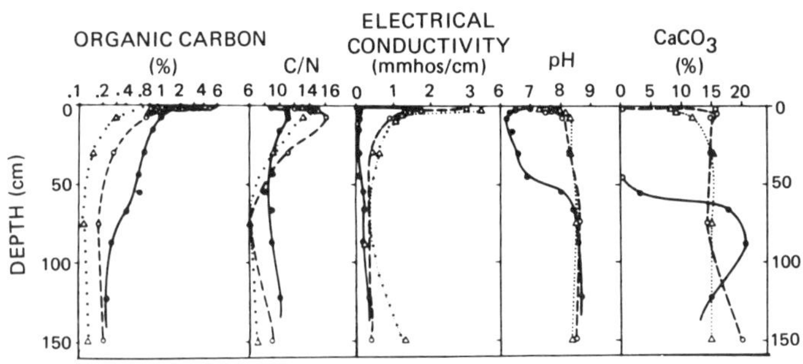

Figure 5 Distribution of organic carbon, C/N ratio, soluble salts, pH, and carbonate in a natural soil and old and new minesoils. Natural soil —●—●—●— old minesoil —○—○—○— new minesoil —△—△—△—. (From Schafer et al., 1980.)

et al., 1980). Few studies have examined the progression of events in the establishment of microbial communities in minespoils, but one point agreed upon is that the rate and extent of development is directly related to nitrogen and organic matter accumulation (Wilson and Hedrick, 1957; Wilson, 1965; Cundell, 1977; Dancer et al., 1977; Hersman, 1977; Reeder and Berg, 1977; Hersman and Temple, 1979; Stroo and Jencks, 1982).

In a study that compared the mineralization of nitrogen in several mine spoils as compared with a nearby undisturbed soil, Reeder and Berg (1977) demonstrated that the rates in unamended materials were far less than in either a vegetated spoil or the undisturbed soil of the area (Table 4). The authors attributed the phenomenon to possible immobilization by heterotrophic microorganisms, because the C:N ratio of the fresh spoil was 35 (near the generally accepted critical value for immobilization of N by soil microbes [Alexander, 1977]) as compared with 22 and 11 for the vegetated spoil and the soil, respectively. However, an additional factor not mentioned was the potential difference of both the carbon and nitrogen availability. The vegetated spoil and soil samples would be expected to contain a higher portion of readily available nitrogen than the freshly deposited spoil, and that fact, combined with a lower abundance of microbes (not evaluated by Reeder and Berg), could also account for the observed differences. Nitrification of added NH_4^+ was also inhibited in the fresh material as opposed to that under a vegetated treatment, which in turn, was slightly reduced as compared with the undisturbed soil. Reeder and Berg's (1977) study generally supports earlier work demonstrating that developed soils mineralize and oxidize nitrogen more rapidly than freshly deposited geologic materials (Hall and Miller, 1908; Wilson and Stewart, 1955). Because they found appreciable CO_2 evolution in all of their treatments (NH_4^+ amendments had no effect, but soil > vegetated spoil >> unvegetated spoil), they concluded that some factor was

Table 4 Rates of N Mineralization in Vegetated and Unvegetated Mine Waste

Material	Incubation period (weeks)				
	0–1.71	0–3	0–6	0–12	0–24
Soil	12.07	10.98	14.69	15.60	16.04
Vegetated spoil	15.27	12.69	14.39	12.11	10.26
Fresh spoil	0.34	0.92	1.25	1.76	1.35

Units are ppm NO_3^--N/weeks$^{1/2}$.
Source: Reeder and Berg (1977).

actually inhibiting the nitrification in the unvegetated spoil. Nevertheless, all of the aforementioned studies point to the fact that the presence of vegetation on a developing mine soil enhances nitrogen mineralization, nitrification, and carbon mineralization.

Although studies of microbial communities undergoing successional recovery from the effects of acid pollution have not been made, implications may be made from data dealing with moderate acidification such as in studies of acid precipitation. Use of deductive reasoning assumes that communities recovering from acidic perturbations undergo an exact reversal of the sequence of events observed during the acidification process, i.e., that there is no hysteresis. That assumption is almost certainly not valid for quantitative evaluations, but changes that are observed to occur early in the acidification process might be expected to be some of the last to return to normal under a regime of neutralization.

Alexander (1980) has reviewed the effects of acidity and acidification on soil microbes and their associated processes with emphasis on acid precipitation. Much of the attention to processes in soils has, like the studies mentioned above, centered on nitrogen transformations. In addition to the effects on ammonification and nitrification, low pH conditions may alter pathways of denitrification, the conversion of NO_3^- to volatile products N_2O and N_2. For example, at pH values below 6, the final reduction of N_2O to N_2 may be inhibited so that N_2O would be emitted to the atmosphere (Wijler and Delwiche, 1954; Bremner and Shaw, 1958; Bollag et al., 1973).

Nitrogen fixation by both non-symbionts (e.g., *Azotobacter, Clostridium*) and by the *Rhizobium*-legume symbiosis is known to be inhibited at low pH (Shriner, 1977; Evans et al., 1980, Francis, 1982). In soil, the reduction may be a direct effect of acidity, or it may be due to increased availability of iron and aluminum.

Non-specific microbial activity may likewise be affected by moderate acidification. Respiration (CO_2 production) and numbers of active bacteria (fluorescein diacetate-active) were reduced in soils receiving artificial acid rain of pH 2 (Baath et al., 1979). Numbers of FDA-active fungi, total fungi, and total bacteria were not significantly different from control treatments.

Extrapolating this information to the more extreme acidities and metal loadings associated with mining wastes, a pattern of recovery might be expected to proceed from almost no activity in any parameter associated with heterotrophic organisms through gradually increasing levels as the soil becomes neutralized, to a state that is virtually indistinguishable from that in an unperturbed site. The rapidity of the recovery will depend largely on the severity of the acidification, the amount of time necessary for the acidic products to leach from the soil, and the amount of organic matter reaching the surface horizons of the profile.

C. Recovery of Aquatic Microbial Communities

1. General Considerations

The effects of acid mine drainage (AMD) on aquatic communities are well described in terms of phytoplankton, benthic and planktonic invertebrates, macrophytes, and fishes (Campbell et al., 1965a,b; Harp and Campbell, 1967; Parsons, 1977; Yan, 1979). Investigations have been made on both acidified streams (Bieseker, 1966; Weaver and Nash, 1968; Nichols and Bulow, 1973; Letterman and Mitsch, 1978) and lakes (Campbell et al., 1965a,b). Other situations have received less attention.

Only a few studies have been attempted on the effects of AMD on heterotrophic bacteria, and those generally dealt with the responses of pure cultures or specific groups of bacteria exposed to full-strength AMD (Joseph and Shay, 1952; Rogers and Wilson, 1966; Karlovic and Wilson, 1969 Hackney and Bissonnette, 1978; Cook and Wilson, 1971). The study of Tuttle et al. (1968) examined changes in the aerobic heterotrophic bacteria in a stream upon confluence with another stream that was exclusively AMD. The data presented in that study were based primarily on enumerations of organisms from various sites above and below the confluence point. The results were descriptive in nature, although informative.

The small amount of information available on the recovery of aquatic microbial communities from the effects of AMD and similar pollutants suggests that a successional pattern of increased heterotrophic biomass is observed along with a concomitant increase in the total activity of that community. Heterotrophs are known to be able to survive immersion in AMD. In early work by Carpenter and Herndon (1933), it was reported that wastewater was sterilized by exposure to AMD, but all later work has demonstrated that viable heterotrophs can be recovered from mixtures of wastewater and AMD (Joseph and Shay, 1952; Hackney and Bissonnette, 1978). Although many of the populations survive the exposure, most are injured or debilitated by the contact, such that attempts at isolation on selective media are often unsuccessful (Hackney and Bissonnette, 1978). Use of a more favorable medium (Roth and Keenan, 1971; Hackney and Bissonnette, 1978) allows the organisms to recover from the effects of the AMD and to regain their former activity.

Natural situations in which heterotrophs encounter AMD undoubtedly work in the same way. Assuming the contact with the AMD is sufficiently brief, pollutant dilution by additional waters or toxicant removal by other mechanism, allows the populations to resume their normal role gradually in that environment. Hackney and Bissonnette (1978) recovered significant numbers of a variety of indicator bacteria from membrane diffusion chambers submerged in an acid mine stream for up to 24-hr. Survival of cells at levels blow the limit of detectability of that study could provide a sufficient inoculum for recovery of the population upon lessening of the pollution stress. Such survival and gradual recovery is consistent with the suggestion that the low levels of heterotrophic activity observed in slightly diluted acid mine streams is due to the presence of transient populations that may be injured but survive the exposure (Tuttle et al., 1969a; Wassel and Mills, 1983).

Because of the low pH of waters contaminated with AMD, it has been suggested that fungi may play an exceptionally important role. In strip mine soils, filamentous fungi are prolific and diverse (Wilson, 1965). Uptake of organic pollutants from AMD-contaminated waters by fungi may result in the removal of iron by the production of organically stabilized colloids, but the observed reduction in acidity in such streams cannot be attributed to those microbes. (Weaver and Nash, 1968). Representatives of the yeasts *Candida, Rhodotorula,* and *Trichosporon* have often been isolated from streams carrying acid mine wastes (Rogers and Wilson, 1966; Weaver and Nash, 1968).

Given a source of inoculum, namely injured or transient heterotrophs, it is to be expected that the general homeostatic model of Fig. 3 will apply for microbial communities, either in time, as in small lakes with no additional input of AMD, or with distance from the pollution source, as in streams or impoundments with a reasonable current. The work of Tuttle et al. (1969a) demonstrated that observation for streams using cultural counts as the quantitative method. Comprehensive study of aquatic microbial community recovery from the effects of acid mine drainage was completed at the Lake Anna-Contrary Creek system of Virginia. Because of the direct relevance to the main thrust of this chapter, the study will be reviewed here in detail.

2. *Lake Anna Case Study*

The area studied is an arm of Lake Anna in east-central Virginia (Fig. 6). Contrary Creek, which flows into the arm, was the site of extensive pyrite mining for 70 years prior to 1920. Although the mining operations have ceased, the flow of AMD has continued to the present. The mean discharge for Contrary Creek for the 1979 water year was 7.02 $ft^3\ s^{-1}$, with a maximum flow rate of 259 $ft^3\ s^{-1}$ and a minimum of 0.45 $ft^3\ s^{-1}$ (U.S.G.S./Va. State Water Control Board, 1979).

Another stream, Freshwater Creek, also flows into the arm, but contains no AMD. This stream and the area into which it flows served as control sites for the study, the object of which was to determine the extent and types of alteration of the microbial communities of the lake due to inputs of acid mine drainage from Contrary Creek. No discharge data were available for Freshwater Creek, but the watershed areas drained by the two streams provide a crude basis for comparing the relative contribution of water from the two streams to the lake. The area drained by the unpolluted stream comprises 2288 ha

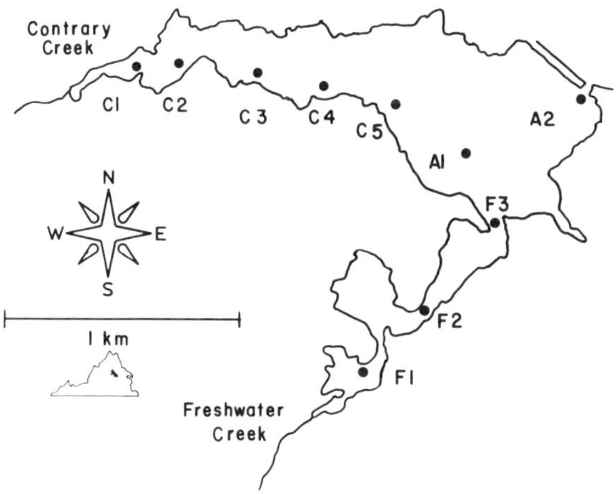

Figure 6 Contrary Creek and the study sites in Lake Anna.

as opposed to 1820 ha for the Contrary Creek drainage basin. These values represent 44% and 35% of the total watershed areas which empty out through the causeway at station A2 (see Fig. 6) for Freshwater and Contrary Creek, respectively.

A series of stations was established in the Contrary Creek and Freshwater Creek areas of the arm and monthly measurements were made of parameters such as temperature, pH, dissolved O_2, conductivity, and microbial biomass and activity in water and sediment. Included in the measurements of biomass were acridine orange counts (AODC), plate counts, and most-probable-number (MPN) determinations of *Thiobacillus ferrooxidans* and sulfate reducing bacteria. In addition, a new method was developed combining fluorescent antibody (FA) counts with the INT (2-(*p*-iodophenyl)-3-(*p*-nitrophenyl)-5-phenyltetrazolium chloride) reduction assay to provide what are termed FAINT determinations of respiring cells of *T. ferrooxidans* (Baker and Mills, 1982).

The major physical parameters associated with some of the stations are presented in Table 5. Data are reported as the annual averages for the 3-year period from June 1978 to May 1981. An obvious gradient of pH and conductivity exists in the Contrary Creek arm with the pH increasing and the conductivity decreasing with increasing distance from the mouth of the stream. In the Freshwater Creek area, the trend is reversed, although the gradient there is not as steep as in the Contrary Creek arm. One noteworthy observation is the magnitude of the standard deviation for each parameter at each station. In this case, the standard deviation is presented to describe the fluctuation in the values obtained over the year. That is, the environ-

Table 5 Values of pH, Conductivity, and Temperature for Selected Stations in Contrary Creek and Freshwater Creek

Station		C1	C4T[a]	C4B[a]	A2T	A2B	F1
depth (m)		0.5	3.0		7.0		1.5
			pH				
78–79	$\bar{x}$	3.5	5.0	4.1	5.9	5.3	6.4
	SD	.29	1.1	.29	.99	1.0	.60
79–80	$\bar{x}$	3.4	5.1	5.4	5.9	5.5	6.3
	SD	.43	.74	1.0	.97	.78	.41
80–81	$\bar{x}$	3.3	5.5	5.2	6.5[b]	6.1	6.3[b]
	SD	.26	.93	1.9	1.0	.83	.78
			Conductivity				
78–79	$\bar{x}$	337.4	117.8	124.0	74.0	83.7	55.4
	SD	87.9	19.9	45.8	10.9	36.2	20.8
79–80	$\bar{x}$	259.3	116.7	77.6	59.4	71.5	52.2
	SD	100.8	130.6	22.8	10.9	22.4	23.9
80–81	$\bar{x}$	252.2	98.5	90.5	81.3	91.0	78.9
	SD	76.3	28.9	25.3	9.0	24.2	28.9
			Temperature				
78–79	$\bar{x}$	17.7	17.8	17.4	18.2	17.6	18.3
	SD	9.9	8.2	8.9	9.6	8.3	9.8
79–80	$\bar{x}$	17.4	19.3	18.9	19.6	17.6	19.3
	SD	10.3	9.2	8.9	8.2	7.9	9.1
80–81	$\bar{x}$	15.0	17.8	16.9	18.5	17.4	15.7
	SD	8.4	8.4	9.0	8.2	8.0	9.2

[a]T, B represent surface samples (0–5 m) and bottom (0.5 m from sediment surface), respectively.
[b]During this year, a pH value of 8.6 was reached in March. That was the highest value ever obtained, and deletion of that reading would lower the average pH for 1980–1981 to 6.3. Similarly, deletion of an extremely low reading of 4.3 at F1 in May 1981 would raise the average pH at that station to 6.5.

ment at the mouth of the acid mine stream was the most extreme with respect to pH and conductivity, but the conditions there were fairly constant over the experimental period, whereas at stations downstream from the pollution source, the environment was more variable.

A pattern of metal concentration (Pb, Cu, Zn, Cd, Fe) consistent with that of pH and conductivity reported above was demonstrated by Nordstrom and Dagenhart (1979) for the Contrary Creek arm. They found high concentrations of all of the metals named at station C1. The concentrations decreased in a downstream direction such that at station A2, the levels of the metals in the surface water were rarely above levels observed in other areas of the lake.

Abundance and diversity of microbes: The densities of bacterial communities were determined using both plate count techniques and AODC. In the 12-month period from June 1978 to May 1979, the sizes of the communities at C1, A2, and F1 as measured by AODC were not significantly different (Fig. 7). Results using the plate count method reveal significant differences among the stations ($p = 0.05$), with F1 > A2 > C1 (Wassel and Mills, 1983). This observation was repeated for the same time period in 1979–1980 and 1980–1981. The usual criticism of cultural count methods notwithstanding, the lack of significant differences among the AODC results is likely due to an abuncance of chemoautotrophic bacteria such as *Thiobacillus* sp. in the acid impacted areas (Baker and Mills, 1982).

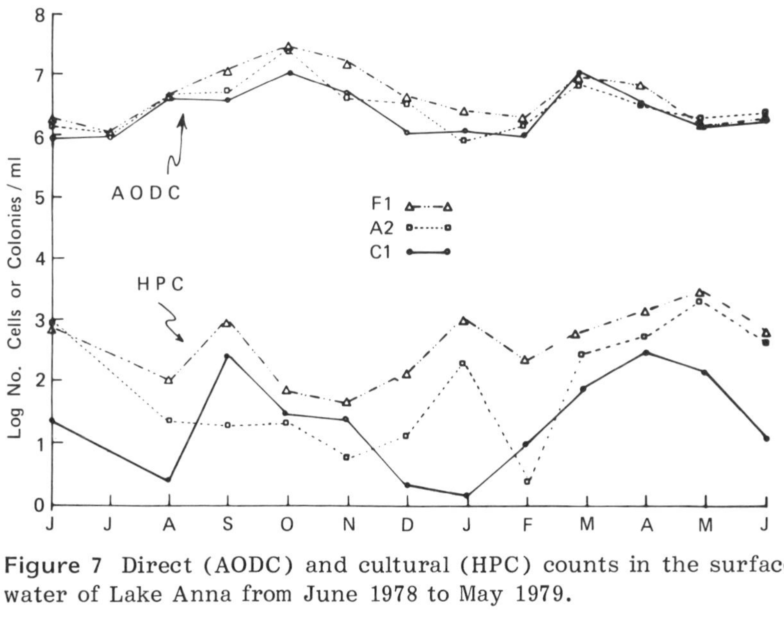

Figure 7 Direct (AODC) and cultural (HPC) counts in the surface water of Lake Anna from June 1978 to May 1979.

If all the stations included in the AODC determinations are examined, a more refined relationship is observed. Whereas statistical significance could not be demonstrated for differences in AODC values among the various stations, an interesting trend was observed (Fig. 8). Bacterial numbers in the unpolluted situation (Freshwater Creek) decreased from a maximum in the shallow productive areas near the stream mouth to a minimum in the deeper, open waters of the lake. However, the presence of acid mine drainage in the Contrary Creek arm has altered the situation to one in which the bacterial densities increase in a downstream direction as the intensity of the pollution damage becomes less and less. The shape of the curve closely resembles that of the general recovery model presented in Fig. 3. The convergence of the AODC curves for the two arms of the lake prior to the point where the waters mix is an important observation indicating essentially complete recovery of the water community within the Contrary Creek arm itself.

Using a FAINT procedure (Baker and Mills, 1982), the contribution of *T. ferrooxidans* to the AODC was assessed. *T. ferrooxidans* accounted for 10% of the AODC obtained at C1, but only 1% at station A2 (Table 6). The numbers of *T. ferrooxidans* and *T. thiooxidans* in Contrary Creek are approximately equal (Scala et al., 1982), and if it is assumed that the relative proportions of organisms hold for the entire year, the points shown at station C1 in Fig. 8 would be lower, demonstrating an even more significant recovery of the heterotrophic fraction in the Contrary Creek arm.

The heterotrophic community diversity was examined on a quarterly basis by isolating organisms from the enumeration plates, subjecting

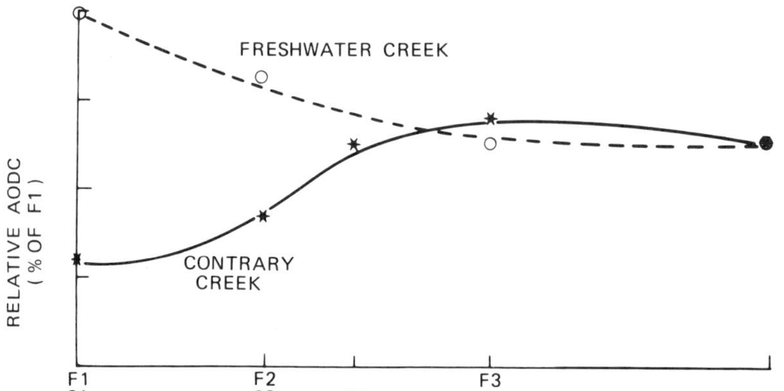

Figure 8 Annual average of the AODC for several stations in the Contrary Creek and Freshwater Creek arms of Lake Anna.

Table 6 Distribution of Respiring Bacteria in the Contrary Creek Arm

Treatment	C1[a]	C3[a]	A2[a]
AODC	370	390	920
INT	74	230	580
FA	38	18	13
FAINT	30	9.2	2.9

Values underscored by the same line are not significantly different ($p = 0.05$). All other values are significantly different.
[a] 10^3 cells/ml.

the cultures to a battery of physiological tests, clustering the isolates into groups using numerical taxonomic techniques, and using these resultant clusters to perform diversity analysis using the rarefaction method (Simberloff, 1978; Wassel and Mills, 1983; Mills and Wassel, 1980).

The results of the diversity analyses obtained by Wassel and Mills (1983) are shown in Fig. 9. Organisms from both sediment and water showed similar diversity patterns, viz., the community at F1 was more diverse than either A2 or C1, which, in turn, were not significantly different. The reduced diversity in the areas receiving the mine wastes could be explained on the basis of absence of the strains found at station F1 but not at the affected sites. That is, a community exists for the entire area (both polluted and unaffected sites) that has, as a large portion of its makeup, a series of common organisms.

At station F1 however, a large number of individuals present had no counterparts at the other stations sampled. Furthermore, on the basis of the patterns of response of the bacterial isolates to the battery of physiological tests, the strains unique to F1 (not present elsewhere, resulting in a reduced diversity at A2 and C1) were different from the general community (common strains) and from the unique strains found at the sites receiving the mine drainage. The F1 unique strains had a lower ability to withstand the metals Pb, Cu, Zn, and Cd in the culture medium. All of those metals are abundant in the Contrary Creek AMD. There was no difference, however, in the ability of the organisms from the various stations to withstand acidity as determined by culturing each isolate at pH 4.0.

In contrast, the sessile bacterial communities at sties C1 and F1 were highly dissimilar in composition (L.A. Mallory and A.L. Mills, unpublished data), and displayed a greater difference in their ability

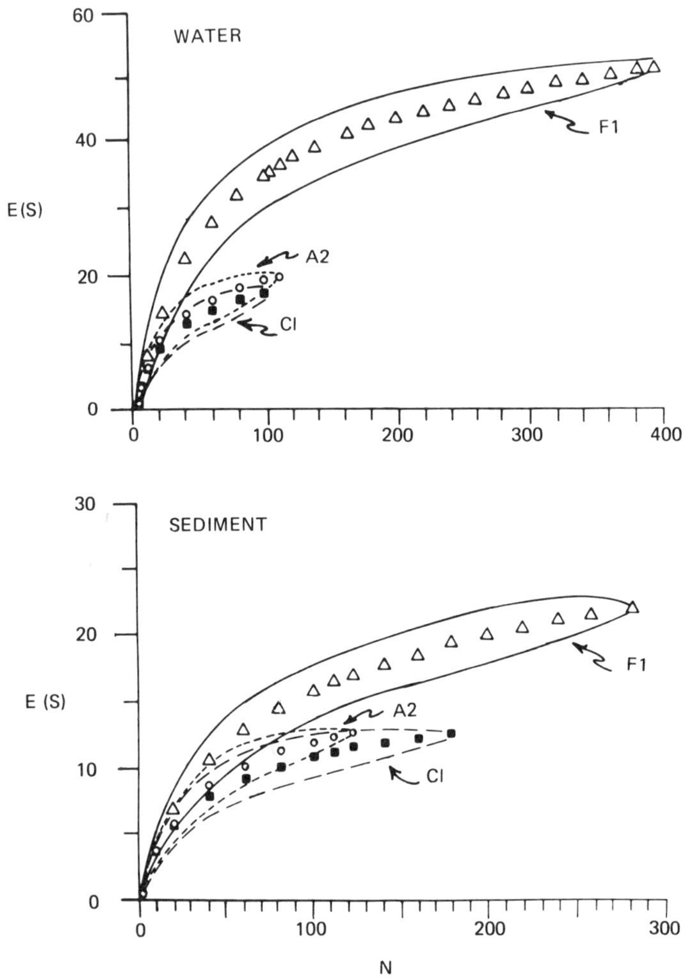

Figure 9 Rarefaction curves for the diversity of the heterotrophic bacterial communities in heavily polluted (C1) and unpolluted (F1), sites, and a site (A2) receiving water from both of the other streams. (Wassel and Mills, 1983. Springer-Verlag, reprinted by permission).

to tolerate the conditions at sites other than their native habitat. These findings imply that, in fact, the planktonic bacteria are transients, whereas the attached forms are able to grow because of an enhanced resistance to the extreme conditions in the acid mine drainage.

Activity of aerobes: Activities of aerobic heterotrophic communities in the water column of the Contrary Creek arm of the lake were examined by two methods. Carpenter et al. (1983) examined the differences in the rates of decomposition of various leaf materials contained in nylon litter bags and placed at various sites in the lake. In addition, frequent assays of the ability of the aerobic heterotrophs to assimilate and/or mineralize [^{14}C]glucose were carried out (Carpenter et al., 1983).

AMD effects on leaf litter decomposition rates were investigated at two sites in the Contrary Creek arm (C1 and C5) in comparison to a control site (F1) in the Freshwater Creek arm. Litterbags containing leaves of yellow birch (*Betula lutea*), flowering dogwood (*Cornus florida*), white oak (*Quercus alba*), and soft rush (*Juncus effusus*) were incubated for up to 149 days in the littoral zone at each site. The decomposition rate was measured by loss of ash-free dry weight of the leaf litter and was expressed as both a linear and an exponential decay rate. Interspecific differences in decomposition rates were found at all sites, with dogwood decaying the most rapidly, followed by birch, oak, and rush (Fig. 10). The leaf decomposition rates also differed significantly among the study sites. For all leaf species, the decay rate coefficients at F1 were at least twice those obtained at C1. At C5, values were generally intermediate between those obtained for the other two sites. Bacterial counts (AODC) in the decomposing material did not differ significantly among the sites at any time. The decomposition process was primarily microbially mediated, as very few direct detritus processors (e.g., shredding and scraping invertebrates) were ever found in the litter bags. Thus, although microbial activity in the Contrary Creek arm was inhibited by the presence of the AMD, some activity was seen even in the most severely affected area.

Glucose assimilation measurements made in various parts of the lake supported the observations of the litter bag incubations. A pattern of assimilation was observed (Fig. 11) that resembled that of the distribution of AODC in the two arms, and the pattern was valid for the 10-month period from spring of 1980 to late winter of 1981 (Table 7). Assimilation of [^{14}C]glucose over the sampling period was lowest at C1 and increased in a downstream direction moving away from the acidic environment. Conversely, the activity was highest at station F1 and decreased in a downstream direction in that arm. The data show that the general relationship observed in Fig. 3 could also be seen in a more integrated examination. The heterotrophic activity results are entirely consistent with those of both the direct and cultural count analyses and show a recovery of the heterotrophic community that is essentially complete before the water from the two arms mix.

Actual rates of heterogrophic activity showed a relationship consistent with all of the other results (Table 8). Values of V_{max} and turnover time (T_t) showed a depression of activity at the acid-impacted

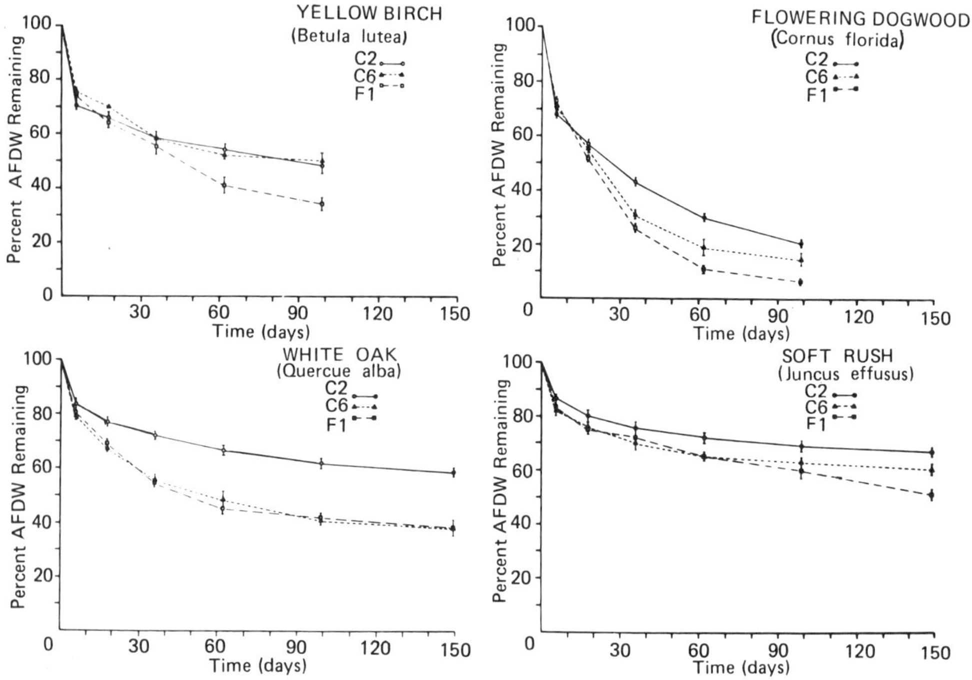

Figure 10 Decomposition of leaf litter at sites in Lake Anna receiving different amounts of the drainage from Contrary Creek. (From Carpenter et al., in press.)

area as compared with the other sites examined. The time for glucose turnover at C1 was an order of magnitude longer than at F1, indicating a much slower rate of activity and supporting the observation of a reduced rate of decomposition of organic matter there.

Anaerobic Activity: Investigation of the effect of AMD from Contrary Creek on heterotrophic activities in sediments revolved around the idea that $SO_4^=$ from the mine drainage might enhance anaerobic carbon mineralization by providing the sediment community with a readily available electron acceptor for the complete oxidation of fermentation products in the absence of oxygen. Most freshwater systems are depleted in $SO_4^=$ as compared with marine situations, such that $SO_4^=$ reduction is not a dominant reaction in the former locations. The levels of $SO_4^=$ in the contaminated water of Lake Anna (Table 9) greatly exceed the average value of 11 mg of $SO_4^=$/liter mentioned by Wetzel (1975), although only at station C1 are the levels much above the normal range given in that reference.

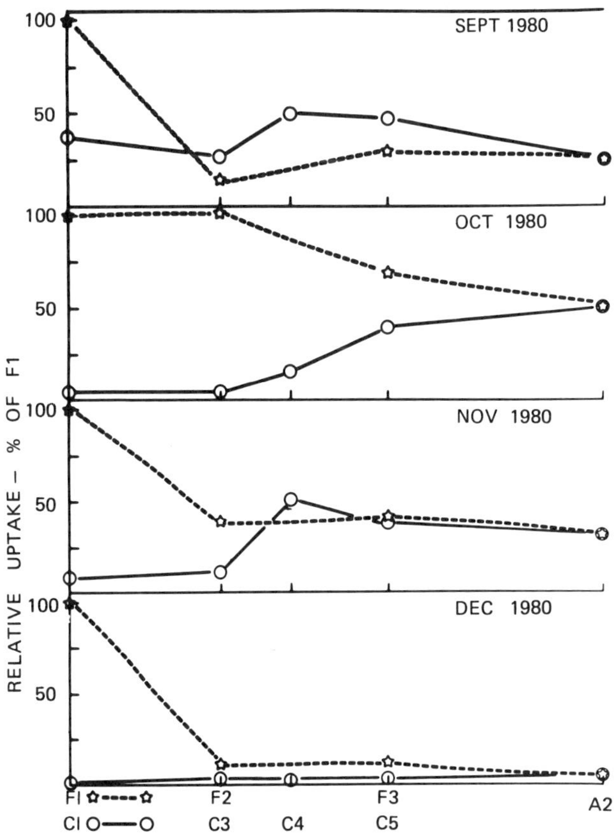

Figure 11 Pattern of [C-14]glucose assimilation by bacteria in Lake Anna during the last 4 months of 1980. Data are normalized by expressing as a percentage of the activity measured at station F1 for the sampling time.

In addition, $SO_4^=$ concentrations in sediments are always measurable, indicating an ample supply of terminal electron acceptor for anaerobic respiration in those areas receiving flow from Contrary Creek (Table 10). Depth profiles of $SO_4^=$ concentrations in areas receiving AMD resemble similar profiles published for marine situations. The importance of the gradient of $SO_4^=$ concentration in these sediments is supported by the distribution of $SO_4^=$-reducing bacteria (Table 11), although the data showed a stronger correlation between the number of organisms and the amount of organic matter present than between the organisms and the amount of $SO_4^=$ present.

Table 7 Comparative Levels of [^{14}C]Glucose Assimilation in the Contrary Creek Area of Lake Anna

Station	Heterotrophic activity: Percent of F1	Heterotrophic activity: SD
C1	8.3	11.9
C3	15.0	14.1
C4	29.0	19.2
C5	31.0	22.0
A2	36.0	22.6
F3	46.1	30.0
F2	49.0	38.2
F1	100	—

To clarify the relationships, all values are expressed as the percentage of the activity observed at station F1. Values presented are the averages of 10 months from November 1979 to August 1980. Comparisons are made on equal-sized water samples.

Table 8 Kinetic Values for Heterotrophic Uptake of [^{14}C]Glucose at Stations in the Contrary Creek Area (April 1981)

Station	V_{max} (nM/h)	T_t (h)
C1	10.2	133
A2	78.2	52.6
F1	124.3	9.38

Source: Carpenter et al. (1983).

Table 9 Concentration of Sulfate in Bottom Waters of the Contrary Creek Area of Lake Anna

Station	mg $SO_4^=$/liter	Maximum	Minimum
C1	101 ± 6.1	110	69
C5	30 ± 1.9	38	23
A2	27 ± 6.1	29	8
F3	18 ± 1.5	28	3
F1	10 ± 2.1	23	2

Values are the means and standard errors of the mean for data collected monthly from July 1979 to April 1980.

Table 10 Concentration of $SO_4^=$ in the Upper 6 cm of Sediment in the Contrary Creek Area, August, 1980.

Station	mg $SO_4^=$/liter pore water	SEM
C1	179.7	9.5
C5	69.3	19.1
A2	21.6	1.1
F1	7.6	1.3

Table 11 Distribution of $SO_4^=$-Reducing Bacteria in the Sediments of the Contrary Creek Area

Station	MPN/ g sediment[a]	Minimum	Maximum
C1	280 ± 110	20	1200
C5	510 ± 130	21	1300
A2	480 ± 360	47	1100
F3	81 ± 16	20	170
F1	86 ± 19	23	220

Values are the means ±1 SEM for a 9-month period from June 1979 to February 1980.

[a]Determinations made using a liquified version of the medium of Postgate (1965).

The physical appearance of collected sediment cores was consistent with the other observations of the $SO_4^=$ concentration gradient and $SO_4^=$ reduction activity. Cores taken from Station C1 have a characteristic surface layer (5–10 cm) of a red iron oxide floc. Beneath that layer is a black zone of sulfide accumulation extending to 20–25 cm below the sediment surface. At Station C5, the sediment surface is most often covered with a thin (1–2 cm) light gray layer underlain by a black layer extending to 25–30 cm. The surface layer is presumed to occur as a result of aerobic conditions that are present due to the shallow water depth (3 m) and abundant rooted macrophytes found there. Station A2 usually has no oxidized zone and is black with precipitated sulfides from the surface to a depth of 30–35 cm.

Stations in the Freshwater Creek arm, show no visible signs of $S^=$ production, viz., black coloration. Cores taken at both F1 and F3 are uniformly light gray at all depths. An oxygen electrode inserted into grab samples of the sediment from those stations always indicated anaerobic conditions. Frequently, large gas bubbles were observed in cores from F1, and, qualitatively, the bubbles were shown to contain CH_4.

Sediment pH values, determined by inserting a combination electrode into freshly collected grab samples taken at each station, indicated no significant differences among the various stations. pH values ranged from 6.0 to 6.5 over most of the year. The high pH of the sediment as compared with the overlying water suggests that sediment underlying acid mine waters may provide a protected environment for decomposition processes. As described later, the high sediment pH is likely due to microbial activities, and that the sediment may provide buffering to the water as a result.

Simultaneous measurements of $SO_4^=$ reduction and glucose mineralization provided strong evidence that $SO_4^=$ respiration plays an important role in the carbon turnover in sediments of the study area. As seen in Table 12, the rate of glucose oxidation did not vary greatly among the sites sampled. This fact supports the view that environmental factors such as pH and metal concentration do not significantly affect total carbon mineralization in these sediments. In an analysis involving creation of a predictive model using stepwise multiple regression, no significant relationship could be found between the rate of glucose oxidation and pH or conductivity (which is proportional to the dissolved ion content) of the sediment or the overlying water. Respiration of $SO_4^=$ in the top 10 cm of the sediment, in contrast, did show a great variation related to the presence of the AMD. On the basis of the results of this experiment, it can be concluded that the inflow of AMD to Lake Anna neither enhanced nor inhibited carbon mineralization, but the pathways whereby organic matter was degraded were dominated by $SO_4^=$ reduction in the Contrary Creek arm and by other means (e.g., the methanogenesis) in the area receiving no AMD.

Table 12 Rates of $SO_4^=$ Reduction and [^{14}C]glucose Oxidation in Sediments in the Contrary Creek Area

Station	μM $SO_4^=$ reduced/cm^3/day	Percent of added [^{14}C]glucose mineralized/24 hr
C1	0.0 ± 0.0[a]	13.4
C5	6.8 ± 7.0	10.2
A2	0.96 ± 0.41	21.2
F1	0.47 ± 0.16	16.4

Values are the means and SEM for the top 10 cm.
[a]All $SO_4^=$ reduction at C1 occurred below the 10-cm depth and was 0.67 ± 0.37.

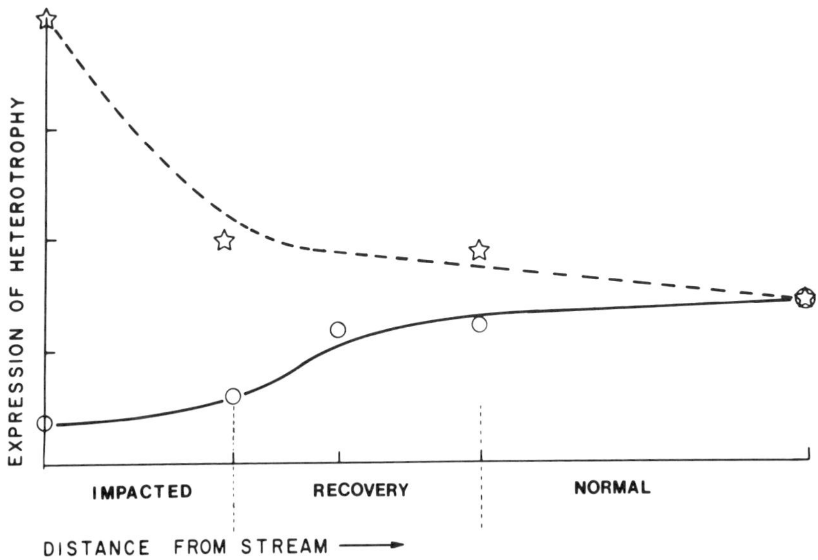

Figure 12 General model of the recovery of the heterotrophic community in the Contrary Creek arm of Lake Anna as compared with the adjacent Freshwater Creek arm. The data points used in the graph are those given in Table 6 for relative heterotrophic activity.

In general, the heterotrophic bacterial community in the Contrary Creek arm of Lake Anna shows a recovery with distance that accurately mimics the general model presented above (Fig. 3). The data from Table 7 have been plotted in Fig. 12 to illustrate the area of severe impaction and the zone of recovery. Prior to the creation of Lake Anna in the early 1970s, effects of the AMD from Contrary Creek were reported as far as 80 km from the present confluence of the stream and the lake (Simmons, 1972). Based on the microbiological studies, along with visual observations made over the period from 1975 to the present, the pollution from Contrary Creek is effectively contained in the first 1–2 km of the lake arm—that area referred to as the zone of recovery.

V. MICROBIOLOGICAL PROCESSES AND THE RECOVERY OF AMD-CONTAMINATED SYSTEMS

Along with the earlier studies of Tuttle et al. (1968), the Lake Anna study demonstrated that, indeed, aquatic communities of microorganisms can rapidly recover from the effects of AMD in cases where the pollution levels are diminished with increasing distance from the source in agreement with the general model of Fig. 3. The key factor permitting that recovery would appear to be the depletion of the pollutants in the water column. Several means exist whereby the pollution may be abated. Some are passive, such as natural dilution with cleaner water, or neutralization by the alkalinity (usually HCO_3^-) contained in the diluting water or the sediment. However, active biological mechanisms also are apparent (Berner et al., 1970; Schindler, et al., 1980; Schindler et al., 1982). Generation of weak organic acids by decomposition will tend to raise the pH if the pK_a of the products is higher than that of H_2SO_4 ($pK_1 = 1.92$). The buffering contributed by organic decomposition products is difficult to predict, however, due to the complexity of the mixture and the spatial and temporal variability of the mixture composition. A survey of alkalinity in 100 lakes of the Experimental Lakes Area in Canada indicated that up to 20% of the alkalinity might be due to such weak acid metabolites (Schindler et al., 1980). Photosynthesis is a net consumer of H^+, but because most of the fixed carbon is subsequently decomposed, the reaction is probably of little overall importance in buffering acidification. The biological reaction that may be of most importance in buffering the effects of both acid precipitation and AMD is the anaerobic reduction of $SO_4^=$ by bacteria (Schindler et al., 1980, 1982). Sulfate reduction is the principal factor affecting carbonate alkalinity in the pore water of organic-rich anoxic marine sediments (Berner et al., 1970). In freshwater environments receiving allochthonous $SO_4^=$ from the atmosphere or in inflow waters, the enhanced $SO_4^=$ reduction may serve to neutralize a significant portion of the acid pollution (Schindler et al., 1980, 1982; Tuttle et al., 1969a,b; Mills and Pien, unpublished data).

Sulfate reduction is carried out under anaerobic conditions by members of the nonsporeforming genus *Desulfovibrio* and by the sporeforming genus *Desulfotomaculum*. Other organisms are mentioned as sulfate reducers, but their importance is probably less than that of the aforementioned genera (Postgate, 1979). Although the sulfate reducers cannot themselves oxidize glucose, they obtain the energy sources needed (primarily low-molecular-weight fatty acids) from fermentors coexisting in the same habitat. The overall reaction of sulfate reduction is:

$$2CH_2O + SO_4^{=} \rightarrow S^{=} + 2CO_2 + 2H_2O \quad (4)$$

$$S^{=} + 2CO_2 + 2H_2O \rightarrow H_2S + 2HCO_3^{-} \quad (5)$$

Bacterial $SO_4^{=}$ reduction has the capacity to help cleanse contaminated water through a series of reactions (Fig. 13). Although the conversion of $SO_4^{=}$ to $S^{=}$ can result in the accumulation of toxic H_2S,

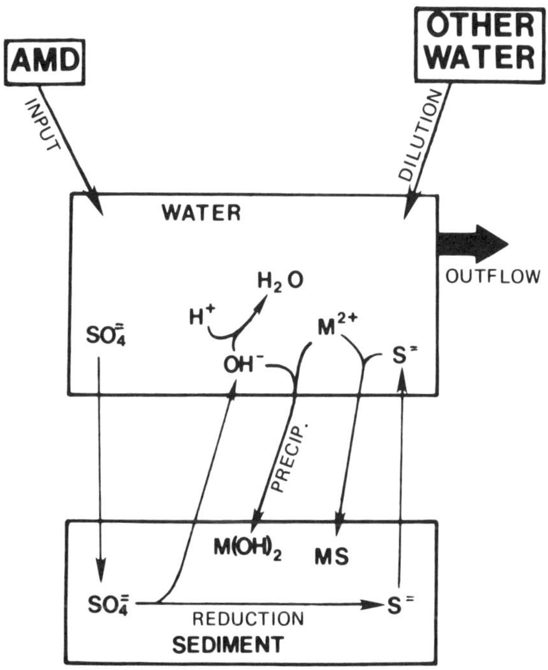

Figure 13 Schematic view of the relationship of $SO_4^{=}$ reduction in sediment and the removal of AMD from the overlying water column. Although $SO_4^{=}$ reduction generates alkalinity as carbonate, OH has been used here for simplicity in depicting the subsequent reactions.

in situations where AMD is the primary source of $SO_4^=$, a variety of metals are often present. The tendency is for the metals to complex with $S^=$, and because most of the metal sulfides are relatively insoluble, precipitation removes both the sulfide and the metal from the water. Furthermore, $SO_4^=$ reduction generates alkalinity by several means, and that alkalinity neutralizes the acid. The concomitant increase in pH forces further precipitation of the heavy metals as oxides and oxyhydroxides. The increase in alkalinity comes about as a result of the products of decomposition, namely CO_2 and $S^=$. The carbonate buffer situation for lakes undergoing acidification by H_2SO_4 has been described by Schindler et al. (1980). The reduction in alkalinity upon input of sulfuric acid may be expressed in terms of the $SO_4^=$ concentration, because equivalent amounts of H^+ and $SO_4^=$ (in mEq/liter) enter simultaneously:

$$\text{Alk}_{\text{equilib.}} = \text{Alk}_{\text{orig}} - SO^{=}_{4\text{added}} = HCO^{-}_{3\text{equilib.}} - H^{+}_{\text{equilib.}} \tag{6}$$

At the pH values found in acidified water bodies, the amount of $CO_3^=$ is negligible so that the dissolved inorganic carbon (DIC) can be expressed as:

$$\text{DIC} = [H_2CO_3] + [HCO_3^-] \tag{7}$$

Using this expression in combination with the equilibrium expression (9) for Reaction (8), Schindler et al. (1980) demonstrated that the carbonate alkalinity (and, therefore the total buffer capacity) in the acidified system is dependent on the concentration of DIC (10):

$$H_2CO_3 \rightleftharpoons HCO_3^- + H^+ \tag{8}$$

$$K_1 = \frac{[HCO_3^-][H^+]}{[H_2CO_3]} \tag{9}$$

$$\text{Alk} = \text{DIC}\,\frac{K_1}{H^+ + K_1} - H^+ \tag{10}$$

Thus, the increase in the DIC brought about as a result of the sulfate reduction reaction (4 and 5) forces an increase in the alkalinity as shown in Eq. (10). After the acidification of Lake 223 in the ELA, Schindler et al. (1980) noted that an increase in alkalinity of 11 μEq/liter (caused by the reduction of 21.5×10^3 Eq of added $SO_4^=$) represented 60% of the total alkalinity present.

In many cases, the $S^=$ generated by reduction of $SO_4^=$, will be precipitated as metal mono- or disulfides. This precipitation is especially important in AMD-contaminated waters and sediments due to the high metal concentrations typical of these systems. Whereas the process is important in removing metals from the water, loss of the $S^{=}$ would appear to remove one source of alkalinity as depicted in Eq. (5). Berner et al. (1970), showed that because of the presence of insoluble $HFeO_2$, the precipitation of metal sulfides will not affect the carbonate alkalinity:

$$6CH_2O + 3SO_4^{=} + 2HFeO_2 \rightarrow FeS + FeS_2 + 6HCO_3^{-} + 4H_2O \qquad (11)$$

Tuttle et al. (1969a) noted that in an acid mine stream which had passed through a debris dam consisting of wood chips and dust, had a higher pH, and a lower $SO_4^{=}$ and Fe concentration than did the water above the dam (Table 13). In an additional paper, Tuttle et al. (1969b) proposed that, in some cases, wood dust might be added to acid mine drainage to force oxygen utilization and to render the water anaerobic, whereupon respiratory processes could shift to $SO_4^{=}$ reduction. When AMD was experimentally amended with wood dust and allowed to incubate, the pH was increased and the metal and $SO_4^{=}$ content of the liquor were drastically reduced (Table 14). Other authors have also suggested that the addition of organic matter to acid mine water might assist in the reclamation of the liquor (Ogg, 1972; King and Simmler, 1973).

On the basis of the results of experimental addition of H_2SO_4 to a Canadian Shield lake, Schindler et al. (1980) proposed that the added sulfate could eventually shift the anaerobic metabolism of the hypolimnion and sediments from a methanogenesis dominated system to a $SO_4^{=}$ reduction-dominated system. In the ELA study, 25--35% of the added $SO_4^{=}$ was precipitated as FeS (Schindler et al., 1982). Furthermore, they postulated that small deep lakes might be more resistant

Table 13 Quality of Acid Mine Drainage Before and After Passing Through a Debris Dam of Wood Dust

Station	pH	$SO_4^{=}$ (μmol/ml)	Total Fe (μmol/ml)	$SO_4^{=}$ reducers (MPN/100 ml)
1	3.1	10.10	1.32	0
2	2.8	9.89	1.47	0
DAM				
3	4.3	3.83	0.68	700

Source: Tuttle et al. (1969a).

Table 14 Comparison of the Effects of Wood Dust Quality (i.e., the extent of wood dust degradation by microorganisms) and Incubation Temperature on the Reduction of Sulfate and Removal of H^+ from Acid Mine Water in Untreated Wood Dust Cultures Containing 400 g of Wood Dust and 2 Liters of Water

	Nondegraded wood dust			Partially degraded wood dust		
Determination	Ambient temp (22 ± 2°C)	37°C	50°C	Ambient temp (22 ± 2°C)	37°C	50°C
Amt of sulfate (μg) removed per ml of water after 14 days	30	480	65	290	590	460
Maximal rate of sulfate removed (μg) per ml of water per day	10	71.7	30.0	19.4	53.2	38.8
pH range, 0–14 days	3.6–4.2	3.7–5.8	3.8–4.2	3.5–4.3	3.5–6.4	3.7–5.8

[a]Each culture contained acid mine water with an initial sulfate concentration of 890 μg of sulfate per ml of water.
Source: Tuttle et al. (1969b).

to the effects of acidification from precipitation inputs due to the alkalinity generated in the anoxic hypolimnion by $SO_4^=$ reduction.

The enhanced $SO_4^=$ reduction in the sediment of the Contrary Creek arm of Lake Anna could provide neutralizing capabilities to the ecosystem that allows for removal of $SO_4^=$ acidity and metals from the overlying water. In the Lake Anna study site, the bottom water is rarely, if ever, anoxic. Thus, it is essential to be able to transport the $SO_4^=$ into the sediments and return the resulting alkalinity to the water column. Such a process is highly feasible if it is assumed that the AMD entering the arm of the reservoir behaves as that described by Koryak et al. (1979). Water containing a high concentration of AMD is often more dense than the receiving water, and, as a result, sinks to a level of neutral density upon entry into the lake. In the Contrary Creek situation, the shallow polluted arm caused water to flow along the sediment, where the $SO_4^=$, acidity, and metals are in close proximity to the surface. Although a complete $SO_4^=$ budget has not been determined for the Contrary Creek area, a crude estimate was made using annual average $SO_4^=$ concentration at the outflow and two inflow sites. These concentrations were coupled with water fluxes based on the known discharge for Contrary Creek, and on the input and output at the other sites estimated on the basis of the relative drainage areas. The results indicated that nearly 55% of the $SO_4^=$ entering the arm was retained, presumably precipitated in the sediment as metal sulfides.

As a means of reducing the effects of AMD, $SO_4^=$ reduction will be effective only in areas that have high levels of organic matter and the potential for anoxia in some portion of the system. These constraints would seem to limit the mechanism to be of importance only in deep, slow-moving water bodies such as lakes and reservoirs. In streams that are normally well aerated and frequently low in organic matter, the presence of debris dams such as those described by Tuttle et al. (1969a,b) may provide sufficient anoxic sites so that $SO_4^=$ reduction may serve to help cleanse those waters as well. In general, however, pollution-induced $SO_4^=$ reduction will be limited to lakes and, primarily, reservoirs.

VI. SUMMARY AND SUGGESTIONS

Microorganisms are intimately involved with the AMD recovery processes occurring in both terrestrial and aquatic habitats, although the roles played by the organisms appear to be somewhat different. In soils, the bacteria force the removal of metals and sulfur species by acidification, then process the organic matter and inorganic nutrients necessary for revegetative growth. In the aquatic systems, the opposite may be true. Through reductive processes, the microbes remove $SO_4^=$, acidity and metals from the water column, and precipitate metal sulfides in the anoxic sediments.

In both cases, much is still unknown about the processes, especially the quantitative aspects of the reactions. Future research needs to be aimed at specific questions concerning the abilities of the microbes to carry out the processes leading to environmental recovery under a variety of field conditions. Coincidentally, appropriate applied research should address the possibilities of enhancing the microbial reactions (such as $SO_4^=$ reduction) in combination with physical or chemical manipulation of the waste to bring about a more rapid rehabilitation of habitats disturbed by mining activity.

REFERENCES

Alexander, M. (1977). *Introduction to Soil Microbiology*, 2nd ed. Wiley, New York.

Alexander, M. (1980). Effects of acid precipitation on biochemical activities in soil. In Proc. Internat. Conf. Ecological Impact of Acid Precipitation. Sandefjord, Norway, D. Drablos and A. Tollan (Ed.), SNSF Project. pp. 47-52.

Ayers, R. W. (1974). *Biological survey of the Contrary Creek arm of Lake Anna, Virginia.* Report No. BS74-003. Virginia State Water Control Board.

Baath, E., Lundgren, B., and Soderstrom, B. (1979). Effects of artificial acid rain on microbial activity and biomass. *Bull. Environ. Contam. Toxicol.* *23*: 737–740.

Babich, H., and Stotzky, G. (1977). Effect of cadmium on fungi and on interactions between fungi and bacteria in soil: Influence of clay minerals and pH. *Appl. Environ. Microbiol.* *33*: 1059–1066.

Baker, K. H., and Mills, A. L. (1982). Determination of the number of respiring *Thiobacillus ferrooxidans* cells in water samples by using combined fluorescent antibody-2-(*p*-iodophenyl)-3-(*p*-nitrophenyl)-5-phenyltetrazolium chloride staining. *Appl. Environ. Microbiol.* *43*: 338–344.

Beers, W. F., Ciolkosz, E. J., and Kardos, L. T. (1974). Soil as a medium for the renovation of acid mine drainage water. Fifth Symposium on Coal Mine Drainage Research. National Coal Association. Washington, D.C., pp. 160–171.

Berner, R. A., Scott, M. R., and Tomlinson, C. (1970). Carbonate alkalinity in the pore waters of anoxic marine sediments. *Limnol. Oceanogr.* *15*: 544–549.

Bhuiya, M. R. H., and Cornfield, A. H. (1974). Incubation study of effect of pH on nitrogen mineralization and nitrification in soils treated with 1000 ppm lead and zinc oxides. *Environ. Pollut.* *7*: 161–164.

Biesecker, J. E. (1966). Stream quality in Appalachia as related to coal mine drainage. *Circ. U.S. Geol. Surv.* No. 256.

Bitton, G., and Freihofer, V. (1978). Influence of extracellular polysaccharides on the toxicity of copper and cadmium toward *Klebsiella aerogenes*. *Microb. Ecol.* *4*: 119–125.

Blevins, R. L., Bailey, H. H., and Ballard, G. E. (1970). The effect of acid mine water on floodplain soils in the western Kentucky coalfields. *Soil Sci.* *110*: 191–196.

Bollag, J.-M., and Barabasz, W. (1979). Effect of heavy metals on the denitrification process in soil. *J. Environ. Qual.* *8*: 196–201.

Bollag, J.-M., Drzymala, S., and Kardos, L. T. (1973). Biological versus chemical nitrite decomposition in soil. *Soil Sci.* *116*: 44–50.

Bremner, J. M., and Shaw, K. (1958). Denitrification in soil. II. Factors affecting denitrification. *J. Agric. Sci.* *51*: 40–52.

Brezina, E. R., Campbell, R. S., and Whitely, J. R. (1970). Effects of acid mine drainage on water quality of a reservoir. *J. Water Pollut. Contr. Fed.* *42*: 1429–1436.

Brierly, C. (1982). Microbiological mining. *Sci. Am.* *27*: 44–53.

Campbell, R. S., Lind, O. T., Harp, G. L., Geiling, W. T., and Letter, J. R. Jr. (1965a). Water Pollution studies in acid strip-mine lakes: Changes in water quality and community structure associated with aging. Symp. on acid mine drainage, Mellon Institute, Pittsburgh, Pennsylvania, pp. 188–198.

Campbell, R. S., Lind, O. T., Harp, G. L., and Geiling, W. T. (1965b). Recovery from acid pollution in shallow acid strip-mine lakes in Missouri, Proc. 19th Indust. Waste Conf., Purdue Univ. Eng. Ext. Ser. No. 117; *Eng. Bull.* *49*: 17–26.

Campbell, R. S., and Lind, O. T. (1969). Water qualtiy and aging of strip mine lakes. *J. Water Pollut. Contr. Fed.* *41*: 1943–1955.

Carpenter, K. E. (1924). A study of the fauna of rivers polluted by lead mining in the Aberystwyth district of Cardiganshire. *Ann. Appl. Biol.* *11*: 1–23.

Carpenter, K. E. (1925). On the biological factors involved in the destruction of river fisheries by pollution due to lead-mining. *Ann. Appl. Biol.* *12*: 1–13.

Carpenter, K. E. (1926). The lead mine as an active agent in river pollution. *Ann. Appl. Biol.* *13*: 395–401.

Carpenter, L. V., and Herndon, L. K. (1933). *Chemical analysis of acid mine drainage*. West Virginia Univ. Eng. Exper. Sta. Res. Bull. No. 10. Morgantown.

Carpenter, J. M., Odum, W. E., and Mills, A. L. (1983). Leaf litter decomposition in a reservoir affected by acid mine drainage. *Oikos.* *41*: 165–172.

Ciolkosz, E., Kardos, L., and Beers, W. F. (1979). The effect of acid mine drainage on two Pennsylvania soils. *Soil Sci.* *127*: 102–107.

Chadderton, R. A., and Shane, R. M. (1978). A model of the behavior of pH determining parameters in impoundments. *Water Resourc. Bull.* *14*: 1357–1373.

Cook, H. A., and Wilson, H. A. (1971). *Influence of acid mine water on the microflora of sewage.* Bull. 3, Water Res. Inst., West Virginia University, Morgantown.

Council on Environmental Qualtiy (1981). *Environmental Trends.* U.S. Government Printing Office, Washington, D.C.

Cundell, A. M. (1977). The role of microorganisms in the revegetation of stripmined land in the western United States. *J. Range Management 30*: 299–305.

Dancer, W. S., Handly, F. J., and Bradshaw, A. D. (1977). Nitrogen accumulation in kaolin mining wastes in Cornwall, I. Natural communities. *Plant Soil 48*: 153–167.

Davis, D. W., Brown, T. S., and Long, B. W. (1972). Dewatering sludge by using rotary vacuum pre-coating filtration. 4th Symp. Coal Mine Drainage Res. Mellon Inst., Pittsburgh, pp. 201–233.

Down, C. G., and Stocks, J. (1977). *Environmental Impact of Mining.* Applied Science Publishers, London, 371 pp.

Ehrlich, H. L. (1981). *Geomicrobiology.* Marcel Dekker, Inc., New York.

Evans, L. S., Lewin, K. F., and Vella, F. A. (1980). Effect of nutrient medium pH on symbiotic nitrogen fixation by *Rhizobium leguminosarum* and *Pisum sativum.* *Plant Soil 56*: 71–80.

Francis, A. J. (1982). Effects of acidic precipitation and acidity on soil microbial processes. *Water Air Soil Pollut.* *18*: 375–394.

Gadd, G. M., and Griffiths, A. J. (1978). Microorganisms and heavy metal toxicity. *Microb. Ecol. 4*: 279–387.

Goodman, G. T. (1974). Ecology and the problems of rehabilitating wastes from mineral extraction. *Proc. R. Soc. Lond. A 339*: 373--387.

Goto, K., Tokichi, T., and Kawamura, S. (1978). Effect of acid mine drainage on the pH of Lake Toya, Japan. *Water Res.* *12*: 735--740.

Griffith, E. D., and Clarks, A. W. (1979). World coal production. *Sci. Am. 24*: 38–47.

Guthrie, R. K., Cherry, D. S., and Singleton, F. L. (1978). Responses of heterotrophic bacterial populations to pH changes in coal ash effluent. *Water Resour. Bull. 4*: 803–808.

Hall, A. D., and Miller, N. H. J. (1908). The nitrogen compounds of fundamental rocks. *J. Agric. Sci. 2*: 343–354.

Hackney, C. R., and Bissonnette, G. K. (1978). Recovery of indicator bacteria in acid mine streams. *J. Water Pollut. Contr. Fed. 50*: 775–780.

Harp, G. L., and Campbell, R. S. (1967). The distribution of *Tendipes plumosus* (Linne) in mineral acid water. *Limnol. Oceanogr. 12*: 260–263.

Harrison, W. and Van Luik, A. (1979). *Suitability of Dredged material for reclamation of Surface-mined land.* NTIS Report No. ANL/ES-73. Nat. Tech. Inform. Serv.

Hersman, L. T. (1977). Microbial activity in strip-mine spoils. Ph.D. Diss. Montana State Univ., Bozeman. *Diss. Abstr. 38*: 4649.

Hersman, L. E., and Temple, K. (1979). Comparison of ATP, phosphatase, pectinolyase, and respiration as indicators of microbial activity in reclaimed strip mine spoils. *Soil Sci. 127*: 70–73.

Hultberg, H., and Andersson, I. B. 91982). Liming of acidified lakes: Induced long-term changes. *Water Air Soil Pollut. 18*: 311–331.

Jensen, V. (1977). Effects of lead on biodegradation of hydrocarbons in soil. *Oikos 28*: 220–224.

Joseph, J. M., and Shay, D. E. (1952). Viability of *Escherichia coli* in acid mine water. *Am. J. Pub. Health 42*: 795.

King, D. L., and Simmler, J. J. (1973). *Organic Wastes as a means of accelerating recovery of acid strip mine lakes.* Univ. of Missouri, Water Resources Center, Completion Rept to U.S. Dept. of Interior, OWRRT, 65 pp.

Klein, L. (1962). *River Pollution. Vol. II. Causes and Effects.* Butterworth's London, pp. 119–122.

Kleinman, R. L., and Crerar, D. A. (1979). *Thiobacillus ferrooxidans* and the formation of acidity in simulated coal mine environments. *Geomicrobiol. J. 1*: 373–387.

Koryak, M., Stafford, L. J., and Montgomery, W. H. (1979). The limnological response of a West Virginia Multipurpose impoundment to acid inflows. *Water Resour. Res. 15*: 929–934.

Kralovic, R. C., and Wilson, H. A. (1969). *Survival activity of sewage microorganisms in acid mine water.* Res. Report No. 1, Water Res. Inst., West Va. Univ., Morgantown.

Lawrey, J. D. (1977a). The relative decomposition potential of habitats variously affected by surface coal mining. *Can. J. Bot. 55*: 1544–1552.

Lawrey, J. D. (1977b). Soil fungal populations and soil respiration in habitats variously influenced by coal strip-mining. *Environ. Pollut. 14*: 195–205.

Letterman, R. D., and Mitsch, W. J. (1978). Impact of mine drainage on a mountain stream in Pennsylvania. *Environ. Pollut. 17*: 53–73.

Madsen, B. W., and Wadsworth, M. E. (1981). *A Mixed Kinetics Dump Leaching Model for Ores Containing a Variety of Copper Sulfide Minerals.* United States Bureau of Mines, Report of Investigations, No. 8547.

McCoy, B., and Dugan, P. R. (1968). The activity of microorganisms in acid mine water. II. The relative influence of iron, sulfate, and hydrogen ions on the microflora of a non-acid stram. p. 64–79. Proc. Symp. on Coal Mine Drainage Res., 2nd. Mellon Inst., Pittsburgh.

Mikkelsen, J. P. (1974). Effect of lead on the microbiological activity in soils. *Tidsskr. Planteavl. 78*: 509–516.

Mills, A. L., and Colwell, R. R. (1977). Microbiological effects of metal ions in Chesapeake Bay water and sediment. *Bull. Environ. Contam. Toxicol. 18*: 99–103.

Mills, A. L., and Wassel, R. A. (1980). Aspects of diversity measurement for microbial communities. *Appl. Environ. Microbiol. 40*: 578–586.

Minshall, G. W., and Andrews, D. A. (1973). An ecological investigation of the Portness River, Idaho: a semi-arid land stream subjected to pollution. *Freshwater. Biol. 3*: 1–30.

Moran, R. E., and Wentz, D. A. (1974). *Effects of Metal-Mine Drainage on Water Quality in Selected Areas of Colorado, 1972–73.* Colorado Water Pollution Control Commission, Colorado Water Resources Circular No. 25.

Mount, D. I. (1973). Chronic effects of low pH on fathead minnow survival, growth and reproduction. *Water Res. 7*: 987–993.

Nichols, L. E., Jr., and Bulow, F. J. (1973). Effects of acid mine drainage on the stream ecosystem of the east fork of the Obey River, Tennessee. *J. Tenn. Acad. Sci. 48*: 30–39.

Nordstrom, D. K., Jenne, E. A., and Averett, R. C. (1977). *Heavy metal discharges into Shasta Lake and Keswick Reservoirs on the upper Sacremento River, California: A reconnaissance during low flow.* United States Geological Survey, Open-File Report 76–49.

Nordstrom, D. K., and Dagenhart, T. V. (1979). *Major Sources of acid and heavy metals which contribute to the acid mine waters of Contrary Creek, Louisa County, Virginia.* Final Report to the Virginia State Water Control Board.

Norris, R. H., Lake, P. S., and Swain, R. (1981a). Ecological effects of mine effluents on the South Esk River, North-eastern Tasmania. I. Study area and basic water characteristics. *Aust. J. Freshwater Res. 31*: 817–827.

Norris, R. H., Swain, R., and Lake, P. S. (1981b). Ecological effects of mine effluents on the South Esk River, North-eastern Tasmania. II. Trace metals. *Aust. J. Freshwater Res. 32*: 165–173.

Ogg, C. W. (1972). *Organic wastes for strip-mine lake recovery.* M. S. Thesis, Univ. of Missouri, Columbia.

Olem, H. (1981). Coal and coal mine drainage. *J. Water Pollut. Contr. Fed. 53*: 814–824.

Olson, J. G., McFeters, G. A., and Temple, K. L. (1981). Occurrence and activity of iron- and sulfur-oxidizing microorganisms in alkaline coal strip mine spoils. *Microb. Ecol. 7*: 39–50.

Paone, J., Morning, J. L., and Giorgetti, L. (1974). *Land utilization and reclamation in the mining industry, 1930–71.* U.S. Bureau of Mines, IC 8642, Washington, D.C.

Parsons, J. D. (1977). Effects of acid mine water on aquatic ecosystems. *Water Air Soil Pollut. 7*: 333–354.

Postgate, J. R. (1965). Recent advances in the study of sulfate reducing bacteria. *'Bacteriol. Rev. 29*: 425–441.

Postgate, J. R. (1979). *The Sulphate Reducing Bacteria.* Cambridge University Press, London.

Reed, L. A. (1980). *Effects of strip mining the abandoned Deep Anna S Mine on the hydrology of Babb Creek, Tioga County, Pennsylvania.* Water Resources Investigations 80–53, U.S. Geol. Sury. NTIS Report No. PB81-121337, Nat. Tech. Infor. Serv.

Reeder, J. D., and Berg, W. A. (1977). Nitrogen mineralization and nitrification in a cretaceous shale and coal mine spoil. *Soil Sci. Soc. Am. J. 41*: 922–927.

Rogers, T. O., and Wilson, H. A. (1966). pH as a selecting mechanism of the microbial flora in wastewater: Polluted acid mine drainage. *J. Water Pollut. Contr. Fed. 38*: 990.

Roth, L. A., and Keenan, D. (1971). Acid injury of *Escherichia coli. Can. J. Microbiol. 17*: 1005–1008.

Scala, G. S., Mills, A. L., Moses, C. O., and Nordstrom, D. K. (1982). Distribution of autotrophic Fe and sulfur oxidizing bacteria in mine drainage from several sulfide deposits measured with the FAINT assay. *Abstr. Ann. Meet. Am. Soc. Microbiol.* N68.

Schafer, W. M., Nielsen, G. A., and Nettleton, W. D. (1980). Minesoil genesis and morphology in a spoil chronosequence in Montana. *Soil Sci. Soc. Am. J. 44*: 802–807.

Schindler, D. W., and Turner, M. A. (1982). Biological, chemical and physical responses of lakes to experimental acidification. *Water Air Soil Pollut. 18*: 259–271.

Schindler, D. W., Wagemann, R., Cook, R. B., Ruszczynski, T., and Prokopowich, J. (1980). Experimental acidification of Lake 223, Experimental Lakes Area: background data and the first three years of acidification. *Can. J. Fish. Aquat. Sci. 37*: 342–354.

Shriner, D. S. (1977). Effects of simulated acid rain acidified with sulfuric acid on host-parasite interactions. *Water Air Soil Pollut. 8*: 9–14.

Simberloff, D. (1978). Use of rarefaction and related methods in ecology. In *Biological Data in Water Pollution Assessment: Quantitative and Statistical Analyses.* K. L. Dickson, J. Cairns Jr., and R. J. Livingston (Ed.). ASTM STP652. American Society for Testing and Materials, Philadelphia, pp. 150–165.

Simmons, G. M., Jr. (1972). *A Preimpoundment Study of the North Anna River, Virginia.* Bulleton 55, Virginia Water Resources Research Center.

Smith, R. M., Tryon, E. H., and Tyner, E. H. (1971). *Soil development on mine spoil.* West Virginia Univ. Exp. Stn. Bull. no. 640T.

Smith, R. W., and Frey, D. G. (1971). *Acid mine pollution effects on lake biology.* Indiana University Water Resources Research Center. Report to USEPA, Water Pollut. Control Res. Ser 18050 EEC, 131 pp.

Sopper, W. E., and Kerr, S. N. (1980). Strip mine reclamation using municipal sludge. Proc. Symp. Surface Mining Hydrology, Sedimentology, and Reclamation. Univ. of Ky., Lexington, p. 179.

Sterritt, R. M., and Lester, J. N. (1980). Interactions of heavy metals with bacteria. *Sci. Tot. Environ. 14*: 5--7.

Stroo, H. F., and Jencks, E. M. (1982). Enzyme activity and respiration in minesoils. *Soil Sci. Soc. Am. J. 46*: 548--553.

Stumm, G. M. W., and Morgan, J. J. (1970). *Aquatic chemistry: An introduction emphasizing chemical equilibria in natural waters.* Wiley and Sons, New York, 583 pp.

Sunda, W., and Gillespie, P. (1979). The response of a marine bacterium to cupric ion and its use to estimate cupric ion activity in seawater. *J. Mar. Res. 37*: 761–777.

Terkeltoub, R. W. (1971). Supplemental irrigation with stream water contaminated by acid mide drainage. *Water Resour. Res. 7*: 704--708.

Thorpe, V. J., and Lake, P. S. (1973). Pollution of a Tasmanian river by mine effluents. II. Distribution of macroinvertebrates. *Int. Rev. Gesamten. Hydrobiol. 58*: 885–892.

Tuttle, J. H., Randles, C. I., and Dugan, P. R. (1968). Activity of microorganisms on acid mine water. I. Influence of acid mine water on aerobic heterogrophs of a normal stream. *J. Bacteriol. 95*: 1495--1503.

Tuttle, J. H., Dugan, P. R., MacMillan, C. B., and Randles, C. I. (1969a). Microbial dissimilatory sulfur cycle in acid mine water. *J. Bacteriol. 97*: 594–602.

Tuttle, J. H., Dugan, P. R., and Randles, C. I. (1969b). Microbial sulfate reduction and its potential utility as an acid mine water pollution abatement procedure. *Appl. Microbiol. 17*: 297–302.

Tyler, P. A., and Buckney, R. T. (1971). Pollution of the South Esk River, by mine effluents. *Bull. Aust. Soc. Limnol.* No. 4. p22.

Tyler, P. A., and Buckney, R. T. (1973). Pollution of a Tasmanian river by mine effluents. I. Chemical evidence. *Int. Rev. Gesamten. Hydrobiol. 58*: 873–883.

United States Environmental Protection Agency (1973). *Methods for identifying the nature and extent of non-point source pollution.* USEPA, Washington D.C.
United States Geological Survey (1979). *Water Resources Data for Virginia.* U.S.G.S. Water Data Report, Va-79-1. Nat. Tech. Inform. Serv., Springfield, Virginia.
Wakao, N., Hanada, K., Sakurai, Y., and Shiota, H. (1978). Seasonal variations in number of acidophilic iron-oxidizing bacteria and iron oxidation in the river contaminated with acid mine water. *Soil Sci. Plant Nutr. 24*: 491–501.
Wassel, R. A., and Mills, A. L. (1983). Changes in water and sediment bacterial community structure in a lake receiving acid mine drainage. *Microb. Ecol. 9*: 155–169.
Weaver, R. H., and Nash, H. D. (1968). The effects of strip mining on the microbiology of a stream free from domestic pollution. 2nd Symp. Coal Mine Drainage Res. Mellon Inst., Pittsburgh, Pennsylvania, pp. 80–97.
Wentz, D. A. (1974). *Effect of mine drainage on the quality of streams in Colorado, 1971–72.* Colorado Water Pollution Control Commission. Colorado Water Resources Circular No. 21.
Wetzel, R. G. (1975). *Limnology.* W. B. Saunders Co., Philadelphia.
Wichlacz, P. L., and Unz, R. F. (1981). Acidophilic, heterotrophic bacteria of acid mine waters. *Appl. Environ. Microbiol. 41*: 1254–1261.
Wijler, J., and Delwiche, C. C. (1954). Investigations on the denitrifying process in soil. *Plant Soil* 5: 155–169.
Wilmoth, R. C., Scott, R. B., and Hill, R. D. (1972). Combination limestone-lime treatment of acid mine drainage. 4th Symp. on Coal Mine Drainage Res. Mellon Inst., Pittsburgh, pp. 244–265.
Wilson, H. A. (1965). *The microbiology of strip-mine spoil.* West Virginia Univ. Agric. Exp. Sta. Bull. no. 506T.
Wilson, H. A., and Hedrick, H. G. (1957). Carbon dioxide evolution from some strip mine spoils. *Appl. Microbiol.* 5: 17–21.
Wilson, H. A., and Stewart, G. (1955). *Ammonification and nitrification in a strip mine spoil.* W. Va. Univ. Agric. Exp. Sta. Bull. 379T, 16 pp.
Wollitz, R. W. (1972). The effect of acid mine drainage on the limnology of a small impoundment in Southwest Virginia. Proc. 26th Ann. Conf. S. E. Assoc. Game Fish Comm.
Yan. N. D. (1979). Phytoplankton community of an acidified, heavy metal-contaminated lake near Sudbury, Ontario: 1973–1977. *Water Air Soil Pollut. 11*: 43–55.
Zevenhuizen, L. P. T. M., Dolfing, J., Eshuis, E. J., and Scholten-Korselman, I. J. (1979). Inhibitory effects of copper on bacteria related to the free ion concentration. *Microb. Ecol.* 5: 139–146.

3

Diagnostic Aspects of Algal Ecology in Disturbed Lands

L. ELLIOT SHUBERT and THOMAS L. STARKS* / *University of North Dakota, Grand Forks, North Dakota*

I. INTRODUCTION

The emphasis of scientific study on the recovery of ecosystems from manmade disturbances has focused on the reestablishment of higher plant communities and associated fauna. If one conceives the restoration of soil fertility to be an essential prerequisite for efficient land reclamation, then soil biological studies must be an integral part of any comprehensive, cohesive research effort.

Curry (1975) emphasized that restoration of mined lands should include the establishment of above-ground vegetation *and* below-ground biota. Higher plant establishment provides a "visible cue" that recovery is taking place, but without the development of appropriate soil microbial communities any initial plant establishment may be short-lived.

The below-ground microbial populations, and in our case, the soil algae appear to offer a rapid and sensitive method of determining the resiliency and stability of a reclaimed surface soil system. As noted by McCann and Cullimore (1979) the soil algae are a more sensitive indicator, than higher plants, of the effects of pesticides on the soil environment. Algae are not a static soil component, rather they are constantly interacting with other biota and the physical and chemical environment. Allen (1977) observed that "For verifying and developing ecological hypotheses in such areas as succession, algae often offer the biological system of choice." Based on these statements, algae can

---

**Current affiliation*: USDA/ARS, Grand Forks Human Nutrition Research Center, Grand Forks, North Dakota

serve as "model" organisms for investigating the recovery of disturbed soil ecosystems. Thus, we have attempted to correlate the patterns of algal colonization and succession on mined lands with soil properties.

Our outlook, knowledge, and understanding of the role soil algae play in the reclamation process has evolved from a simple floristic approach to an integrated investigation of the recovery process using algae as ecological indicators that has been well documented in previous studies (Pipe and Shubert, 1984).

This chapter summarizes our present knowledge of soil algae and their recovery processes in semiarid, and arid lands. It is hoped that this discussion will stimulate other workers to view algae not as just another soil component, but as a valuable tool in investigating and managing reclamation of mined lands.

II. FLORISTIC STUDIES

Classical studies of soil algae (Metting, 1981; Starks et al., 1981) have involved collections of soil samples and subsequent enrichment cultures to produce species lists. Unfortunately, the media used in these studies are almost as varied as the resulting species lists, making comparisons virtually impossible.

Since 1 m^2 of soil may consist of many different microenvironments, each of which may possess different algal populations, the importance of thorough random sampling during the growing season of the study area, with many collection dates cannot be overemphasized. Proper planning of the sampling protocol will allow the use of valid statistical tests. Consideration must be used when reporting the results as "flora of an area" when it may be the "flora of a specific soil layer or profile."

After collection, culturing should follow immediately, since storage under anything but ideal conditions would change the floristic composition. Whatever medium is employed, an uninoculated medium should be used as a control to ensure that the medium does not contribute any algal contaminants. Comparisons of enrichment cultures, using sterile water, soil extract medium, defined inorganic media, and/or selective media, would provide a more accurate list. In several studies, diatoms were not reported even though they inhabited comparable soils. The use of a selective medium may have altered these results.

The resulting enrichment cultures must be sampled periodically after inoculation, since competition or antagonism among species, or the changes in the nutritional and physical conditions of these cultures, may favor one particular alga while inhibiting or eliminating another species before it can be identified. The physical conditions of culturing may also alter the final species list. The temperature, quality and quantity of light, type of culture container, and other parameters may influence algal growth and reproduction. Considering the above points, it seems obvious that there is a real need to standardize methods of collection and culturing of soil algae to allow for valid comparisons similar to those that can be obtained from the standard aquatic algal bioassay prodecure (APHA, 1980).

In addition to these limitations, morphological variation in algae may produce false differences in species lists. The phenomenon of morphological variation in algae is well established, since we know that many environmental factors affect the morphology of various algae (Trainor et al., 1971). Some algal morphology is influenced by physical conditions, light, temperature, cell density, etc., whereas others are affected by nutritional changes. Unfortunately, morphs of one alga are often misidentified or labeled as another alga. The occurrence of ecotypes has been recognized for higher plants, but their identification in algae has been slow and erratic. Although the cosmopolitan blue-green alga, *Nostoc commune* Vaucher, occurs worldwide, its basic growth requirements may be as varied as its habitats. This problem can be solved by use of standardized methods, to the extent possible.

III. EDAPHIC STUDIES

Our initial study of soil algae from mined lands involved a comparison of species identified from enrichment cultures collected in North Dakota, New Mexico, and Arizona (Table 1), (Shubert and Starks, 1980). The paucity of species and the lack of any visible relationships, coupled with the numerous limitations noted previously, made species list comparisons of limited value, a point that was emphasized in reviewing species lists from other studies (Shubert and Starks, 1979; Starks and Shubert, 1982), as summarized in Tables 2 and 3. Apart from these concerns, it was apparent that environmental factors and in particular, soil properties influenced the quantity and types of algae present, and that these factors were site specific. Our work then focused on identifying soil-algal relationships in an attempt to view the ecology of soil algae from the standpoint of site characteristics and changes that might occur in the course of plant-soil system development.

A. Cluster Analysis

Previous studies of soil-algal relationships focused on pH, soil moisture, particle size, and general soil type (sandy, loam, clay, etc.) without any further characterization of the soil environment (Starks et al., 1981), resulting in a poor understanding of soil-algal interactions. Cluster analysis or similarity (indices) of the North Dakota, New Mexico, and Arizona data presented in Table 1 demonstrated certain relationships between soil types and algae present (Fig. 1), but the groupings indicated that many algal species were found in all the soils tested. Therefore, it was assumed that the differences or similarities in algal communities would be related to more detailed soil characteristics. A multilinear regression comparing the soil properties of the various

Table 1 Algae Identified from Surface-Mined Soils in North Dakota, New Mexico, and Arizona

Alga	New Mexico	Arizona	North Dakota
Chlamydomonas sp.		X	
Chlorella sp.	X	X	
Chlorococcum sp.	X	X	X
Chlorosarcinopsis pseudominor	X	X	X
Golenkinia sp.		X	X
Anabaena sp.	X	X	
Lyngbya aestuarii			X
Nostoc commune			X
Oscillatoria brevis	X	X	X
Oscillatoria sp.	X	X	X
Nitzschia palea			X

Source: Shubert and Starks (1980).

sites showed only limited differences among the soils. We postulated that these differences affected the algae at each site and determined the species present (Shubert and Starks, 1980).

B. Successional Analysis

The application of the concept of a "climax" community to soil algae is difficult. Rapid generation times, environmental fluctuations and other external forces can result in significant short-term population changes. In spite of this, succession follows a predictable course toward a climax state in many disturbed systems. Based on this premise, Shubert and Starks (1979) initiated a study of soil algal succession on abandoned mined lands. The study area offered a unique situation. Abandoned spoil piles of several age groups (1, 7, 17, and 45 years after mining) were located in the same mining area and consisted of similar parent material. In addition, native sites possessing similar aspects and slopes were located adjacent to the abandoned sites. These sites provided an assumed "climax" successional stage and a data base with which the various aged spoil banks could be compared. The resuting species list provided evidence that succession was occurring on the abandoned spoil banks, but even after 45 years, the algal community

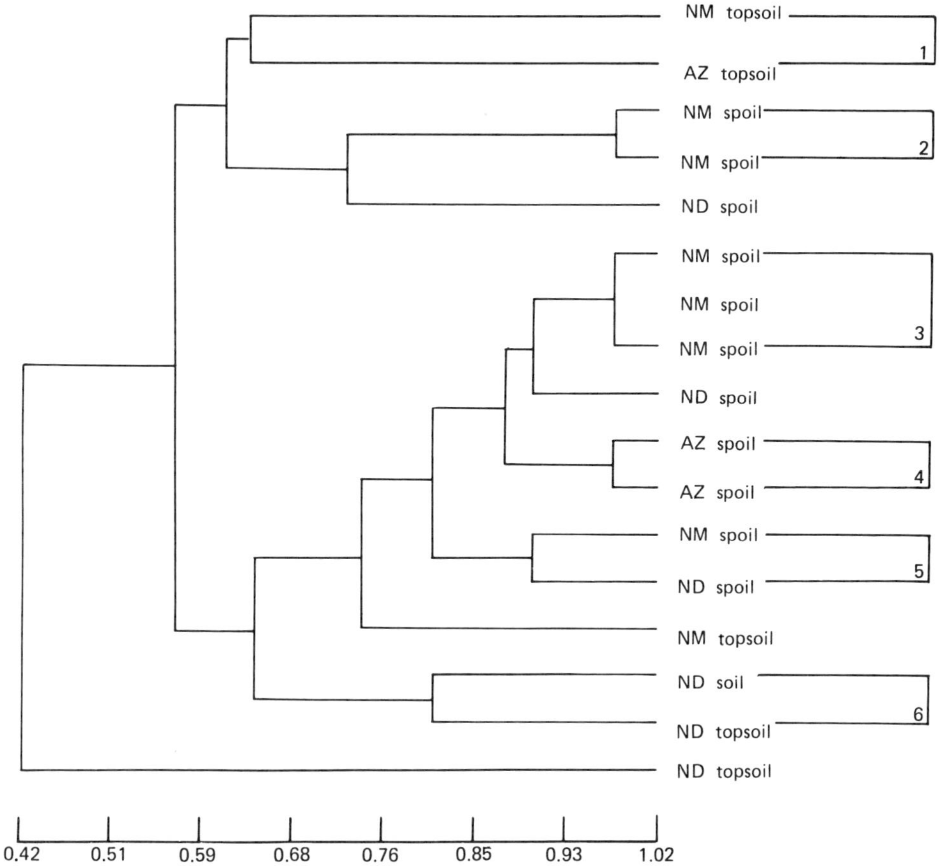

Figure 1 Cluster analysis of soil algae compared with soil types from Arizona, New Mexico, and North Dakota. Numerical axis represents a similarity index. The vertical numbers represent algal groupings. (From Shubert and Starks, 1980.)

had not reached the "climax" state represented by the unmined areas (Table 2). Comparisons of species list from one area obtained from cultures grown under the same environmental conditions are useful and valid when used to evaluate ecological trends such as succession.

C. Slope and Aspect Analysis

A comparison of the algal species at the varied aged sites produced distinct distribution patterns, but this method did not address the in-

Table 2 Algal Species Present at the Abandoned Spoil Banks, Near Velva, North Dakota

Species	Age of mined sites (years) 1	7	17	30	45	Unmined sites
CHLOROPHYCEAE						
Apatococcus lobatus	X	X	X	X	X	X
Chlorococcum humicola	X	X	X	X	X	X
Chlorosarcinopsis pseudominor	X	X	X	X	X	X
Sphaerocystis schroeteri	X	X	X	X	X	X
Stichococcus subtilis		X	X	X	X	X
Eremosphaera viridis	X	X		X	X	X
Chlorococcum ellipsoideum	X	X			X	X
Oocystis crassus	X				X	X
Klebsormidium sp.			X	X		
Chlorella vulgaris			X			
Ulothrix sp.		X	X			
Chlorochytrium biennis	X					X
Coelastrum sp.						X
Chlorococcum minicula						X
Chlorosarcina sp.						X
Desmococcus viridis						X
Gonium sp.						X
Hazenia mirabilis						X
Oocystis apiculata						X
CYANOPHYCEAE						
Nostoc commune	X	X	X	X	X	X
Oscillatoria brevis	X	X	X	X	X	X
Wollea saccata	X	X	X	X	X	X
Anabaenopsis sp.		X	X	X	X	X
Lyngbya aestuarii			X	X	X	X
Oscillatoria sp.			X	X	X	X
Cylindrospermum licheniforme	X	X		X	X	X
Tolypothrix distorta	X			X	X	X
Calothrix parietina	X	X			X	X
Gloeocystis sp.	X		X			X
Anabaena sp.	X		X	X		
Cylindrospermum sp.	X					
Schizothrix macbridei				X		
Plectonema nostocorum					X	
Arthrospira sp.						X

Table 2 (Continued)

Species	Age of mined sites (years) 1	7	17	30	45	Unmined sites
CYANOPHYCEAE						
Nodularia spumigena						X
Scytonema hofmannii						X
BACILLARIOPHYCEAE						
Nitzschia palea		X	X	X	X	X
Navicula minima			X	X	X	X
Pinnularia saxicola			X	X	X	X
EUGLENOPHYCEAE						
Euglena sp.						X
XANTHOPHYCEAE						
Monocilia sp.		X		X	X	

Source: Shubert and Starks (1979).

fluence of aspect or slope or the algal communities. We shall define "aspect" as the compass direction a stand faces, and "position" as its locale relative to elevation on a particular hillside. Differences among aspects in soil moisture, temperature, soil development, incoming solar radiation, air temperature, and various microenvironmental factors and their effect on vegetation have been reported (Aayad and Dix, 1964). As a general rule, in the Northern hemisphere northerly aspects are cooler and moister, while southerly aspects are warmer and drier.

Although the differences among higher vegetation with respect to aspect and position are well documented (Aayed and Dix, 1964; Jonescu, 1979), the effect of these factors on soil algae were not known. After investigating the successional patterns on these abandoned areas (Shubert and Starks, 1979), relationship of the aspects, positions, and the algal community were investigated. Therefore, target diagrams were used to assess the differences in algal community structure by age, aspect, and slope, (Fig. 2) using the techniques of Aayad and Dix (1964) where the influence of aspect and position on higher plant growth and distribution were studied. By the use of this technique,

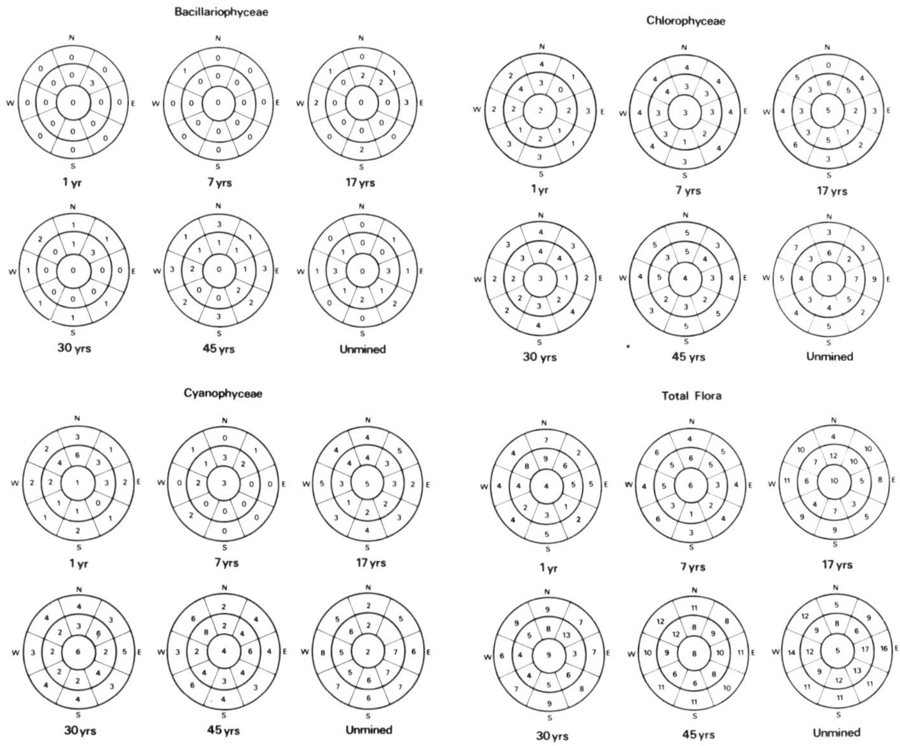

Figure 2 Target diagrams of algae present (numbers represent species) at abandoned spoil banks near Velva, North Dakota in relation time, aspect, and slope.

a trend toward increasing species variety, with little significant differences among the aspects or slopes, became evident. It is interesting to note that diatoms (Bacillariophyceae) were essentially absent from the 1-year and 7-year-old sites and were never found on all aspects or slopes for any age group including the native site, whereas the Cyanophyceae and Chlorophyceae were found throughout the study area. The difficulty of evaluating these data based only on species numbers or community composition became very apparent. Although the northern aspects were higher in moisture, the number of species present was not greater as one would postulate, and in many cases more species were found on the drier, warmer aspects. Quantification of the algal biomass appeared to be essential to interpret and identify possible soil-algal interactions. It was hypothesized that variations in algal abundance would reflect soil property variation (Starks and Shubert, 1979). To provide information on this point, soil algae were quantified by chlorophyll *a* extraction, a procedure that is simple, reliable, reproducible, and allows for comparison of varying soil types (Shubert and Starks, 1979). This procedure has been discussed in detail by Starks et al. (1981).

IV. STABILITY

Reclamation of surface-mined lands is directed toward the establishment of a vegetation complex that resembles the premining ecosystem in plant cover and productivity. The premise has been made that as species diversity increases over time so will the stability of the system. This diversity-stability relationship is a point of continuous discussion and Goodman (1975) expressed an alternate viewpoint stating that "clearly the belief that more diverse communities are more stable is without support." Even so, many workers continue to use the former assumption in ecosystem analysis.

Stability is the ability of a system to return to an equilibrium state after a temporary disturbance (Hollings, 1973). Since stability is a dynamic property of ecosystems, systems that return more rapidly to the original state after a perturbation with fewer fluctuations, are considered to be more stable (Hollings, 1973; Van Voris et al., 1980). Ecosystem stability is also used in the sense of resistance and resilience. Resistance is the capacity of the system to return to prestress equilibrium (Holling, 1973). The emphasis on the disturbance is important since an ecosystem may appear to be stable simply because it seldom experiences a perturbation.

A study was initiated to compare the stability of abandoned spoil banks to managed areas using soil microbial measurements and calcium export levels rather than above-ground vegetation analysis (Starks and Shubert, 1982). The use of above-ground vegetation (diversity, productivity, cover, etc.) would be limited, since managed areas are planted with a limited diversity (usually five species), and variously aged areas have been seeded with different planting mixtures, which would also invalidate comparisons to the abandoned sites. Based on this type of limitation, the soil microbiota were evaluated as indicators of stability with primary emphasis on the soil algae.

Study areas located southwest of Velva, North Dakota and near Zap, North Dakota, were selected as typical managed areas. These areas represented three age classes (3, 5, and 7 years after mining). The managed areas were topsoiled according to existing laws and seeded with various planting mixtures. In addition, native sites near Velva and Zap were also sampled.

Three intact soil cores approximately 15 cm in diameter and 20 cm deep were extracted from each sampling area for a total of 24 cores. This size microcosm was selected as the smallest soil section that could maintain autotrophic components for the period study (Van Voris et al.,

1980). Each microcosm was equilibrated for 90 days. Microcosms were selected for this study for several reasons. Manipulation of entire ecosystems is costly and difficult, and microcosms (artificially bounded replicable experimental units) designed to exhibit processes occurring in an entire system can provide valuable data at reduced costs and with less difficulty. In addition, environmental interactions are scaled down so that complex natural processes can be studied under more controlled conditions. Microcosms offer replicability and can aid in a better understanding of fundamental properties of ecosystems, in this case, of their reactions to a perturbation in terms of system stability.

Prior to perturbation, biweekly measurements of algal populations were determined by chlorophyll *a* extraction of soil cores (Shubert and Starks, 1979). Bacterial population were estimated from soil cores by serial dilutions and plating methods. Leachates were made by adding approximately 200 ml of distilled water to each microcosm. The resulting leachate was collected through the bottom port, filtered, and calcium analyzed on a Perkin-Elmer Atomic Absorption Spectrophotometer, Model 503.

Cadmium, as $CdCl_2$, was used to stress the microcosms. Cadmium was chosen because it would cause the desired perturbation and since previous studies had provided background information on the effect of cadmium on an ecosystem (Van Voris et al., 1980). In addition, earlier work has shown that cadmium is an important limiting factor for soil algal growth in North Dakota (Starks and Shubert, 1982). Based on literature values, a dose per unit area of 0.424 mg/cm^2 was used. This was applied by misting the cadmium solution after a 90-day equilibration experiment. Sampling of the stressed microcosms commenced 2 weeks after the cadmium addition and was conducted in the same manner as prior to the perturbation.

The amount of chlorophyll *a* in the soil, as an indicator of algal abundance, exhibited a trend that may make this parameter a useful tool in studying the stability of systems (Fig. 3). Although only the 35-year and native site data are presented, all managed and abandoned sites, showed similar trends with the Cd stress producing a delayed response in the algal populations. The increase in abundance during the 8–12-week poststress period was probably due to the elimination of competitive species. As a result, the soil algae could reproduce rapidly. Later as the systems showed signs of recovery, based on the calcium export data, competition among the algae probably reduced their abundance to the prestress levels. Although algae are highly sensitive to perturbation, the chlorophyll *a* values followed patterns similar to the calcium export data and therefore may be a useful tool in studies of this type.

In contrast to the algal data, bacterial populations were not a useful parameter for measuring ecosystem stability due to population

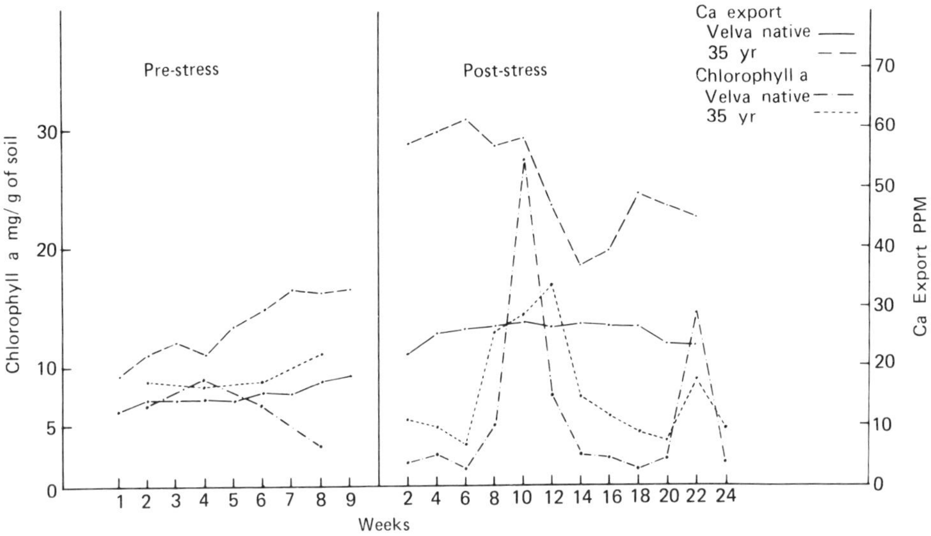

Figure 3 Comparison of calcium (Ca) export values and chlorophyll *a* levels at two sites near Velva, North Dakota with and without cadmium (Cd) stress. (From Starks, 1982.)

density fluctuation, (Starks and Shubert, 1982). Rapid generation times of bacteria, and the influence of environmental factors on these organisms, leading to large population fluctuations, added convincing evidence to this statement (Atlas and Bartha, 1981). The fluctuation of bacterial numbers among the individual microcosms from the same sites and the biweekly changes in the numbers showed the unreliability of the parameter as an indicator of resistance or stability. Thus, inclusion of this data would only confuse the issue, not clarify it.

The abandoned 12-, 35-, and 50-year Velva sites had varying degrees of resistance to the cadmium perturbation based on the calcium export data. A comparison of these sites showed that the 35-year site had the least resistance, whereas the 12- and 50-year sites had similar resistance responses. In contrast, the native site showed a high level of resistance so that the amount of calcium export in the pre- and poststress leachates were similar.

The managed sites (Zap: 3, 5, and 7 year) showed variability in resistance to the cadmium perturbation. The 3- and 5-year sites had the most resistance to the cadmium stress but did not exhibit signs of recovery, since the poststress calcium export levels were significantly higher than the prestress values. In contrast, the 7-year site showed very little resistance to the cadmium stress, but appeared to be recovering with the calcium export levels approaching prestress values.

Calcium exports, as analyzed in leacheates, have been demonstrated to be a reliable indicator of the stability of a system (Van Voris, et al., 1980). The increased calcium loss following the perturbation and the subsequent recovery observed on this study were not correlated with any changes in the bacterial populations of the soil. Therefore, the behavior of the calcium has been shown to be a measure of total ecosystem response based on this study and others (Van Voris, et al., 1980). Combination of chlorophyll *a* data with calcium values may prove to be more reliable than calcium export data alone.

Comparison of the results from this study to that of Van Voris et al. (1980) demonstrated the importance of topsoiling in the reclamation of mined lands. In the latter study, an abandoned field, previously used for agriculture, was fully recovered from a cadmium perturbation after 9 weeks. In the former study, none of the abandoned sites at Velva had recovered after 22 weeks. Only the 50-year site was approaching prestress algal community levels. The managed sites, although they showed limited resistance to the perturbation, appeared to be recovering more rapidly than the abandoned sites. This faster recovery of the managed sites was probably a direct result of reclamation practices, such as topsoiling, fertilizing, etc.

Linear regression equations were calculated for all sites. Only the Zap native and Velva 12-year sites had negative slopes (Table 3). A negative slope indicates a movement toward recovery. It is difficult to predict when the other sites will show signs of recovery, however, based on the managed area data, 10 years or less is a realistic estimate of the time needed for managed areas to reach a state of stability similar to the native sites.

V. SOIL-ALGAL INTERACTIONS

Information describing possible interactions among soil algae and soil components was developed from statistical analysis of the data. Such statistical correlations may not necessarily indicate the existence of biological relationships, but in this relatively new area of soil-algal ecology this was used to provide information on algal distribution and abundance using more basic ecological concepts.

Much of the available information on soil factor relationship to algal community development has been based on liquid cultures, including information on algal nutritional requirements or tolerance ranges. Therefore, the application of these studies to the soil environment should be interpreted with this in mind. Although the soil solution in which the algae interact would resemble a culture medium with a constant replenishment of nutrients similar to soil water bottles, differences in the manner in which algae interact with the soil solution and liquid culture probably exist. In addition, the concentrations of soil nutrients reported and the actual amount available for algae may differ. We must

Table 3 Regression Equations Using Algal Abundance (Chlorophyll *a*) as the Dependent Variable Where $y = mx + b$

Zap (managed site)		Velva (abandoned site)	
3 year		*12 year*	
$y = 1.15x - .17$	$r = -2.08$	$y = -2.86x + 4.0$	$r = -0.23$
5 year		*35 year*	
$y = 1.96x - .92$	$r = 1.89$	$y = 4.57x - 3.64$	$r = 2.58$
7 year		*50 year*	
$y = .79x + .33$	$r = 0.0$	$y = .75x + .36$	$r = 0.00$
Native		*Native*	
$y = -.21x + 1.19$	$r = -0.23$	$y = -32.66x + 34.44$	$r = 0.00$

Source: Starks (1982).

assume that the nutrient concentration in soil extracts would be similar to the actual levels available.

Nutrient requirements or tolerance ranges vary among algal species, therefore a change in one soil parameter may inhibit or eliminate one species while enhancing the growth of another. The species composition may change but the productivity or abundance may remain relatively constant. A factor that alters the abundance would be of greater importance to the stabilization of the algal community. Further support for utilizing algal abundance came from a study of various aged abandoned spoils and native sites (Shubert and Starks, 1979). Across this range of sites, a direct comparison of algal species variety to abundance showed a significant positive linear response, in that as the species variety increased, the algal abundance also increased.

From statistical analysis of the data (Shubert and Starks, 1979; Starks and Shubert, 1979, 1982) linear regressions using algal abundance (chlorophyll *a*) as the dependent variable and soil properties as the independent variables generated results that were developed into conceptual models. These models were used as a nonmathematical visual explanation to interpret the results of linear regression analysis. A positive linear relationship indicated that, as algal abundance increased over time, the concentration of a specific soil parameter also increased. A negative linear relationship denoted an inverse where the algal abundance increased as the levels of a specific soil property decreased over time. The results of this analysis (Fig. 4) indicated that no chemical property of the topsoiled site correlated significantly with algal abundance. Although topsoil storage has been shown to affect adversely, soil microorganisms, the chemical nature of the soil remains basically the same (Miller and Cameron, 1976, 1977; Visser et al., 1979). Therefore, the increase in algal abundance over time that was observed at the topsoiled site was not greatly influenced by soil chemical variation since the native algae had adapted to this material From this we assumed that abundance increased due to colonization and growth of new species. This premise, that topsoil represented a favorable algal environment, was substantiated from the Velva study. The native sites had very few negative factors affecting soil algal abundance and these were trace elements and calcium (Fig. 4).

In direct contrast, the spoil material at the test sites and the abandoned spoil banks was basically parent material and presented an environment quite different from the topsoiled and native sites. As many of the chemical properties changed via leaching; plant uptake, percolation, or other factors, they directly affected algal abundance (Fig. 4). Most of the parameters were at inhibitory levels to some algae, as reflected by the negative correlations.

The possible influence of the various factors that showed significant correlations with algal abundance could then be analyzed. Iron, one of the first factors to be considered, is important for growth, pigment synthesis, nitrogen metabolism, and photosynthesis in algae (Wiessner, 1962). The levels of iron in the spoil material at the test site (42.5–57.6 mg/liter) were below inhibitory concentrations and above critical levels reported for *Nostoc muscorum,* a common soil inhabitant (Eyster, 1958). Lindsay (1977) reported that inorganic iron in soil solution reaches a minimum at pHs of 6.5–8.0, which was similar to the pH reported for this site (7.1–7.7). This pH factor may have caused the amount of available iron to be below optimal or critical growth levels and as this level increased over time the algae responded in a positive manner.

The high levels of calcium, sodium, and potassium (236.5–510.0 mg/liter, 670–1100 mEq/100 g, 50–90 mEq/100 g) showed a negative correlation, which resulted from the calcareous spoil used at the test site. Although these cations are required for growth by most algae (O'Kelley, 1974), the levels at the test site greatly exceeded these values. As these levels were reduced through the process of soil development, other algal species were able to colonize and propagate on this site. Raju (1972) reported better development of blue-green algae in soil rich in calcium and sodium. Also, sodium has long been considered essential to certain cyanophytes (O'Kelley, 1974). This information may aid in explaining the abundance of algae reported for this site in 1976 (Table 4). The members of the Cyanophyceae found in 1975 and 1976 may have been able to grow and reproduce in these conditions, but as the calcium, sodium, and potassium levels were reduced, other algae could grow and compete with the blue-green algae.

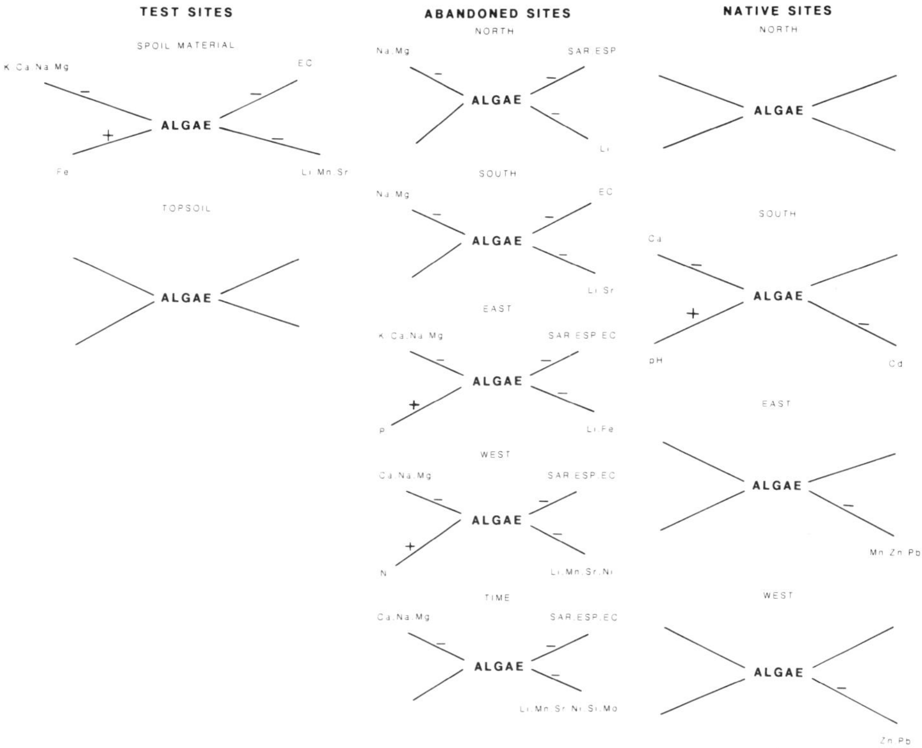

Figure 4 Conceptual models of soil-algal interactions in surface-mined lands.

The electrical conductivity (EC) is directly related to the cation concentration of the soil and is a measure of salinity. The negative correlation of this parameter with abundance further supported the possible reasons suggested for the effects of potassium, calcium, and sodium (Fig. 4) (Starks and Shubert, 1982).

All forms of magnesium had significant negative correlations with algal abundance in this study. Hunt et al. (1979) reported a negative correlation of magnesium on soil blue-green algal abundance and a positive correlation with eukaryotic soil algae at levels that ranged from approximately 50 mg/g to 180 mg/g of soil. The levels in our study (Mg, 85–250 ppm) exceeded reported critical levels for growth (Eyster, 1972; Fogg et al., 1973). Although no inhibitory magnesium levels for soil algae have been published, the concentrations for the spoil material sites were probably sufficient to inhibit algal growth. As the levels were reduced over time, algal abundance increased (Starks and Shubert, 1982).

Table 4 Algae Identified from Test Sites near Beulah, North Dakota (8/75–9/77)

Species	8/75	5/76	6/76	7/76	8/76	9/76	5/77	6/77	7/77	8/77	9/77
	Date										
Chloroccoccum sp.	1,2[a]	1,2	1,2	1,2	1,2	1,2	1,2	1,2	1,2	1,2	1,2
Chlorosarcinopsis pseudominor	1	1,2	1,2	1,2	1,2	1,2	1,2	1,2	1,2	1,2	1,2
Navicula minima	1,2	1,2	1,2	1,2	1	1,2	1	1,2	1,2	1,2	1
Apatococcus lobatus		1,2	1	1,2	1,2	1,2	1	1		2	1
Anabaena sp. (1)		1	1	1	1,2	1	1,2	1			1
Nostoc commune	1,2	1,2	1,2	1,2	1,2	1,2	1,2	1,2	1,2	1,2	1,2
Oscillatoria brevis	2	1,2	1,2	1,2	1,2	1,2	1,2	1,2	1,2	1,2	1,2
Nitzchia palea	2	1,2	1,2	1,2	1,2	1,2	1,2	1,2	1,2	1,2	1,2
Stichococcus subtilis		1	1	1	1	1	1,2	1,2	1,2	1,2	1,2
Klebsormidium sp.		1	1								
Pinnularia saxicola		1	1	1	1	1	1,2	1,2	1	1	1,2
Microcoleus sp.	1				1				1	1	1
Oscillatoria sp. (1)	2	1	1,2	1,2	1	1,2	1,2	1,2	1,2	1,2	1,2
Monocilia viridis		1	1	1	1	1	1,2	1	1	1,2	1
Anabaena sp. (2)							1,2	2	1	1	1,2
Eremosphaera viridis							1	1	1	1	1

Lyngbya aestuarii	1		1	1	1	1
Gloeocystis sp.				1	1	
Sphaerocystis schroeteri		1,2	1,2	1,2	1,2	1,2
Cylindrospermum lichenforme			1	1		1
Oscillatoria sp. (2)					1	1
Chlorosarcinopsis minuta		1			1	1
Nostoc sp.			1	1		
Chlorochytrium biennis		1				
Nitzschia hantzxchiana						1
Mougeotia sp.	1					
Scytonema hofmanii				1		
Botrydium sp.						1
Chlorococcum minutum						1
Plectonema sp.				2		2

Site 1 was constructed with recontoured spoil material. Site 2 was constructed with topsoil.
[a]1 = Present at site 1; 2 = present at site 2.
Source: Starks and Shubert (1982).

Trace elements and their effects on algae have received increasing attention in the past few years, but their influence on soil algae has not been investigated. In soils, most heavy metals occur as inorganic compounds or are bound to organic matter, clays, or hydrous oxides of iron, manganese, and aluminum (Foy et al., 1978). For these reasons only zinc, copper, and nickel toxicities have occurred frequently in higher plants, and lead, cobalt, and cadmium toxicities occur only under unusual conditions. Other elements may be toxic in liquid cultures but are not phytotoxic in soils, even at extremely high levels such as cadmium for which availability is strongly affected by soil pH, chlorine, and chelation (Foy, Channey and White, 1978). Rosko and Rachlin (1977) reported depression of cell division in *Chlorella vulgaris* at 0.06 ppm of cadmium. Toxicity at levels greater than 1.0 ppm was demonstrated for *Scenedesmus obliquus* (Cain, et al., 1980). Cain and Allen (1980) showed some growth of *Chlamydomonas reinhardtii* at levels of 10.0 ppm of cadmium. Such variations in findings among various species added to the difficulty of deducing any possible inhibitory levels for these metals in the soils. The observed relationships suggested that the elements cadmium, lithium, manganese, and strontium may have inhibited algal growth. Inhibition of algal growth by high levels of manganese has been reported by Velichko (1968) and Eyster (1972). The trace elements may be interacting, or replacing major cations in metabolic pathways as strontium does with calcium (O'Kelley, 1974). Unfortunately, very little work has been done to identify which essential metabolic processes are sensitive to heavy metals, but it is known that trace metals are cofactors for enzymes. Therefore, trace elements, including heavy metals, may play an important role in soil algal distribution and succession, however the exact mechanisms are still unknown (Starks and Shubert, 1982).

VI. CONCEPTUAL MODELS

The conceptual models generated from our studies illustrated the complexity of the soil system, but also presented strong evidence that on topsoiled or native areas the algae are in or close to a state of equilibrium (Fig. 4). The absence of any factors correlating with algal abundance at the topsoiled reclamation site and few factors on the native test sites supported this finding. Therefore, it was apparent that climatic factors (moisture, temperature) were most important in determining distribution and abundance of algae at undisturbed sites.

As stated earlier, the influence of aspect and position on higher plant growth and distribution has been well documented (Aayad and Dix, 1964). Differences among aspects in soil moisture, temperature, and various other microenvironmental factors and their resulting effects on vegetation have been reported.

Based on the models generated for the native sites, the influence of aspect was also apparent for the soil algae. Comparison of the chemical factors of the northern and southern aspects supported the data reported by Aayad and Dix (1964) and indicated that these chemical factors affected the algae in the same manner as higher plants. The southern, eastern, and western aspects showed varying degrees of disequilibrium, compared with the northern aspect which was in equilibrium (Fig. 4). This same trend was found at the reclamation test sites, since topsoiled areas were in a state of equilibrium. However, the nontopsoiled areas (spoil material) were unstable biologically and chemically.

The models allowed identification of several common factors in spoil areas: calcium, sodium, magnesium, EC, lithium, and manganese. Based on this finding, it appears that these negative factors contributed to the harsh environment and were inhibiting the growth and or the colonization of algae. This conclusion has been substantiated by our previous studies (Shubert and Starks, 1979; Starks and Shubert, 1982). Decreases in these soil factors led to increasing algal abundance and the number of species.

The factors that were found to limit or enhance algal abundance are quite different from these that limit higher plant growth in native grasslands. Native grassland productivity, prior to any disturbance, is limited by low levels of nitrogen and phosphorus and various micronutrients, such as manganese and zinc (Woodmansee and Duncan, 1980; Malakondaiah et al., 1980). Typically nitrogen and phosphorus are found to be the major limiting algal growth factors in aquatic systems. The difference in these findings may be due to the low levels of nitrogen and phosphorus required by algae. The nitrogen and phosphorus levels in soils and spoil material exceed these levels and since algae are probably in a state of equilibrium with the soil solution, these nutrients are present in sufficient quantities to insure growth. In addition the blue-green algae can contribute to the soil nitrogen pools through biological fixation processes. In contrast, higher plants require greater quantities of these growth-stimulating elements, in many cases in concentrations above those present in native or disturbed soils.

The various models generated from our studies showed the similarity in factors influencing algal abundance in semiarid and arid environments. Although the studies were conducted at different times and sites, by using the same techniques for sampling and analysis, it has been possible to develop a reliable and reproducible model format.

The model concept has limitations since a simple causal effect may not exist in the complex soil environment. Algae vary from species to species in their requirements or tolerance for most elements, therefore a change in one soil parameter may inhibit or eliminate one species but may enhance the growth of another.

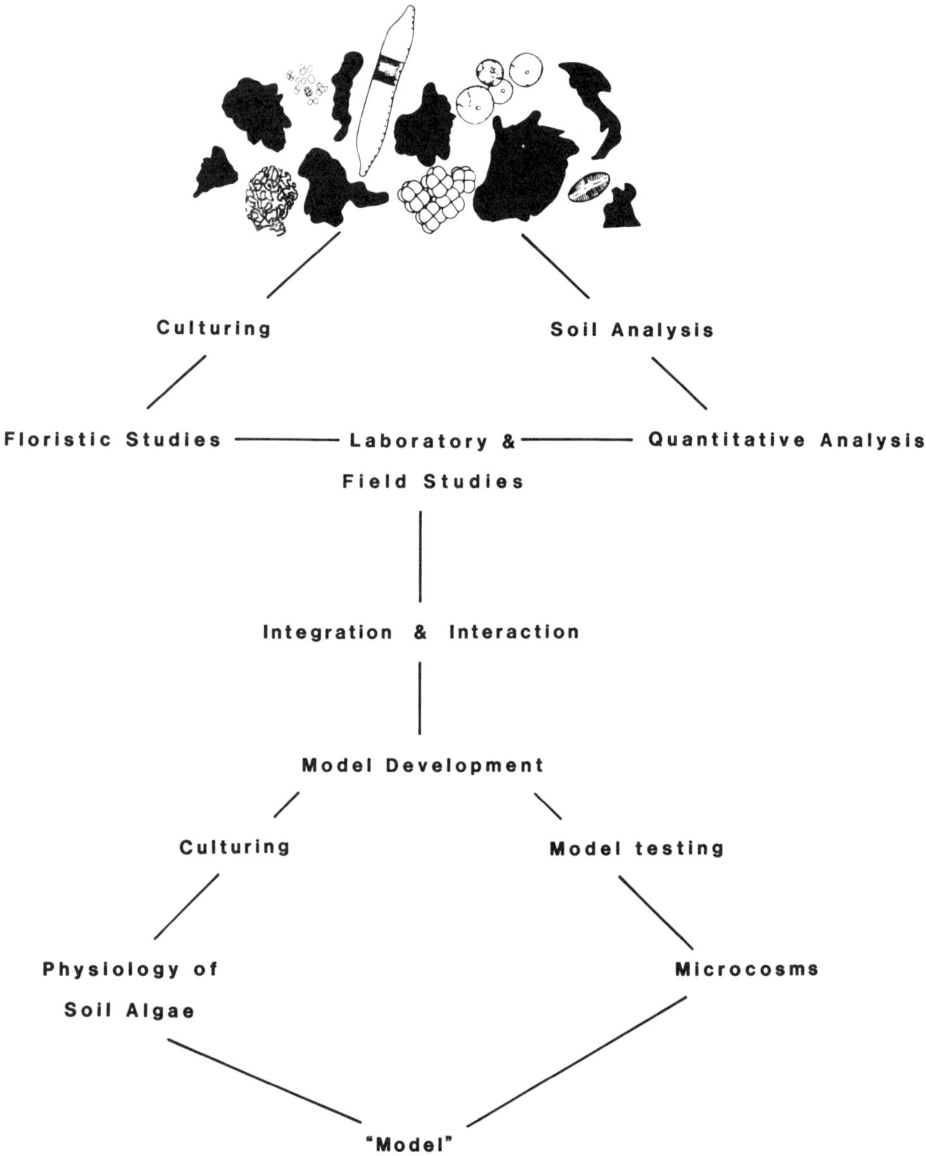

Figure 5 Present and future approaches to soil-algal experimentation and research.

Comparison of models generated by the same statistical method and based on data formulated in the same manner can produce useful information. But if data from different ecosystems are combined and less extensive soil analysis are used, this will result in models that tend to produce less reliable results and have a lesser chance of improving our understanding of the soil-algal environment. This is not to be confused with the comparison of models generated from different ecosystems, which is valid and does provide useful concepts.

VII. CONCLUSIONS

The development of a model can involve analysis at many levels (Fig. 5). We initiated our studies with two simple approaches: culturing (floristics) and soil analysis. To deal with the limitations of this approach, the data were integrated statistically to identify possible soil-algal interactions. On the basis of this approach, we generated the conceptual models discussed in this chapter.

These models suggest that it has been possible to identify limiting factors in spoil material and topsoils and to monitor changes in algal community development in the course of reclamation. Further model development needs to include research in the areas of physiology and nutrition, ecosystem stability and other pertinent ecological premises (Fig. 5), to allow more advanced models of soil-algal interactions to be developed. It should be emphasized that model development and interpretation should be focused on specific ecological situations, such as semiarid, arid, and temperate environments where specific environmentally independent variables can be identified and quantified for use in these types of analyses. This concept could be advanced to encompass a better understanding of the effects of acidification, toxic wastes, pesticides, and other perturbations on the algal component of specific ecosystems.

REFERENCES

Aayad, M. A. G., and Dix, R. L. (1964). An analysis of a vegetation—microenvironmental complex on prairie slopes in Saskatchewan. *Ecol. Monogr. 34*: 421–442.

Allen, T. F. H. (1977). Scale in microscopic algal ecology: A neglected dimension. *Phycologia 16*: 253–257.

APHA (1980). *Standard Methods for the Examination of Water and Waste-Water,* 15th ed. Amer. Public Health Assoc., Washington, D.C., 1134 pp.

Atlas, R. M., and Bartha, R. (1981). *Microbial Ecology: Fundamentals and Applications.* Addison-Wesley Publ. Co., Reading, Massachusetts, 560 pp.

Cain, J. R., and Allen, R. K. (1980). Use of a cell wall-less mutant strain to assess the role of the cell in cadmium and mercury tolerance by *Chlamydomonas reinhardtii*. *Bull. Contam. Toxicol.* *25*: 797–801.

Cain, J. R., Paschal, D. C., and Hayden, C. M. (1980). Toxicity and bioaccumulation of cadmium in the colonial green alga, *Scenedesmus obliquus*. *Arch. Environ. Contam. Toxicol.* *9*: 9–16.

Curry, R. R. (1975). Biogeochemical limitations on western reclamation: The high northern Great Plains example. In *Practices and Problems of Land Reclamation in Western North America*, M. K. Wali (Ed.). The Univ. N. Dak. Press, Grand Forks, North Dakota, pp. 18–47.

Eyster, C. (1958). The micro-element nutrition of *Nostoc muscorum*. *Ohio J. Sci.* *58*: 25–33.

Eyster, C. (1972). *Nostoc muscorum*, its mineral nutrition and its use as a bioassay organism. In *Taxonomy and Biology of Blue-green Algae*, T. V. Desikachary (Ed.). Symposium, Univ. Madras, Bangalore Press, Bangalore, India, pp. 480–520.

Fogg, G. C., Stewart, W. D. P., and Walsby, A. E. (Ed.) (1973). *The Blue-green Algae*, Academic Press, New York, 459 pp.

Foy, C. D., Channey, R. L., and White, M. C. (1978). The physiology of metal toxicity in plants. *Ann. Rev. Plant Physiol.* *29*: 511–566.

Goodman, D. (1975). The theory of diversity-stability relationships in ecology. *Q. Ref. Biol.* *50*: 237–266.

Hollings, C. S. (1973). Resilience and stability of ecological systems. *Ann. Rev. Ecol. System.* *4*: 1–23.

Hunt, M. E., Floyd, G. C., and Stout, B. B. (1979). Soil algae in field and forest environments. *Ecology* *60*: 362–375.

Jonescu, M. E. (1979). Natural revegetation of strip-mined lands in the lignite coal fields of southeastern Saskatchewan. In *Ecology and Coal Resource Development*, M. K. Wali (Ed.). Pergamon Press, New York, pp. 592–608.

Lindsay, W. L. (1977). Inorganic phase equilibrium of micronutrients in soils. In *Micronutrients in Agriculture*, J. J. Mortvedt, P. M. Giordano, and W. L. Lindsay (Ed.). Soil Sci. Soc. Am. Inc., Madison, Wisconsin, pp. 41–57.

Malakondaiah, N., Safaya, N. M., and Wali, M. K. (1980). Productivity potential of some stockpiled topsoils from coal mine lands in western North Dakota. 72nd Ann. Meeting Amer. Soc. Agron., *Agron. Abst.*, p. 171.

Metting, B. (1981). The systematics and ecology of soil algae. *Bot. Rev.* *47*: 195–312.

McCann, A. E., and Cullimore, D. R. (1979). Influence of pesticides on the soil algal flora. *Residue Rev.* *72*: 1–31.

Miller, R. M., and Cameron, R. E. (1976). Some effects of soil microbiota ontop soil storage during surface mining. *Fourth Symp. Surface Mining Reclam. NCA/BCR,* Louisville, Kentucky, pp. 131–139.

Miller, R. M., and Cameron, R. E. (1977). *Microbial Ecology Studies at Two Coal Mine Refuse Sites in Illinois.* NTIS, Springfield, Virginia, 41 pp.

O'Kelley, J. C. (1974). Inorganic nutrients. In *Algal Physiology and Biochemistry,* W. D. P. Stewart (Ed.). Univ. Calif. Press, Berkeley, California, pp. 610–635.

Pipe, A. E., and Shubert, L. E. (1984). The use of algae as indicators of soil fertility. In *Algae as Ecological Indicators,* L. E. Shubert (Ed.). Academic Press, London, pp. 213-233.

Raju, N. S. (1972). The blue-green algae from the soil of India. In *Taxonomy and Biology of Blue-Green Algae,* T. V. Desikachary (Ed.). Symposium, Univ. Madras, Bangalore Press, Bangalore, India, pp. 332–353.

Rosko, J. J., and Rachlin, V. W. (1977). The effect of cadmium, copper, mercury, zinc, and lead on cell division, growth and chlorophyll *a* content of the chlorophyte *Chlorella vulgaris. Bull. Torrey Bot. Club 104*: 226–233.

Shubert, L. E., and Starks, T. L. (1979). Algal succession on orphaned coal mine spoils. In *Ecology and Coal Resource Development,* M. K. Wali, (Ed.). Pergamon Press, New York, pp. 661–669.

Shubert, L. E., and Starks, T. L. (1980). Soil-algal relationships from surface mined soils. *Br. Phycol. J. 51*: 417–428.

Starks, T. L. (1982). *Ecosystem stability of surface mined lands.* Final Report, Mining & Miner. Resources Res. Inst., Univ. of North Dakota, Grand Forks, North Dakota.

Starks, T. L., and Shubert, L. E. (1979). Algal colonization on a reclaimed surface mined area in western North Dakota. In *Ecology and Coal Resource Development,* M. K. Wali (Ed.). Pergamon Press, New York, pp. 652–660.

Starks, T. L., and Shubert, L. E. (1982). Colonization and succession of algae and soil-algal interactions associated with disturbed areas. *J. Phycol. 18*: 99–107.

Starks, T. L., Shubert, L. E., and Trainor, F. R. (1981). Ecology of soil algae: a review. *Phycologia 20*: 65–80.

Trainor, F. R., Rowland, H. L., Lylis, J. C., Winter, P. A., and Bonanomi, P. L. (1971). Some examples of polymorphism in algae. *Phycologia 10*: 113–119.

Van Voris, P., O'Neill, R. V., Emanuel, W. R., and Shugart, H. H., Jr. (1980). Functional complexity and stability. *Ecology 61*: 1352–1360.

Velichko, I. M. (1968). The role of iron and manganese in the vital activities of blue-green algae of the genus *Microcystis*. *Mikroelem. Sel. Khoz. Med. Resp. Mezhved. Sb. (Kiev) 4*: 11–17.
Visser, S., Zak, J., and Parkinson, D. (1979). Effects of surface mining on soil microbial communities and processes. In *Ecology and Coal Resource Development*, M. K. Wali (Ed.). Pergamon Press, New York, pp. 643–651.
Wiessner, W. (1962). Inorganic micronutrients. In *Physiology and Biochemistry of Algae*, R. A. Lewin (Ed.). Academic Press, New York, pp. 267–288.
Woodmansee, R. G., and Duncan, D. A. (1980). Nitrogen and phosphorus dynamics and budgets in annual grasslands. *Ecology 61*: 893–904.

4

Microbial Development on Oil Shale Wastes: Influence on Geochemistry

RAYMOND E. WILDUNG and THOMAS R. GARLAND / *Battelle Pacific Northwest Laboratories, Richland, Washington*

I. INTRODUCTION

Soil microbial activity has long been established as a key factor influencing soil genesis and productivity and playing a major role in the chemical and physical aspects of soil fertility, soil structure, and water relationships. Microbial processes may therefore markedly influence the type of surface reclamation required for the oil shale industry as well as the necessity for, and type of, technology used in control of leachate migration to ground and surface waters. In a commercial oil shale industry, retorted oil shales would be produced in large volume and, after disposal to the ground, may be expected to become a permanent part of the landscape in many parts of the United States. In this sense, retorted shales may be viewed as nascent soils, initially at geochemical disequilibrium with their environment and subject to most of the processes important in soil formation.

This review will focus on the development of microbial processes in retorted shales and examine the possible effects of these phenomena on the form and solubility of carbon, nitrogen, sulfur, and selected trace elements. The limited information available will be discussed within the context of our understanding of the properties of the wastes, methods of disposal, and the role of microflora in soil formation and function as it may apply to these wastes.

II. RETORTED SHALE AND ASSOCIATED WASTES

A. Waste Sources

Most of the known U.S. oil shale reserves are contained in the Green River Formation of northwestern Colorado, northeastern Utah, and southwestern Wyoming (Donnell, 1964; Donnell and Blair, 1970). A number of in situ, modified in situ, and above-ground processes have been proposed for extraction of this resource. However, microbial processes are likely to play the most significant role in the behavior of wastes from surface retorting facilities. The locations of above-ground retorting facilities, the technologies to be employed for extraction, the likely disposal procedures as well as the geochemistry of wastes, and the potential impacts of disposal on water quality have been reviewed (Wildung and Zachara, 1980).

Conceptually, most of the above-ground retorting facilities will be similar in design. After mining, higher-grade oil shales will be crushed and heated (retorted) at 500–750°C until most of the organic matter is vaporized. The products (Shih et al., 1977; Routson et al., 1979) are shale oil and some low BTU gases; and the wastes are primarily retorted shale and retort water (the aqueous phase that condenses with the shale oil). Secondary products from the extraction and retorting process include overburden, low grade or off-size oil shale, and wastes from prerefinery distillation, gas stream purification, and retort water treatment (Hand, 1969; Cook, 1974a; DOI, 1977; Routson et al., 1979). However, due to their high volume, retorted shale and retort water will represent the primary problems in ground disposal.

B. Waste Characteristics

1. *Retorted Shale*

Retorted shale would be produced in quantities of up to 2 tons/barrel of oil produced. The mineralogy and chemical composition of the raw shale, in conjunction with the specific retorting process, govern the quantity and chemical and physical characteristics of the retorted shale and retort water produced.

Surface retorting technologies can be divided into two general categories based on their method of heat exchange to the raw shale. These have been the subject of several reviews (Cook, 1974a; Shih et al., 1977; Cameron Engineers, 1978) and include direct processes (combustion of part of the carbonaceous material as a heat source) and indirect processes (external heat source, heat carrier solids). Basic differences between the technologies include particle size of the raw shale feed, retorting temperatures, heating rates, use and composition of recycle gas, and degree of combustion of carbon. The direct heating processes must generally discard raw shale fines with particle sizes less than their feed requirements (Sladek, 1975).

Few published studies have investigated the redistribution of elements during the retorting process. However, 50–60% of the mercury (original weight percent), nitrogen and organic carbon, 20% of the sulfur, and 5% of the arsenic and selenium were shown to be redistributed to shale oil product, retort water, or off gases during the Paraho retorting process (Shendriker and Faudel, 1978; Fruchter et al., 1979). Most of the elements analyzed, however, were conserved in the retorted shale.

The physical and chemical properties of the Paraho, TOSCO II, and Lurgi retorted shales represent a range in process conditions likely to be encountered from aboveground retorting processes (Table 1). Yet, they are quite similar in their overall elemental composition. Differences arise principally in their aqueous chemistry. The Paraho process uses the direct, though partially anoxic, combustion of the carbonaceous residual as a heat source, and as a result produces a retorted shale with less residual organic carbon (Table 1) than those produced by indirect processes (Pfeffer and Kerr, 1974; Routson et al., 1979). The higher retorting temperatures (600–700°C) reached in the Paraho process induce the decarbonation of a portion of the carbonate minerals to their respective oxides, resulting in pH values of from 11 to 12 when contacted with water (Wildung et al., 1979a). Due to the minimum size required for raw shale feedstock, retorted shale produced by the Paraho process is considerably coarser than that produced by some indirect processes. Approximately 92% of the materials are sand size (0.05 mm or greater); 34% are greater than 8 mm (Striffler et al., 1974). TOSCO II is an indirect process that retorts at approximately 500°C and produces a residue with 3–6% organic carbon; 63–75% of the materials are <0.05 mm in size. Temperatures reached are probably adequate to induce the decomposition of at least a portion of the nahcolite, dawsonite, and analcite. The pH of the TOSCO II retorted shale, when mixed with water (1:1), is approximately 8–9 (Stollenwerk and Runnells, 1977).

The Lurgi-Ruhrgas is another indirect process that will likely produce several types of retorted shale that have been exposed to differing temperatures and extent of combustion. This will result from the use of a hot recirculating shale residue, heated by combustion to approximately 630°C, as the heat-exchange media. In this process, minus 6-mm raw shale is mixed with six to eight times its volume of hot circulating shale residue, raising its temperature to 530°C where the evolution of gas, shale oil vapor, and water vapor occurs (Schmalfeld, 1975; Shih et al., 1977).

The Union B process currently under most active commercial development, will likely be used in the initial commercial scale retorts (Cloninger, 1978). This indirect process produces a black carbonized

Table 1 Elemental Composition of Retorted Oil Shales

Component	Process: Paraho	TOSCO II	Lurgi (Overhead fines)
		Percent	
Organic C	3.0	3.2	0.3
Inorganic C	4.0	6.6	2.6
Total N	0.4	0.5	0.2
Na_2O	3.0	2.1	2.3
MgO	6.8	6.3	7.1
Al_2O_3	8.1	6.8	10.0
SiO_2	35.7	28.8	39.4
Total P	0.07	0.06	0.04
Total S	1.2	1.0	2.1
$SO_4(S)$	0.9	0.3	4.9
K_2O	2.5	1.6	3.1
CaO	16.1	20.6	14.8
Fe_2O_3	3.3	2.7	4.1
		μg/g	
B	120	—	—
NH_4	22	52	33
F	670	58	372
Cl	44	37	<20
Cr	34	30	41
Mn	365	325	395
Co	12	14	29
Ni	27	24	27
Cu	58	85	71
Zn	94	73	114
As	60	<30	<30
Se	3.1	—	—
Sr	814	945	800
Mo	42	26	38
Cd	0.6	<0.6	<0.6
Ba	576	530	526
Hg	0.03	—	—
Pb	19	17	<10

retorted shale residue similar to that produced by the TOSCO II process. Detailed chemical analyses of retorted shales from this process have not been published. The pH of saturated extracts of the Union B process is 8.5; the particle sizes are relatively coarse with 32–97% greater than 2 mm and 18–13% less than 2 mm (Cloninger, 1978).

2. *Retort Waters*

Estimates of retort water production vary with the site and process from 5 gal to 40 gal/barrel of oil, but even the lowest estimates will require major treatment and reuse-disposal efforts. The concentrations of nitrogen, sulfur, and soluble organic carbon are highly elevated in retort waters relative to retorted shales or natural waters (Table 2). The nitrogen and certain volatile sulfur and organic carbon compounds may be removed effectively by steam stripping. However, in view of the potential volumes that may be produced, the retort water, if codisposed with the retorted shales, may represent a major source of nonvolatile organic materials and reduced sulfur disposed to the ground.

C. Waste Disposal

General plans for the ground disposal of wastes from surface retorting operations have been published (e.g., DOI, 1977; Cloninger, 1978; Rio Blanco, 1981). The expansion in shale volume (10–40%) during mining, crushing, and retorting as well as technical and economical barriers have precluded the use of mines as primary disposal locations (Cook, 1974b; Pfeffer and Kerr, 1974; Routson et al., 1979). Therefore, a solid waste embankment will be required. An embankment of commercial dimensions has not been constructed, and it is clear that the designs will evolve as new information becomes available and is used by industry and regulatory agencies. Current knowledge has been summarized by Wildung and Zachara (1980, 1981) and Wildung et al. (1982) and illustrated in Fig. 1. Basically, in canyon-fill operations, catchment basins would be constructed and possibly lined with a sealant, perhaps highly compacted retorted shale, in an attempt to minimize or prevent water movement to subsoil and underlying alluvial and bedrock aquifers. The basin would be filled with retorted shale, possibly moistened with treated retort water to aid in dust control and facilitate compaction, and laid down and compacted in lifts. Drainage systems would divert runoff and leachates to evaporation ponds. Seasonal stream flows would be impounded above the disposal site and bypassed using conduits, flumes, and diversions. The latter would not be necessary in mesa, head-of-hollow, or basin-type fills, which appear to be favored by regulatory agencies, but the general design would be similar to canyon-fill operations.

Table 2 Elemental Composition of Paraho Retort Water

Component	Concentration
	g/liter
Organic C	42.0
Inorganic C	2.0
Total N	31.4
$NH_3(N)$	29.6
Total P	0.002
SCN	$\leqslant$1.0
SO_3	15.0
SO_4^+	6.65
S_2O_3	33.1
	μg/ml
B	2.1
F	52
Na	420
Mg	950
Cl	83
K	59.2
Ca	46.6
Cr	0.49
Mn	0.25
Fe	3.1
Co	0.30
Ni	0.68
Cu	22.7
Zn	5.5
As	10.7
Se	6.2
Sr	1.37
Mo	0.60
Cd	<0.10
Ba	0.25
Hg	0.001
Pb	0.32

Source: Adapted from Cataldo et al. (1983).

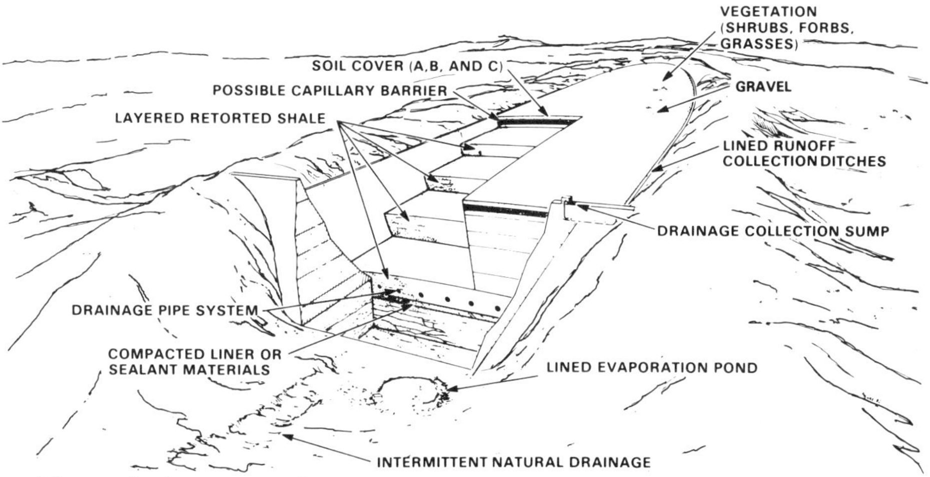

Figure 1 A conceptual retorted shale embankment. (From Wildung et al., 1982.)

Due to the high salt content and dark color (increased surface temperatures) of retorted shale, it may be recommended that the embankment be covered with a minimum of 60 cm of soil to facilitate revegetation (Redente et al., 1981). However, the magnitude of the proposed disposal operation and some early indication of favorable changes in retorted shale as a plant growth medium with time and weathering (Wildung and Zachara, 1981) would suggest that alternative treatments, such as direct revegetation in conjunction with gravel on the surface, might also be appropriate for steeper slopes subject to soil erosion.

Like natural terrain, the waste embankment would be subject to precipitation in the form of rainfall or snowmelt. The extent of moisture penetration of the surface, as well as the net direction of unsaturated water movement in the embankment, will depend on the surface and subsurface characteristics of the embankment and the balance between incoming precipitation, infiltration, runoff, evapotranspirational processes, and hydraulic properties of the retorted shale. The extent of erosion, moisture penetration and movement, and the possible accompanying mobilization and transport of potential pollutants are major uncertainties in disposal.

III. ESTABLISHMENT OF MICROBIAL COMMUNITIES IN RETORTED SHALE EMBANKMENTS

The temperatures of retorted shales on disposal (250–300°C) are sufficient to preclude microbial growth. However, as cooling occurs and the waste embankment is constructed, airborne particles from the surrounding soils will serve as potential inocula. This process, in conjunction with construction of the embankment in lifts over the 20–30-year lifetime of the plant (Fig. 1), will insure that the entire embankment will be exposed to the broad range of microorganisms associated with soil. If a soil overburden is used in reclamation, leachates from the soil will also serve as a continual source of microorganisms to the depth of water penetration.

Situations in which environments initially free of microbial life are subsequently colonized by microorganisms are not uncommon (Alexander, 1971). Where an initial microbial inoculum is available, colonization will be governed primarily by the physical and chemical properties of the material and the associated environment (Harrison, 1978). On the basis of chemical properties and general microbial growth requirements, the primary factors likely limiting initial microbial colonization in retorted shale would be pH in the case of the direct-heated retorting processes, e.g., Paraho, and high levels of soluble salts resulting from decomposition of minerals in both direct and indirect retorting processes. Although the upper limit of the optimum pH range for most microorganisms seldom exceeds pH 9.0, growth may occur up to pH 10–11 for strains of a number of common soil bacteria, e.g., *Thiobacillus* and *Nitrobacter*, and fungi, e.g., *Penicillium* (Thimann, 1967). Bacterial sensitivity to sodium salts appears to occur at concentrations approximating 0.01 M ± a factor of 10 (Thimann, 1967). Initial leachates from retorted shales are included in this range. Therefore, if other growth factors were not limiting (temperature, nutrient supplies, water potential, oxygen levels, other toxic substances), it would appear that pH or salt concentrations would not wholly limit microbial colonization.

The pioneering organisms would be expected to proliferate to a point where either nutrients become limiting or the physical and chemical composition of the environment changes (Alexander, 1971). In environments of low nutrient value, the death of pioneering organisms will provide nutrients for other species. Successive microbial populations will develop to accommodate the new substrates. Ultimately, a self-supporting, dynamic heterogeneous population should develop within the constraints of the physicochemical environment. This phenomenon is exemplified in the process of soil development, i.e., the formation of soil types with unique properties reflecting the influence of soil-forming factors such as parent material and climate on texture, structure, the weathering of minerals, and the formation of humus (Webley et al., 1963).

The processes of soil development in retorted shale, particularly microbial development, and the relationships to geochemistry will be discussed in the context of observations over a 5-year period on a lysimeter containing Paraho retorted shale alone and retorted shale with a 60-cm soil overburden. The lysimeter is located in the field at the Department of Energy Rifle Oil Shale Facility, Anvil Points, Colorado, and details of its design, construction and management have been described by Garland et al. (1979) and Harbert (1978).

By analogy to surface soils, the composition of retorted shale (Table 1) suggests that carbon, nitrogen, and phosphorus are available to support initial microbial growth provided these elements are sufficiently soluble and in forms available to the microorganisms. This is supported by examination of bacterial and fungal colony forming units (CFU) in the surface (0--5 cm) of the Paraho retorted shale after placement in the field lysimeter (Table 3). There was little difference with time or between the retorted shale and soil treatments. However, several factors might lead to differences in the distribution of organisms within these classes. The organic residue was formed under unique pyrolytic processes in the retort and, although largely uncharacterized, might represent a more selective nutrient source than native soil organic matter. Furthermore, as discussed above, the high concentrations of salts and elevated pH of retorted shales might be expected to result in some selective effect on initially colonizing species. Then effects may be illustrated by examination of the properties of organisms isolated from the surface of the lysimeter (Rogers et al., 1982). These studies showed that a single bacterial genus, *Micrococcus*, constituted 30% of the aerobic, heterotrophic bacterial population in June 1977. This bacteria was not

Table 3 Total Bacteria and Fungi in Retorted Shale and Soil, with Time, After Placement in a Field Lysimeter

Time of sampling[a]	Retorted shale[b]	Soil
	Bacteria (CFU)	
June 1977	2×10^6	2×10^6
April 1978	1×10^7	1×10^7
April 1979	2×10^7	1×10^7
	Fungi (CFU)	
April 1978	5×10^4	5×10^4
April 1979	1×10^4	6×10^4

[a]Retorted shale and soil added to lysimeter in April 1977.
[b]*Micrococcus* sp. constituted 30% of the bacterial population in retorted shale in June 1977; <1.0% in soils or in retorted shales thereafter.
Source: Adapted from Rogers et al. (1981).

present (<1.0%) in associated soils or in the retorted shales at later dates.

The development of a diverse, aerobic heterotrophic bacterial population in the retorted shale lysimeter after initial colonization appeared to be a successional process in which *Micrococcus* sp., uniquely adapted to use of the residual organic carbon under high salt conditions, dominated initially and were followed by other organisms that subsequently took advantage of the organic substrates provided by the initial colonizing species. Cultures containing retorted shale, a soil inoculum, and readily available organic nutrients exhibited a diverse, aerobic heterotrophic population after 7 days of growth with <1% *Micrococcus* sp. present, whereas cultures that did not receive organic amendment consisted of 100% *Micrococcus* sp. (Table 4).

Addition of phosphate provided selective advantage to *Micrococcus* sp. However, the nutrients sulfur and nitrogen were not limiting to microbial growth on fresh, retorted shales (Rogers et al., 1982). This might be expected from the substantial levels of sulfur (as sulfate) present (Table 1). Soluble nitrate concentrations in the Paraho shale were, however, quite low initially (next section) and little NH_4^+ might be expected at the pH levels encountered, suggesting that organic forms may have been the prime initial source of nitrogen.

Microbial development with depth will be a function of the same parameters operational on the surfaces, i.e., temperature, water content, oxygen supply, pH, organic and inorganic nutrients, and the presence of toxic substances (Alexander, 1967). These parameters will have complex interdependent effects. However, oxygen supply is a key factor governing the distribution of microorganisms and their activity with depth of an embankment. Oxygen diffusion to lower reaches of a retorted shale embankment may be quite limited as evidenced by the reduced oxygen levels in many groundwaters. Similar effects on the microbial population of a reported shale lysimeter are illustrated in Table 5.

Table 4 Effect of Complex Organic Substrates on the Growth of Microorganisms in Retorted Shale

Composition of organic medium[a]	Bacteria CFU (Log_{10})	*Micrococcus* sp. (%)
Complete	8	<1
No organic amendment	7	100

[a]Glucose (2 mg/ml), yeast extract (1 mg/ml), and casamino acids (5 mg/ml).
Source: Adapted from Rogers et al. (1981).

Table 5 Distribution of Aerobic and Anaerobic Heterotrophic Bacteria and Fungi with Depth in a Retorted Shale Lysimeter

	Bacteria[a]			
Depth (cm)	Aerobic/ faculative aerobic (A)	Anaerobic/ faculative anaerobic (B)	Ratio A/B[a]	Fungi[a]
0--30	1×10^6	2×10^4	50	5×10^4
30--60	3×10^5	7×10^3	42	8×10^2
60–90	1×10^4	$<10^2$	<40	3×10^2
90–120	4×10^3	$<10^2$	<40	10^2
120–150 (Near saturated zone)	9×10^5	2×10^5	<5	$<10^2$

[a]cfu/g soil.
Source: Adapted from Rogers et al. (1981).

Measurements were conducted a year after construction of the lysimeters when the total microbial population in the surface horizon (Table 3) of the retorted shale lysimeter had largely stabilized. The fungi, which are strictly aerobic, decreased with depth. This was likely in response to reduced diffusion of oxygen with depth. Reflecting the reduced oxygen levels, the ratio of aerobic to anaerobic organisms also decreased with depth. Although oxygen diffusion is reduced with depth, this effect may be accentuated if the pore spaces become saturated with water. Saturation of soils with water has been known to dramatically alter the soil microbial population (Parr and Papendick, 1971; Griffin, 1981). At the 120–150-cm depth, where water accumulated above a compacted retorted shale zone, the ratio of aerobic to anaerobic organisms was markedly reduced relative to the upper reaches of the lysimeter. An increase in the total number of organisms in both classes relative to the 60–120-cm depth likely resulted primarily from increases in the facultative anaerobic organisms in both classes. The water that accumulated at the 120–150-cm depth was enriched in organic carbon and other nutrients leached from the upper reaches of the lysimeter.

Although the lysimeter served as a microcosm reflecting many of the phenomena that would occur in a retorted shale embankment, it represented only a fraction of the depth of a commercial scale embankment (estimated to exceed 200 m at certain sites). A commercial scale embankment would also be much more heterogenous in composition, physical properties and water content with depth (Wildung et al., 1982). However, it is clear that successional changes in microbial populations will occur in retorted shales with time, and responses will vary with soil depth. The growth and metabolism of these populations, as well as their influence on biogeochemical processes, will also be a function of time and conditions at different locations in the embankment. The overall importance of the microflora in influencing the quality of leachates from retorted shales remains to be elucidated. However, a sufficient body of knowledge is available to define the nature of these processes and assign some priority to their importance.

IV. EFFECT OF MICROORGANISMS ON GEOCHEMICAL PROCESSES

There is wealth of information on the role of microflora in organic matter degradation and nutrient cycling in soils, but much remains to be learned about nonnutrient elements and specialized processes under unique or stressed environments, such as oil shale wastes. If recently disposed retorted shales may be viewed as in the first stages of soil development, offering initially a unique environment for microbial growth but progressing with time to conditions approximating native surface soils, examination of the key microbial processes in soil in the context of retorted shale properties and initial behavior may provide some indication of the nature and direction, if not the rate, of these processes in a future disposal site.

By analogy to soil systems, microorganisms must play a significant role in transformations of carbon, nitrogen, sulfur, and a number of trace elements (Routson and Wildung, 1969). This may occur through: (1) immobilization/mineralization reactions resulting from incorporation of these elements into cells and/or release on decomposition of cellular or other organic residues; (2) indirect mechanisms such as changes in pH or Eh (redox potential), which affect mineral solubilities, surface reactions, or solution speciation; and production of exocellular ligands, which may affect mineral weathering, surface reactions or solubility; and (3) direct transformations such as valence transformations or alkylation (Wildung and Garland, 1979; Wildung et al., 1979b). Ultimately, all of these processes have potential for operating to some extent under the conditions projected for retorted shale disposal. The probability of occurrence of soil-associated processes would increase as the properties approach those of adjacent soils, as a result either of retorting conditions or weathering processes (mineralogical changes, leaching, microbial development) after disposal. Unequivocal evidence for the operation of any of the microbial processes in retorted shale alone is very limited, but some circumstantial evidence is at least consistent with a role for the microbes in (1) and (2), and direct evidence is available for their participation in (3). All of the processes have potential for influencing the composition of waters percolating through retorted shales.

In field studies, the compositions of percolates from surface-retorted shales over periods of several years under natural weathering conditions have been examined (Ward and Reinecke, 1972; Harbert and Berg, 1978; Garland et al., 1979). The maximum observed concentrations of several major elements in percolates from Paraho retorted shale and a representative adjacent soil during the first year in the field lysimeter referred to above are compared with Colorado River water in Table 6. The retorted shale leachates were elevated in lithium, boron organic carbon, fluorine, sodium, magnesium, aluminum, sulfur (10% was in the reduced $S_2O_3^{2-}$ form), chlorine, potassium, calcium, chromium, iron, arsenic, molybdenum, and mercury. In initial leachates at least, these are the elements with greatest potential for entering natural waters in the vicinity of a waste disposal site, assuming minimal attenuation by substrata. Of the soluble elements elevated in concentration, carbon, sulfur, iron, arsenic, and selenium are most likely to be influenced by known microbial processes. The NO_3 levels are also of interest because nitrogen is one of the few elements that is initially present in the retorted shale leachate at lower concentrations than the irrigation water applied to the lysimeter and because its chemistry is closely linked to microbial processes in soils. These elements are also present in significant quantities in the retorted shale (Table 1) and retort water which might be codisposed with the retorted shale (Table 2). Therefore, the remainder of this discussion will focus on these elements.

A. Carbon

Of the many possible metabolic processes occurring in oil shale wastes, decomposition of aliphatic hydrocarbons and aromatic compounds likely represent two major pathways through which substances containing organic carbon are converted to carbon dioxide in aerobic systems or to methane under anoxic conditions. Most hydrocarbons, including compounds exotic to the soil systems, such as gasoline, tars, and rubbers, are decomposed relatively rapidly in soil. Hydrocarbons unsaturated with respect to hydrogen, branched, and of high molecular weight are generally more susceptible to degradation than their saturated, unbranched, low-molecular-weight analogs (ZoBell, 1948).

Aromatic materials are generally considered quite resistant to microbial and chemical degradation in soil, and carbon in aromatic forms constitutes a major portion of the relatively stable soil organic fraction. However, it is likely that most naturally occurring aromatic hydrocarbons can be metabolized by soil microorganisms, although the rate may vary depending on the environmental constraints. Once the aromatic ring is cleaved, the resulting straight-chain organic acids

Table 6 Composition of Leachates from Paraho Retorted Shale and Soil in a Field Lysimeter Relative to the Composition of Colorado River Water Used in Lysimeter Irrigation

	Lysimeter leachates[a]		
Component	Paraho retorted shale	Soil	Colorado River water used for irrigation
pH	11.1	7.9	8.3
Conductivity (mmhos/cm)	30	6.2	3.2
	mg/l		
Li	21.6	0.09	0.09
B	2.2	1.7	0.3
Organic C	241	82	32
Inorganic C	39	167	60
NO_3	<0.1	41	20
F	12.5	1.4	0.5
Na	9,050	840	400
Mg	6.8	256	150
Al	0.1	0.025	0.25
SiO_2	27	27	2.6
SO_4^+	20,000	2,050	1,070
$S_2O_3^{2-}$	2,100	<10	<10
Cl	1,280	520	140
K	1,460	90	8.2
Ca	513	170	180
Cr	0.001	0.014	0.001
Fe	0.3	<0.1	0.02
As	<0.07	0.006	<0.001
Se	0.010	0.004	0.007
Sr	12	3.0	2.5
Mo	5.6	0.6	0.04
Hg	<0.001	<0.001	<0.001

[a]Maximum value attained during the first year of weathering in the field.

Source: Adapted from Garland et al. (1979).

are subject to relatively rapid degradation and oxidation to the lowest energy states, carbon dioxide and water.

Degradation rates for organic compounds associated with oil shale wastes will be dependent upon the suitability of the compounds as energy or nutrient sources and on retorted shale environmental conditions that can influence microbial growth rates.

1. Oxygen Supply

Organic matter decomposition is essentially an oxidative process with oxygen serving as the primary electron acceptor. Certain microorganisms may oxidize carbon under anaerobic conditions with an alternate electron acceptor; however, this is limited to a fairly narrow range of compounds. Factors limiting the supply of oxygen would, therefore, tend to lower decomposition rates, except in the surface horizons of a retorted shale embankment.

As illustrated previously, saturation of a portion of the embankment with water or construction of an embankment with depth sufficient to limit oxygen diffusion to lower reaches of the pile would effectively alter the microbial population (Table 5) and dramatically limit decomposition. A succession of microorganisms has been observed as oxygen levels decrease with time (Takai and Kamura, 1966). Under largely aerobic conditions (>-0.3 bar water tension), oxidative decomposition of organic substrates proceeds with the formation of carbon dioxide from the activity of aerobic and facultative anaerobic microorganisms. Little or no accumulation of incompletely oxidized compounds occurs. As the level of oxygen decreases, the microbial community becomes dominated by facultative anaerobes and some accumulation of incompletely oxidized organic compounds, such as organic acids and alcohols, will occur. Correspondingly, nitrate, manganic, and ferric ions will be reduced successively. As the redox potential drops further to a level where oxygen is present in minimal quantities, obligate anaerobes begin to dominate the microbial community. Decomposition rates decrease because energy yield is lower and alternate electron acceptors are present in lower concentrations. The formation of sulfide ion, molecular hydrogen, and methane will occur as sulfate, hydrogen, and carbon dioxide are progressively used as alternate electron acceptors. Such changes in microbial populations will have a marked effect on the rate and direction of decomposition and solubility of nitrogen and trace metals, with possible corresponding effects on leachate water quality.

2. Retorted Shales/Retort Waters

Weathered retorted shale has been shown to contain extractable alkanes, fatty acids, alkylbenzenes, and polycyclic aromatic hydrocarbons (Pereira et al., 1981). Although the alkanes and fatty acids may have originated from biological activity after retorting, the aromatic compounds likely originated from the pyrolysis of kerogen and may be expected in most retorted shales. A portion (20%) of the benzene-extractable residual carbon (~0.2 weight %) is water soluble (Schmidt-Collerus, 1974).

Thus, although the residual organic carbon in retorted shales seldom exceeds 6% (Table 1) and contains aromatic forms that are more resistant to degradation, a fraction is soluble and available for sole support of initially colonizing species (Table 4). Furthermore, the carbon in retort water (Table 2), if codisposed, is mostly water soluble with a major fraction consisting of aliphatic acids (Riley et al., 1979; Leenheer et al., 1982), which are readily degradable by microorganisms (Rogers et al., 1981). The benzoic acids are also degraded, but at a slower rate. Thus, there would appear to be sufficient biologically available carbon to support an initially limited microbial population under most disposal scenarios.

The concentration of organic carbon in leachates from retorted shale over a 5-year period is shown in Fig. 2. An initial pulse of leached organic carbon during the first year (0.1 column volume of water) was followed by several years in which organic carbon levels

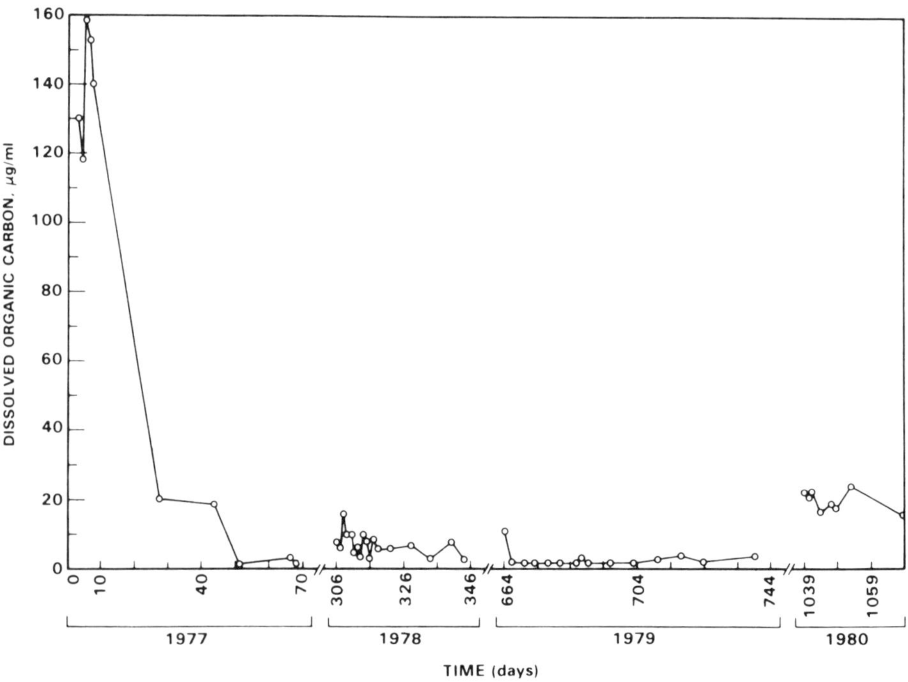

Figure 2 Concentration of dissolved organic carbon in leachates from retorted shale in field lysimeters. (From Wildung and Zachara, 1981.)

were minimal. The pulse of organic carbon in percolates during the first year likely reflected simple displacement of soluble residuals with minimal microbial influence. Noteworthy, however, was the initial small increase in organic carbon eluted with the first percolate of the year after weathering over winter. This effect, which occurred each year and was most pronounced after 4 years, may have resulted from the microbial breakdown of residual organic carbon into more soluble, lower-molecular-weight components. The relationship of this process to nitrogen behavior and possible effects on the solubility of other elements will be discussed in subsequent sections.

B. Nitrogen

Nitrogen exists predominantly in organic combinations in surface soils with no more than 2–3% of the total being present in inorganic forms (Bremner, 1965). The rate of assimilation or incorporation into new protoplasm is dependent upon the availability of organic carbon. The carbon:nitrogen ratio of most soils is in the range of 8–12. In general, if a soil additive contains a carbon:nitrogen ratio >38, the inorganic nitrogen reserves of the soil will be depleted rapidly. If the soil additive contains a carbon:nitrogen ratio <25, decomposition of the material results in an immediate increase in inorganic nitrogen. Decomposition of materials containing carbon:nitrogen ratios between 25 and 38 typically results in temporary net immobilization of nitrogen followed by net mineralization. The carbon:total nitrogen ratio of freshly retorted shales and retort waters range from 1.4 to 8.2 (Tables 1 and 2). Sufficient nitrogen should therefore be present for organic carbon degradation in retorted shales, for incorporation into cellular material, and ultimately for humus formation provided other conditions (pH, salt content, and Eh) become favorable with weathering. Gradual increases in the soluble organic carbon content of retorted shale leachates in the second and subsequent years of residence in the environment (Fig. 2) may reflect microbial degradation of organic carbon with subsequent utilization of residual nitrogen after weathering to reduce soluble salts and pH to levels commensurate with significant microbial activity.

The concentration of nitrate in percolates from the retorted shale lysimeter is given in Fig. 3. Initially, nitrogen entering retorted shales as nitrate in irrigation and rain waters (Table 6) was not present in the leachates. However, beginning in the second year, nitrate began to be eluted (Fig. 3). After the fourth year, nitrate was present in sufficient quantities to degrade natural waters.

These phenomena appear to reflect a unique combination of chemical and microbiological controls. The initial absence of nitrate likely reflects the chemical conversion of oxidized forms to ammonia and its loss to the atmosphere at the high pH (pH 11) of the freshly retorted shale. However, at lower pH values (pH 9–10), significant amounts of

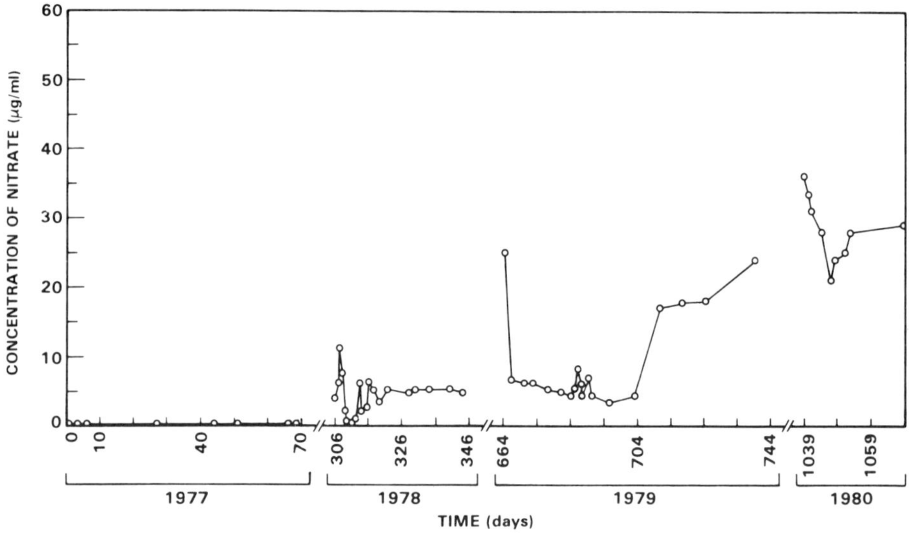

Figure 3 Concentration of nitrate in leachates from retorted shale in field lysimeters. (From Wildung et al., 1982.)

ammonium ion would be expected in most aqueous solutions, reducing losses due to volatility. Reductions in pH did occur with weathering in the lysimeter over a two year period (Fig. 4, discussed below), but these changes did not result in significant concentrations of nitrate in leachates from the retorted shale. This suggests that any ammonium formed during the first year was either sorbed by the retorted shale or was volatilized.

Nitrogen can be lost from soil by volatilization either as NH_3, N_2, or N_2O produced through denitrification processes carried out by a select group of facultative anaerobic organisms. Facultative anaerobic organisms were present in the retorted shale lysimeter, increasing with depth and reduced oxygen supplies (Table 5). Organisms such as *Thiobacillus denitrificans* are known to grow at pH values up to 10.7 (Thimann, 1967). However, the denitrification rate is a direct function of aeration, with maximum denitrification occurring under anaerobic conditions (Alexander, 1967). The depth of the lysimeter and the limited thickness of the water saturated region in which anaerobic conditions may have occurred (Table 5), in conjunction with the marked increase of NO_3^- in leachates beginning in the second year, suggest that microbial denitrification in the first year did not result in significant losses of nitrogen from the retorted shale lysimeter.

If the nitrogen entering the retorted shale in irrigation waters was not volatilized, but chemically converted to ammonium and sorbed,

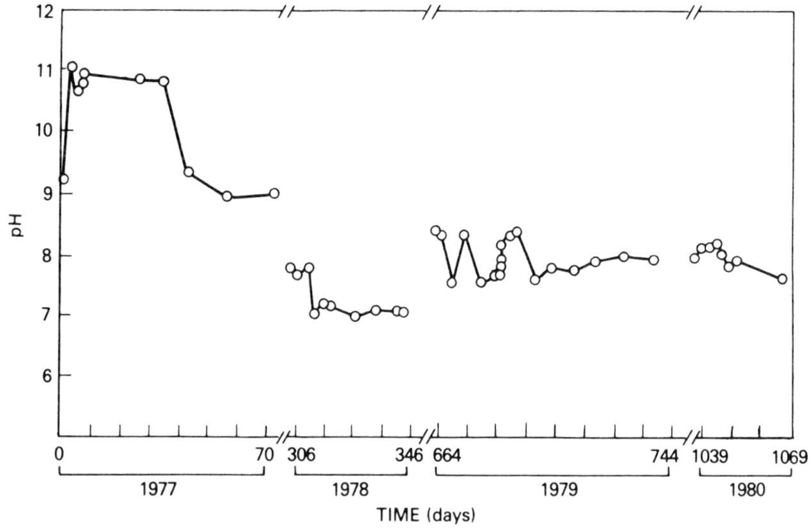

Figure 4 The pH of leachates from retorted shale in field lysimeters. (From Wildung and Zachara, 1981.)

a significant reservoir of nitrogen may be expected to develop in the retorted shales over the first few years under an irrigated reclamation regime. The microflora may then be expected to play an important role in governing its subsequent behavior through mineralization processes.

Mineralization of soil nitrogen involves conversion of the organic nitrogen forms to mineral forms. Organic forms of nitrogen exist in shale oil (Riley et al., 1979) and retort waters (Table 2). It is therefore possible that they may also contribute to the small quantities of nitrogen in retorted shale (Table 1). However, it is likely that mineralization would be most significant in the latter stages of weathering after biological activity results in increased levels of nitrogen in organic forms.

The long-term behavior of nitrogen in retorted shale leachates could be explained on the basis of the nitrification process. A viable nitrifier population may have developed in response to reductions in pH during the latter half of the first year (Fig. 4). The nitrifying organisms have pH optima in the range of 8.5–9.0 with minimal growth occurring in solution above pH 9.5 (Thimann, 1967). These organisms may have been responsible for the progressively increased levels of nitrate observed in subsequent years.

Under most conditions, nitrite does not accumulate in the soil; however, organisms involved in the conversion of nitrite to nitrate appear to be quite sensitive to the ammonium that predominates under alkaline conditions. This ammonium does not significantly influence the ammonium oxidizers (Stojanovic and Alexander, 1958). Thus, conditions in retorted shale might restrict conversion of nitrite to nitrate, resulting in an accumulation of nitrite, which is highly toxic. Nitrite was present at levels up to 4 μg/ml in the retorted shale leachates during 1978, equalling the nitrate levels at certain time periods (Fig. 3). However, nitrite remained at concentrations less than 0.1 μg/ml throughout 1979 and 1980, suggesting that nitrate may indeed accumulate in the early stages of leaching but that, as weathering progresses, the conversion of nitrite to nitrate is not the rate-limiting step in nitrification.

C. Sulfur

Oil shale is a relatively complex marlstone. The high temperatures of most surface retorting processes induce decarbonization of a portion of the dolomite ($CaMg[CO_3]_2$), nahcolite ($NaHCO_3$), and some of the compound carbonates (e.g., shortite, $Na_2Ca_2[CO_3]_3$). The hydration of the resulting oxides (Table 1) upon wetting results in leachates with high levels of soluble calcium and sodium and high pH (10–12) relative to most natural waters of the region (Table 6). With the low oxygen fugacity maintained during retorting, pyrite (FeS_2), and perhaps organically bound sulfur, is liberated and reacts with hydrogen present in the kerogen or generated during retorting to produce pyrrhotite (FeS) and H_2S (Burnham et al., 1981). The H_2S produced may be sorbed by or react with the calcite remaining (Smith et al., 1978), possibly forming a CaS. It may be this intermediate that upon hydration and oxidation results in the high levels of sulfate and thiosulfate observed in the initial leachate (Table 6).

In most laboratory and field studies, soluble chemical species (salts, organic carbon, sulfur compounds) are released from retorted shale in a concentrate pulse in the first 0.1–1.0 pore volume of leachate. The concentrations of these components generally decrease rapidly with leaching (water volume) suggesting removal of readily soluble amorphous by-products of retorting. Accompanying these changes, the pH of the leachates from retorted shales weathered in the field lysimeter decreases markedly the first year (to as low as pH 7; Fig. 4) and remains in the range of local soils in subsequent years (Garland et al., 1979). This pH reduction is highly significant as it, alone, markedly increases leachate quality, but it is also important because of possible accompanying chemical and biological changes that tend toward the properties of local soils with time. Associated with this pH change, the concentration of the reduced sulfur compound, thiosulfate, is also depleted (not shown). Noteworthy, it is depleted in the field lysimeter at rates exceeding those in laboratory column leaching studies. However, if retorted shale is subjected to short-term, wet-dry/warm-cold cycles in the laboratory, the reductions in leachate pH and thiosulfate concentrations can be accelerated and the change is accompanied by an increase in Eh (Silviera et al., 1978). These data suggest that the pH and Eh regimes of the leachates are regulated by complex solubility associations with the relatively soluble $MgO/CaO/Na_2O/CaS$ residues remaining after retorting. The oxidation of reduced sulfur species such as sulfide and thiosulfate as well as recarbonization would lower leachate pH.

What role could the microbes play in this process? The oxidation of inorganic sulfur in soils is quite complex, and not all the reactions involved can be considered strictly enzymatic or biological. Elemental sulfur, sulfide, and thiosulfate may be oxidized chemically, but oxidation may also be accomplished through microbial processes. The microbial sulfur oxidation process is analogous to nitrification, and the similarity extends to physiology of the organisms involved as well as to the essential environmental conditions. A general pathway for the chemoautotrophic oxidation of sulfur in soils may involve the conversions: sulfide to thiosulfate to tetrathionate to trithionate to sulfite to sulfate, with sulfite also being capable of direct conversion to thiosulfate (Alexander, 1967). However, the chemical instability of many of the proposed intermediates makes the process difficult to study.

Although a number of soil microorganisms may be involved in sulfur oxidation, members of the reportedly chemoautotrophic genus *Thiobacillus* are probably the most important. With one exception, these organisms are obligate autotrophs deriving energy from sulfur oxidation, utilizing carbon dioxide or bicarbonate as sources of carbon for growth. From the standpoint of predicting the fate of sulfur in wastes, knowledge of the requirements of the individual organisms participating in the various phases of sulfur oxidation is important, since individual organisms of different environmental requirements participate in unique phases of sulfur oxidation. For example, *T. thiooxidans* is a chemoautotroph oxidizing elemental sulfur and having optimum growth at pH 3, whereas *T. thioparus*, a chemoautotroph capable of oxidizing thiosulfate and tetrathionate, is acid sensitive. Thus, it is entirely feasible that sulfide and thiosulfate oxidation was conducted, at least in part, through chemoautotrophic processes.

Indirectly, the heterotrophic microbial population (Table 3) may also play a role in pH reduction by oxidation of reduced forms of carbon (discussed above) and by increasing the partial pressure of carbon dioxide in the system. Increased carbon dioxide levels would accelerate the rate of solid-phase recarbonization and provide the necessary sources of carbon dioxide and bicarbonate for chemoautotrophic oxidation processes. Commensurate with this process, the ratio of sulfate to carbonate-bicarbonate in the leachate gradually decreases (Fig. 5) as water percolating through retorted shale removes sulfate and recarbonization proceeds. After 4 years, the ratio more closely approximates that of local soils.

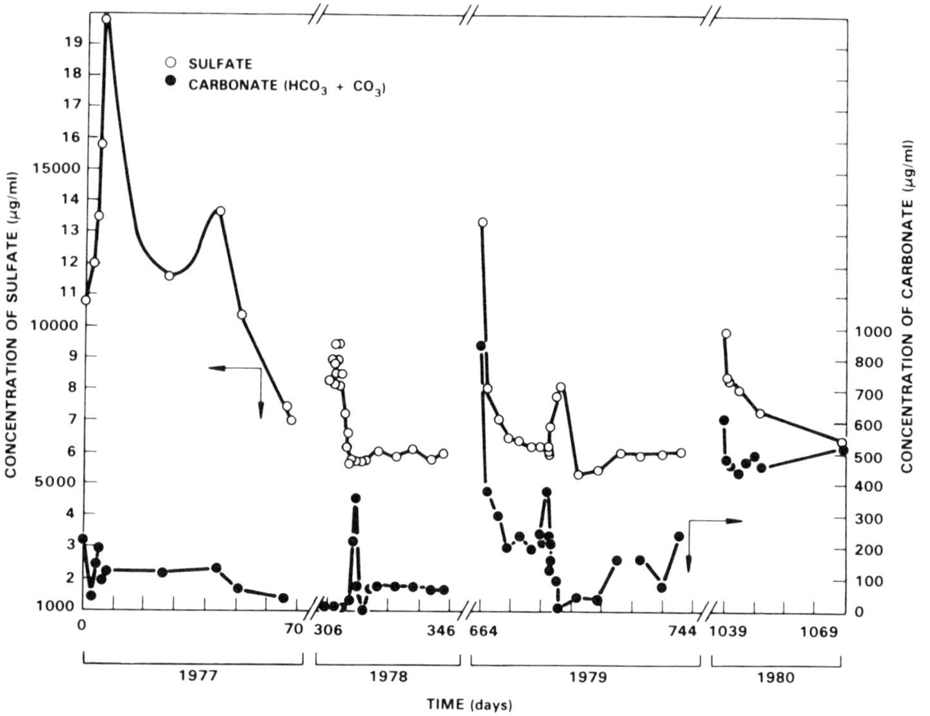

Figure 5 Concentrations of sulfate and carbonate-bicarbonate in leachates from retorted shale in field lysimeters. (From Wildung et al., 1981.)

Sulfate may be reduced in soils under anaerobic conditions but not by a reversal of the above oxidation pathways. The bacteria that produce sulfide from sulfate, sulfite, thiosulfate, tetrathionate, or sulfur are obligate anaerobes. Probably the most important of these organisms is *Desulfovibrio desulfuricans*, which utilizes sulfate, sulfite, and thiosulfate and grows best in media below pH 5.5. Sulfide was never detected in leachates from the retorted shale lysimeters. This may result from the limited area in the lysimeter (120–150-cm depth; Table 5) in which significant numbers of strict anaerobes may have existed, from the relatively low pH requirements for sulfur-reducing bacteria compared with the retorted shale, or from other nutritional factors. In deeper embankments, the opportunity for growth of strict anaerobes and effects on geochemistry may be greater.

D. Trace Elements

All of the possible mechanisms for microbial modification of the geochemistry of the retorted shales outlines above have potential for impacting indirectly or directly on the solubility and speciation of trace elements. The possible formation of CaS intermediates during retorting may result in coprecipitation of a number of trace elements as sulfides. Oxidation of metal sulfides by microorganisms has been demonstrated (Silver and Torma, 1974), and this process may represent a long-term source of a number of trace elements in retorted shale leachates. However, of the trace elements known to be subject to modifications in behavior by microbial processes, only arsenic, iron, and selenium exhibited elevated concentrations in initial retorted shale leachates (Table 6). These elements have been shown to be subject to valence alteration by microorganisms. Additionally, arsenic and selenium are also subject to alkylation by microorganisms, leading to increased toxicity. It is unlikely that elevated iron levels would pose a significant problem in water quality. The levels of iron (Table 6) do exceed what might be expected from inorganic solubility considerations. It is possible therefore that the elevated iron solubility in initial leachates that accompanied high soluble organic carbon levels may be due to organic complexation or organic carbon-related redox reactions. By analogy to soils, ligands responsible for iron complexation might also impact copper, zinc, cadmium, and nickel with time. Although none of these elements is elevated in total concentration in retorted shales (Table 1) relative to most soils, differences in mineralogy and chemistry of the retorted shales compared with soils, both initially and with residence time in the environment, suggest that some examination of possible complexation reactions is appropriate. Therefore, this discussion will focus on microbial valence transformations, alkylation, and complexation as they may apply to arsenic, selenium, and the transition elements.

1. Valence State

Alteration of valence state by microorganisms is closely related to the oxidation state of the system. As discussed above, a succession of microorganisms will occur in response to decreased oxygen, and this will be accompanied by utilization of a series of alternate electron acceptors. Facultative anaerobes are known to reduce iron and manganese in this process increasing solubility by orders of magntiude.

Arsenic is not reduced to the elemental form. However, a number of soil microorganisms are capable of reducing arsenate to arsenite (Bautista and Alexander, 1972). Microorganisms are also capable of reversing these reactions under appropriate conditions. Arsenite oxidation to arsenate has been observed for a number of *Pseudomonas* sp. (Turner, 1981). Oxidation occurred early during the growth of these organisms in the presence of arsenite and was strongly inducible; noninduced cells were inactive.

Unlike arsenic, inorganic selenium can be reduced biologically to its elemental form (Bautista and Alexander, 1972). It has been estimated that 22% and 50% of the bacteria in soil are capable of reducing selenate and selenite, respectively, to elemental selenium, indicating high probability for these transformations in the terrestrial environment.

It is clear that the chemical speciation of arsenic and selenium may vary markedly, depending upon oxygen supplies and the microbial population. However, little is known of the importance of these processes in wastes or in strata beneath wastes. Increases in soluble arsenic and selenium did not occur with time after the initial solute pulse in the retorted shale lysimeter. This is consistent with microbial assays (Table 5) which indicated that the potential for highly reduced conditions occurred only in a limited area of the lysimeter and with the behavior of carbon, nitrogen, and sulfur, which appeared to reflect overall oxidative conditions.

2. *Alkylation*

The methylation of arsenite and arsenate has been observed in studies with fungi, bacteria, and algae in defined growth media amended with arsenic. Methylation has been observed in both aerobic and anaerobic studies. A culture of the strict anaerobe *Methanobacterium* has been found to generate dimethylarsine from arsenite under anaerobic conditions (McBride and Edwards, 1977). This organism is known to be present in reduced groundwaters. Furthermore, a number of aerobic microorganisms have been shown to form methyl arsenicals. Two fungi, *Scopulariopsis brevicaulis* and *Aspergillus niger*, were shown to produce trimethyl arsine from various arsenic compounds (Challenger et al., 1954); *Candida humicola* produced trimethyl arsine from the precursors, arsenite, arsenate, monoethylarsenate, and dimethylarsenate (Cox and Alexander, 1973); and the formation of mono- and dimethylarsenate has been observed with a number of marine algae (Andreae and Klumpp, 1979).

Methylarsenic and dimethylarsenic acids are known to be present in surface, ground, and estuarine waters (Braman and Foreback, 1973). In over half the waters, the concentration of the methylated arsenic components totaled greater than 50%, and, in a few cases, approached 75% of the total soluble arsenic concentration. These observations in conjunction with the presence of methanobacteria and methylarsenicals in groundwater suggest that methylation of arsenic may occur under reduced conditions in retorted shale embankments.

Demethylation of arsenic may also occur, and the rate of demethylation of methylarsenic acid and dimethylarsenate in sediments is dependent on their distribution between soluble and bound forms (Holm et al., 1980). Similar studies (Klein et al., 1979) suggest that arsenic, methylarsenic acid, and dimethylarsenate may be microbiologically converted to volatile forms (e.g., methyl- and dimethylarsine) in aqueous suspensions of retorted oil shale and soil. In these studies, the addition of organic nutrients to retorted oil shale-oil suspensions markedly enhanced volatile arsenic formation, and the transformations occurred under both aerobic and anaerobic conditions, but were slightly more rapid under aerobic conditions. In the retorted shale field lysimeter, optimum conditions for this type of reaction occurred at the 120–150-cm depth (Table 5).

The formation of several methylated forms of selenium has been reported to occur in sewage sludge and soil (Reamer and Zollner, 1980). Methylation occurred with both selenite and selenate amendments and was two- to three-fold greater under aerobic than anaerobic conditions. The formation of dimethyl selenide has been demonstrated for 11 soil fungi (Barkes and Fleming, 1974).

It is clear that there is potential for alkylation of arsenic and selenium in retorted shale disposal sites under both aerobic and anaerobic conditions. Methylated derivatives of arsenic or selenium have not been measured in retorted shale leachates; however, specialized sampling would be required due to the high volatility of these chemical species. Given the possible increased mobility of alkylated species in retorted shales and substrata that might occur and the increased toxicity associated with these compounds, further investigation of alkylation phenomena may be warranted, particularly with respect to their effects in deeper embankments where oxygen is restricted.

3. *Complexation*

Transition metals form complexes with a number of organic ligands, many of which are known microbial metabolites (Siegle, 1973, 1974; Neilands, 1979). Much evidence supports the strong relationship between soluble organic carbon and trace metal solubility in soils and sediments (Jenne, 1977), and bacteria have been implicated in the solubilization of some minerals (Duff et al., 1963). However, little is known of the role of microbial extracellular ligands on the behavior of metals in soils and wastes (Cataldo and Wildung, 1983). Bacteria and fungi isolated from soil on the basis of metal tolerance and carbon source have been shown to produce ligands capable of complexing cadmium, nickel, chromium, and plutonium (Drucker et al., 1978; Wildung and Garland, 1979; Wildung et al., 1979b). Several fungi were shown to produce low-molecular-weight, mixed ligand, neutral, and anionic extracellular complexes that markedly increased plutonium and nickel solubility in soil (Wildung and Garland, 1979; Wildung et al., 1979b). It is likely that these processes would play a role in trace element solubility in retorted shales as a viable microbial population develops, geochemical solubility controls permitting.

Microbial complexes would likely contribute most to the maintenance of the aqueous concentration of some elements at levels considerably higher than those maintained by precipitation-dissolution reactions alone. There is some presumptive evidence for the participation of microbes in these processes in retorted shales. Comparison of the solubility of chromium and nickel (Fig. 6) with the behavior of carbon (Fig. 2) in the retorted shale lysimeter over time shows a major increase in chromium and nickel solubility after 4 years of weathering. This corresponds to a similar pulse in organic carbon, which may have been derived from microbial metabolism as discussed previously. It is possible that the increased solubility of organic carbon was due to the formation of lower-molecular-weight metabolites from residual organic carbon associated with the retorted shales. These compounds may have included ligands capable of complexing Cr(III) and nickel, increasing their solubility and mobility in the retorted shales. However, absolute evidence for the importance of these processes in soils or retorted shales remains to be developed.

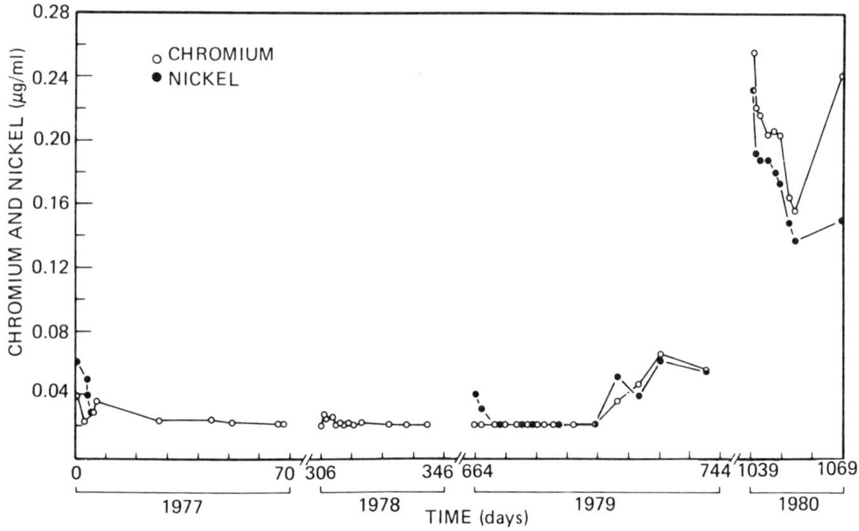

Figure 6 Concentrations of chromium and nickel in leachates from retorted shale in field lysimeters. (From Wildung and Zachara, 1981.)

V. SUMMARY AND RESEARCH NEEDS

Biologically mediated phenomena will likely play a role in the geochemistry of disposed oil shale wastes. The significance of these effects in altering the mobility of important elements in solid wastes and in moderating the attenuation of waste solutes in geologic materials surrounding a disposal site will be mutually dependent on the waste characteristics and the nature of the microbial community that develops in the surface and subsurface environment.

Microbial activity will be controlled to a large extent by inoculum availability, organism density, nutrient and oxygen status, and the presence of toxic substances. The planned construction of solid waste disposal sites in lifts would provide ample opportunity for inoculation of the entire waste embankment by organisms associated with windborne soil particles. Oil shale wastes represent an initially stressed environment for microorganisms because they contain chemically reduced nitrogen and sulfur species, have high soluble salt concentrations, and exhibit high pH values. However, microbial colonization and succession does occur in this environment, initially by a select group of organisms. Analogous to soil-forming processes, geochemical weathering reactions with time and leaching in the field tend to improve the conditions for microbial growth and succession. Laboratory and field evidence is presented, which is consistent with, but not proof of, the participation of microorganisms in waste geochemical reactions, including organic matter decomposition, recarbonation, nitrogen and sulfur, oxidation and the subsequent effects of these processes on the pH and chemical composition of leachates.

The rate and duration of these processes over extended time (>10 years) and their quantitative significance in wastes disposed on a commercial scale remain to be elucidated. There is a particular need to examine possible microbial transformations in water-saturated zones and deeper embankments where limited oxygen supplies after construction may result in changes in the solubility and speciation of trace elements.

To estimate the importance of particular microbial phenomena, it is necessary to establish kinetic rate constants and/or the stoichiometry of specific transformations in relation to organic carbon utilization over a range of environmental conditions. Field studies must then be conducted at selected disposal sites as the industry develops to establish whether biological transformations, which alter the form and mobility of waste components, occur in waste and substrata at environmentally significant levels.

ACKNOWLEDGMENTS

This work was supported by the U.S. Department of Energy under contract DE-AC06-76RLO 1830 with Pacific Northwest Laboratory, Richland, Washington, operated by Battelle Memorial Institute.

REFERENCES

Alexander, M. (1967). *Introduction to Soil Microbiology.* John Wiley and Sons, New York.

Alexander, M. (1971). *Microbial Ecology.* John Wiley and Sons, Inc., New York.

Andreae, M. O., and Klumpp, D. (1979). Biosynthesis and release of organoarsenic compounds by marine algae. *Environ. Sci. Technol. 13*: 738–741.

Barkes, L., and Fleming, R. W. (1974). Production of dimethylselenide gas from inorganic selenium by eleven soil fungi. *Bull Environ. Contam. Tox. 12*: 308–311.

Bautista, E. M., and Alexander, M. (1972). Reduction of inorganic compounds by soil microorganisms. *Soil Sci. Soc. Am. Proc. 36*: 918--920.

Braman, R. S., and Foreback, C. C. (1973). Methylated forms of arsenic in the environment. *Science 182*: 1247–1249.

Bremner, J. M. (1965). Organic nitrogen in soils. In *Soil Nitrogen,* W. V. Bartholmew and F. E. Clark (Ed.). American Society of Agronomy, Madison, Wisconsin, pp. 93–149.

Burnham, A. K., Bey, N. K., and Koskinas, G. J. (1981). Hydrogen sulfide evolution from Colorado Oil Shale. In *Oil Shale Tar Sands and Related Materials,* ACS Symposium Series 163. American Chemical Society, Washington, D.C.

Cameron Engineers. (1978). *Synthetic Fuels Data Handbook,* 2nd Ed. Cameron Engineers, Inc. Denver, Colorado.

Cataldo, D. A., and Wildung, R. E. (1983). The role of soil and plant metabolic processes in controlling trace element behavior and bioavailability to animals. In *Biological Availability of Trace Metals,* E. A. Jenne and R. E. Wildung (Ed.). Elsevier, Amsterdam, pp. 159–168.

Cataldo, D. A., Fallon, W. E., Riley, R. G., Garland, T. R., and Wildung, R. E. (1983). Phytotoxicity of oil shale retort water and its constituents. PNL-SA-11367, Pacific Northwest Laboratory, Richland, Washington.

Challenger, F., Lisle, D. B., and Dransfield, P. B. (1954). Studies on Biological Methylation. Part 14: The Formation of Trimethylarsine and Dimethyl Selenide in Mould Culture from Methyl Sources Containing ^{14}C. *J. Chem. Soc.* (London) Part 2: 1760–1771.

Cloninger, J. S. (1978). Revegetation of retorted shale. Presented at American Nuclear Society, Denver, Colorado, Sept. 26, 1978. Union Oil Co. of California, Grand Junction, Colorado.

Cook, C. W. (1974a). Oil shale technology in the USA—A review of current problems and prospects. *Fuel 53*: 146–151.

Cook, C. W. (Project Coordinator). (1974b). Surface rehabilitation of land disturbances resulting from oil shale development. *Tech. Rep. Ser. No. 1* (Final Report, Phase 1), Environ. Resource Ctr. Colorado State University, Fort Collins, Colorado.

Cox, D. P., and Alexander, M. (1973). Effect of phosphate and other anions on trimethylarsine formation by *Candida humicola*. *Appl. Microbiol. 25*: 408--413.

DOI (U.S. Department of Interior). (1977). *Final Environmental Impact Statement. Proposed Development of the Oil Shale Resources.* Colony Development Operation, Atlantic Richfield Co., Denver, Colorado.

Donnell, J. R. (1964). Mineral fuels and associated resources—oil shales. In *Mineral and Water Resources of Colorado,* U.S. Congress, 2nd Session 83rd Congress, pp. 67–73.

Donnell, J. R., and Blair, R. W. (1970). Resource appraisal of three oil shale zones in the Green River Formation, Piceance Creek Basin, Colorado. *Colorado School of Mines Quart. 65*: 73–87.

Drucker, H., Garland, T. R., and Wildung, R. E. (1978). Metabolic response of microbiota to chromium and other metals. In *Trace Metals in Health and Disease,* N. Kharasch, (Ed.). Raven Press, New York.

Duff, R. P., Webley, D. M., and Scott, R. O. (1963). Solubilization of minerals and related materials by 2-ketogluconic acid-producing bacteria. *Soil Sci. 95*: 105–114.

Fruchter, J. S., Wilkerson, C. L., Evans, J. C., Sanders, R. W., and Abel, K. H. (1979). *Source Characterization Studies at the Paraho Semiworks Oil Shale Retort.* PNL-2945, Pacific Northwest Laboratory, Richland, Washington.

Garland, T. R., Wildung, R. E., and Harbert, H. P. (1979). Influence of irrigation and weathering reactions on the composition of percolates from retorted shale in field lysimeters. In *Proc. of 12th Annual Oil Shale Symposium,* Colorado School of Mines, Golden, Colorado.

Griffin, D. M. (1981). Water potential as a selective factor in the microbial ecology of soils. In *Water Potential Relations in Soil Microbiology,* Soil Science Society of America Special Pub. No. 9., pp. 141–151. Madison, Wisconsin.

Hand, J. W. (1969). Planning for disposal of oil shale chemical and mine wastes. *Colorado Geol. Surv. Special Publ. 1*: 33–38.

Harbert III, H. P. (1978). *Vegetative stabilization of Paraho spent oil shales, lysimeter study.* Thesis, Colorado State University. Fort Collins, Colorado.

Harbert, H. P., and Berg, W. A. (1978). *Vegetative stabilization of spent oil shales: vegetation, moisture, salinity, and runoff—1973–1976.* EPA-600/7-78-021, National Technical Information Service, Springfield, Virginia.

Harrison, Jr., A. P. (1978). Microbial succession and mineral leaching in an artificial coal spoil. *Appl. and Environ. Microbiol. 36*: 861–869.

Holm, T. R., Anderson, M. A., Stanforth, R. R., and Iverson, D. G. (1980). The influence of adsorption on the rates of microbial degradation of arsenic species in sediments. *Limnol. Oceanogr. 25*: 23–30.

Jenne, E. A. (1977). Trace element sorption by sediments and soils—sites and processes. In *Proc. Symp. Molybdenum in the Environment.* Marcel Dekker, New York, Vol 2, Ch. 5.

Klein, D. A., Hassler, R. A., Meglen, R. R., and Sistko, R. (1979). Environmental effects on microbial mobilization of arsenic from retorted oil shale. In *Oil Shale Environmental Research and Coordination.* Progress Report, Department of Energy, DOE-10298-3.

Leenheer, J. A., Noyes, T. I., and Stuber, H. A. (1982). Determination of polar organic solutes in oil-shale retort water. *Environ. Sci. Technol. 16*: 714–723.

McBride, B. C., and Edwards, T. L. (1977). Role of the methanogenic bacteria in the alkylation of arsenic and mercury. In *Biological Implications of Metals in the Environment,* H. Drucker and R. E. Wildung, (Ed.). Technical Information Center, Oak Ridge, Tennessee, pp. 1–19.

Neilands, J. B. (1979). Biomedical and environmental significance of siderophores. In *Trace Metals in Health and Disease,* N. Kharasch (Ed.). Raven Press, New York, pp. 27–41.

Parr, J. F., and Papendick, R. I. (1971). Interactions of microbial metabolism and soil physical properties and their significance in some hydrologic processes. In *Biological Effects in the Hydrological Cycle,* E. J. Monke (Ed.), PB-246308, National Technical Information Service, Springfield, Virginia; pp. 148–162.

Pereira, W. E., Rostad, C. E., Steinheimer, T. R., and Johnson, S. M. (1981). *J. Environ. Sci. Health A16*: 571–587.

Pfeffer, F. M., and Kerr, R. S. (1974). *Pollutional problems and research needs for an oil shale industry.* EPA-660/2-74-067, National Technical Information Service, Springfield, Virginia.

Reamer, D. C., and Zoller, W. H. (1980). Selenium biomethylation products from soil and sewage sludge. *Science 208*: 500–502.

Redente, E. F., Doerr, T. V., Mount, C. E., Grygiel, C. E., Sievers, T. B., and Brondini, M. (1981). Effects of plant species, soil material, and cultural practices on plant establishment and succession. In *Revegetation Research in Oil*

Shale Lands in the Piceance Basin, Progress Report, Dept. Range Science, Colorado State University, Fort Collins, Colorado, pp. 1–31.

Riley, R. G., Shiosaki, K., Bean, R. M., and Schoengold, D. M. (1979). Solvent solubilization, characterization, and quantitation of aliphatic carboxylic acids in oil shale retort water following chemical derivatization with boron trifluoride in methanol. *Anal. Chem. 51*: 1995–1998.

Rogers, J. E., Riley, R. G., Li, S. W., Mann, D. C., and Wildung, R. E. (1981). Microbiological degradation of organic components in oil shale retort water: Organic acids. *Appl. Environ. Microbiology 42(5)*: 830–837.

Rogers, J. E., McNair, V. M., Le, S. W., Garland, T. R., and Wildung, R. E. (1982). *Microbial Colonization of Retorted Shale in Field and Laboratory Studies.* PNL-4311, Pacific Northwest Laboratory, Richland, Washington.

Routson, R. C., and Wildung, R. E. (1969). Ultimate disposal of wastes to soil. *Chem. Eng. Prog. 65*: 19–25.

Routson, R. C., Wildung, R. E., and Bean, R. M. (1979). A review of the environmental impact of ground disposal of oil shale wastes. *J. Environ. Qual. 8*: 14–19.

Schmalfeld, P. (1975). The use of the Lurgi-Ruhrgas process for the distillation of oil shale. *Colorado School of Mines Quart. 70*: 129–145.

Schmidt-Collerus, J. J. (1974). Polycondensed aromatic compounds (PCA) and carcinogens in carbonaceous spent shale from retorting of oil shale of the Green River Formation. *Am. Chem. Soc., Div. Fuel Chem. 19*: 115–122.

Shendrikar, A. D., and Faudel, G. B. (1978). Distribution of trace metals during oil shale retorting. *Environ. Sci. Technol. 12*: 332–334.

Shih, C. C., Prien, C. H., Nevens, T. D., and Cotter, J. E. (1977). *Technological overview reports for eight shale oil recovery processes.* Environmental Protection Agency, Industrial Research Lab, Cincinnati, Ohio.

Sigel, H. (1973). Metal ions in biological systems. *Mixed-Ligand Complexes.* Marcel-Dekker, New York, Vol. 2.

Sigel, H. (1974). Metal ions in biological systems. *Simple Complexes,* Marcel-Dekker, New York, Vol. 1.

Silver, M., and Torma, A. E. (1974). Oxidation of metal sulfides by *Thiobacillus ferrooxidans* grown on different substrates. *Can. J. Microbiol. 20*: 141–147.

Silviera, D. J., Garland, T. R., and Wildung, R. E. (1978). Characterization of retorted oil shale leachate in laboratory studies. In *Agronomy Abstracts,* ASA, SSSA, and CSSA Annual Meetings, Chicago, Illinois, December 5–9, 1978.

Sladek, T. A. (1975). Recent trends in oil shale—part 2. Mining and shale oil extraction processes. Colorado School of Mines Miner. *Ind. Bull. 18*: 21.

Smith, J. W., Robb, W. A., and Young, B. (1978). High temperature reactions of oil shale minerals and their benefit to oil shale processing. In *Proc. 11th Oil Shale Symposium.* Colorado School of Mines, Golden, Colorado.

Stojanovic, B. J., and Alexander, M. (1958). Effect of inorganic nitrogen or nitrification. *Soil Sci. 86*: 208–215.

Stollenwerk, K. G., and Runnells, D. D. (1977). Leachability of arsenic, selenium, molybdenum, boron and fluoride from retorted shale. In *Proc. Second Annual Pacific Chem. Eng. Congress,* American Institute of Chemical Engineers, New York, pp. 1023.

Striffler, W. C., Wymore, I. F., and Berg, W. A. (1974). Characteristics of spent shale which influence water quality, sedimentation and plant growth medium. In *Surface Rehabilitation of Land Disturbances Resulting from Oil Shale Development*—Tech. Rep. Ser. No. 1, C. W. Cook, Coordinator, Colorado State University, Fort Collins, Colorado.

Takai, A., and Kamura, T. (1966). The mechanism of reduction in water logged paddy soil. *Folia Microbiol. 11*: 304.

Thimann, K. V. (1967). *The Life of Bacteria.* The Macmillan Company, New York.

Turner, R. R. (1981). Oxidation state of arsenic in coal ash leachate. *Environ. Sci. Tech. 15*: 1062–1066.

Ward, J. C., and Reinecke, S. E. (1972). *Water pollution potential of snowfall on spent oil shale residues.* Environ. Eng. Progr., Dept. of Civil Eng., Colorado State University, Fort Collins, Colorado.

Webley, D. M., Henderson, M. E. K., and Taylor, I. F. (1963). The microbiology of rocks and weathered stones. *J. Soil. Sci. 4*: 102–112.

Wildung, R. E., and Garland, T. R. (1979). The relationship of microbial processes to the fate and behavior of transuranic elements in soils, plants, and animals. In *Transuranic Elements in the Environment,* DOE/TIC 22800, Technical Information Center, Oak Ridge, Tennessee, pp. 300–335.

Wildung, R. E., and Zachara, J. M. (1980). Geochemistry of oil shale solid waste disposal. In *Oil Shale—The Environmental Challenges.* Colorado School of Mines Press, Golden, Colorado, pp. 201–240.

Wildung, R. E., and Zachara, J. M. (1981). Effects of oil shale solid waste disposal on water quality: Current knowledge, information requirements and research strategy. In *Oil Shale—the Environmental Challenges* II. Colorado School of Mines Press, Golden, Colorado.

Wildung, R. E., Garland, T. R., Bean, R. M., Cataldo, D. A., Fallon, W. E., Li, S. W., McFadden, K. M., McNair, V. M., Neil, L. C., Riley, R. G., Rogers, J. E., Shiosaki, K., and Silviera, D. J. (1979a). Terrestrial effects of oil shale development. In *Pacific Northwest Laboratory Annual Report for 1978,* PNL-2850, PT2, National Technical Information Service, Springfield, Virginia, pp. 2.1–2.8.

Wildung, R. E., Garland, T. R., and Drucker, H. (1979b). Nickel complexes with soil microbial metabolites—mobility and speciation in soils. In *Chemical Modeling in Aqueous Systems.* American Chemical Society Symposium Series 93, Washington, D.C., pp. 181–200.

Wildung, R. E., Garland, T. R., Zachara, J. M., Riley, R. G., Rogers, J. E., Cataldo, D. A., Jenne, E. A., Ames, L. L., Li, S. W., McFadden, K. M., and Neil, L. C. (1981). Terrestrial Effects of Oil Shale Development. In *Pacific Northwest Laboratory Annual Report for 1980,* National Technical Information Service, Springfield, Virginia.

Wildung, R. E., Bond, F. W., Gee, G. W., and Nelson, J. M. (1982). Oil shale waste disposal: Estimation of embankment physical stability and movements of water solutes. In *Oil Shale—the Environmental Challenges III.* Colorado School of Mines Press, Golden, Colorado.

ZoBell, C. E. (1948). Microbial transformation of hydrocarbons. *J. Gen. Microbiol. 2*: viii–x.

5

Soil Enzymes: A Predictor of Reclamation Potential and Progress

DONALD A. KLEIN, DARWIN L. SORENSEN,* and EDWARD F. REDENTE / *Colorado State University, Fort Collins, Colorado*

I. INTRODUCTION

A major concern in reclamation is to be able to predict potential effects of waste materials on biological processes upon which plant-soil systems are dependent. In most work carried out to the present time, chemical methods have been used to assess the quality of materials that will be used as a plant growth medium. In the last two decades, however, there has been an increased interest in the use of soil enzymes as a diagnostic and monitoring approach in reclamation.

In this chapter, the use of soil enzymatic methods to evaluate factors influencing the reclamation of retorted oil shales will be discussed as an example of the use of these approaches. Potentially inhibitory components in waste materials can interact with soil enzymes in direct and indirect ways:

1. Inhibition of soil enzymes and plant growth can result if toxic wastes are used as a plant growth medium. These inhibitory characteristics can be overcome or avoided, but this usually requires extensive management (leaching, fertilization) or the use of topsoil covering to provide a more suitable plant growth environment.
2. If a soil layer is used to cover a known inhibitory waste, soluble or volatile waste components can move or be transported to the surface soil zone, resulting in negative effects on the plant-soil system.

**Current affiliation*: Utah Water Research Laboratory, Utah State University, Logan, Utah

3. Plant growth from a surface soil into an underlying waste material can result in uptake of potentially toxic components by plants. After translocation, decomposition of plant litter can lead to accumulation of these components in surface soils. Microbial activity may be stimulated in the plant root zone, and lead to a wide range of transformation and mobilization/immobilization reactions that will influence these plant uptake processes.

II. HISTORICAL BACKGROUND

The monitoring of belowground processes in relation to reclamation has only developed as a serious area of research in the last 20 years, as summarized by Parkinson (1976). Surveys of microbial populations in mine spoils were initiated in the 1955–1965 period (Wilson and Stewart, 1956; Wilson and Hedrick, 1957b; Wilson, 1961, 1965); and during this time soil enzymatic processes also began to be considered in relation to reclamation (Wilson and Stewart, 1955; Wilson and Hedrick, 1957a; Brüning, et al., 1965). In the study by Brüning et al., management of brown coal spoils to allow "biological activation" was discussed and cellulose decomposition measurements were completed to evaluate the effectiveness of various reclamation procedures. Schuman et al. (1976), provided general information on oil shale development in the United States and discussed important reclamation concerns. Their review contained no specific discussions of microbiological processes or their potential relationships to reclamation, although trace elements and their implications for biological processes were discussed. Cundell (1977) discussed soil microbiological processes in relation to western reclamation, and although no specific data from reclamation studies were available, it was suggested that soil enzymatic approaches could play a significant role in reclamation management. In 1978, in a summary volume of reclamation of disturbed lands (Schaller and Sutton, 1978), mycorrhizal fungi and rhizobia were discussed; however, no further emphasis on microbiological processes was provided. Work carried out at the Argonne National Laboratory (Miller, 1977; Miller and Cameron, 1977, 1978) has provided additional information on soil enzymatic and microbial population responses during reclamation in a series of ecosystems, indicating an increased interest in this area.

In the last several years, more specific emphasis has been given to soil enzymes in relation to reclamation management. Stroo and Jencks (1982) measured microbial respiration and soil enzyme activities in mine soils in relation to time and fertilizer amendments. With added fertilizer, higher microbial activities were observed, but organic matter formation was retarded. Based on their usefulness in this study, soil enzymatic assays were felt to have an increased potential for better understanding reclamation management effects on plant-soil system development. Ross et al. (1982) studied soil enzymatic responses after simulated soil disturbance, and followed changes in seven hydrolytic enzyme activities in relation to reestablishment of a pasture plant community over a 3-year period. Biochemical activities were correlated with herbage yields on the disturbed plots, and invertase and sulfatase activities appeared to be the best indicators of the soil fertility status.

This trend has continued, with studies of retorted oil shale effects on soil microbial populations and enzymatic activities (Hersman and Klein, 1979) and of retorted oil shale interactions with more complex plant-soil systems being completed (Klein et al., 1979a; Sorensen et al., 1981). Microbial processes in the plant rhizosphere have also been emphasized. Fresquez and Lindemann (1982), in a study of coal spoil reclamation, noted different enzymatic responses of control and stored soils and of soil after plant growth. These workers felt that an available carbon source was more important for belowground process development than the use of a particular microbial inoculum.

Together with an increasing awareness of microorganisms and their potential contributions to reclamation, there have been continuing advances in the area of system ecology. Plant-soil system development, succession, system stability and resilience, nutrient cycling, and energy flow processes are increasingly being analyzed on both theoretical and applied levels, based on studies of Jenny (1980), Lindemann (1942), and Odum 1969). Recently O'Neill and Reichle (1979), Reichle, et al. (1975), Clark and Paul (1970), and MacMahon (1979) have developed a strong conceptual framework within which to view microorganisms and their roles in ecosystem functioning. Increasingly, the microbes are being considered as an important labile nutrient pool.

These trends suggest that a convergence is occurring, involving ecosystem theory and our understanding of the role of microorganisms in the functioning of plant-soil systems.

III. SOIL ENZYMES AND RECLAMATION

Although there has been a continuing interest in the study of soil enzymes, there has not been an equivalent effort to link enzymes and enzymatic processes to ecosystem functioning, or to extend these considerations to disturbed ecosystem reconstruction.

A. Soil Enzymes as Ecosystem Components

There are three major system components (i.e., plant, energy waste, and soil) that interact through the soil enzymes in reclamation, as noted in Figure 1. The enzymes that function as critical intermediates in these interactions can exist both free in the soil environment or

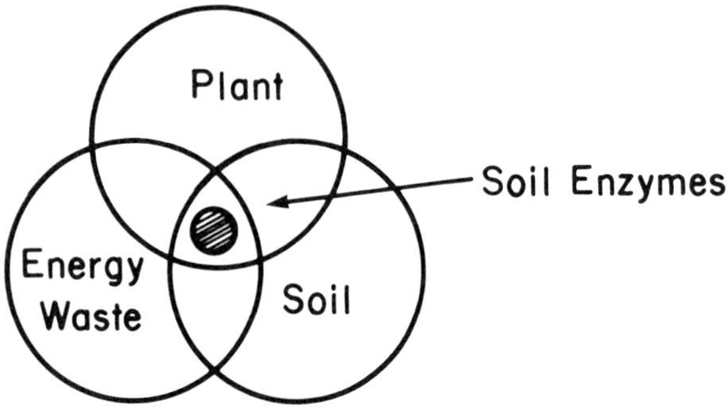

Figure 1 Conceptualized role of soil enzymes as intermediates in reclamation systems.

associated with living organisms. The extracellular or abiontic enzymes (Skujins, 1978) can be protected by soil colloids and clays which limit the degradation and denaturation of these soluble proteins. Such abiontic enzymes primarily carry out hydrolytic reactions, whereas oxido-reductase or dehydrogenase enzymes are associated with living cells.

Reclamation can be initiated without consideration of soil and its attendant soil enzymes. Under these conditions, without a nutrient cycling system being available, fertilizers can be added to provide inorganic plant-utilizable nutrients that would normally be made available through extracellular and organism associated soil enzymes. As the plant-soil-energy waste system develops and soil organic matter accumulates, closer relationships would be expected between soil enzymes and the other system components.

In comparison, if soil is present prior to the development of a plant community, there will be an excess of soil organic matter and soil enzymes when reclamation is initiated. Under these conditions, soil enzymes may be less useful for monitoring development of the plant-soil system, at least in the beginning stages of plant development at a particular site.

Skujins (1967) noted that many workers have attempted to use soil enzymatic measurements to provide a "biological index" of soil fertility. In this research summary, he stated that, "The correlations have not been successful." The observation that free enzymes are present in a series of sites in soils and from a wide range of sources was felt to make it difficult to use enzymatic measurements to predict soil fertility. It is important to note that Skujins did assume that some such correlation might exist, but that "new and improved methods for its demonstration are needed." In another review, Skujins (1976) suggested that a fertility index might be able to be developed based on the recognition of key species or physiological groups and their possible role in nutrient cycling processes.

Although there has been a general feeling that soil enzymes will be less applicable for evaluating soil fertility in intensive agriculture, there is evidence that these approaches may have a greater range of applications in arid or semiarid environments (Skujins, 1973). This may be the result of a closer inherent linkage between plant communities and soil organic matter accumulation in these areas, and the greater influence that minor differences in parent material and abiotic conditions would have on soil organic matter and soil enzymatic processes.

B. Reclamation and Soil Enzymatic Methods

Earlier work on methods for analyses of soil enzymes was initiated by Hoffmann and his co-workers in the mid-1950s. A series of reviews has been completed in the last 20 years, and a multiauthored book edited by Burns (1978) is also available on the subject, together with a recent review on substrate interactions with cell-free enzymes in soils (Burns, 1983). The presence of cell-associated as well as extracellular enzymes, discussed earlier, is important in the design and interpretation of soil enzymatic assays. In addition, actual versus potential activities of soil enzymes can be measured. To provide this type of information, assays can be carried out with or without added substrates or energy sources.

Enzyme stability is also a point of concern. With the known ability of soil clays and colloids to stabilize free enzymes and to protect them from degradation (Sorensen, 1969; Skujins et al., 1974; Burns, 1978), it is possible to accumulate extracellular enzymes in soils. This potential activity may not be expressed due to a lack of interaction with substrates or inappropriate abiotic conditions. There is a need for more detailed information regarding soil enzymes, their functions in plant-soil systems, and their responses to the presence of protecting clays and colloids, especially in relation to inhibitory materials that may be present in particular materials.

Soil enzymes can play important roles in the design of reclamation systems, based on the following characteristics:

1. Enzymatic activities can be measured using specific substrates and test conditions, with or without allowing enzyme synthesis or microbial proliferation to occur. Microbial growth or enzyme synthesis has usually been controlled by use of toluene (Hoffmann and Seegerer, 1951) or irradiation (Skujins et al., 1962), although there is continuing concern regarding the usefulness of toluene (Kaplan and Hartenstein, 1979).
2. Short-term measurements can be used where reaction linearity can be maintained, even without the use of inhibitors. This can involve the use of radiolabel-type assays (Nakas and Klein, 1980) or other sensitive enzymatic measurements that can be completed with shorter (5–15 min) reaction times.

Field sampling and soil handling techniques also can affect the quality of information derived by enzymatic procedures. In contrast to soil chemical analyses, which are frequently carried out using dried soils, preparatory steps plus soil storage can have major effects on soil enzymatic processes (Birch, 1958; Speir and Ross, 1981).

The soil sampling depth is an important consideration, especially with a developing plant-soil system. Based on the recommendations of Hersman (Hersman and Temple, 1979), sampling carried out in this laboratory has been primarily at the 5–10-cm depth, considered the zone where plant growth would have a maximum influence on soil development, and where shorter-term variations is soil temperature and moisture would have lesser effects on soil enzymatic activities.

Soil storage is also critical. Soil samples can be held at the field water content and temperature or at some set of arbitrarily established conditions that may allow more optimum enzyme expression and functioning. A method that allows extended storage of soils involves holding of soils at their field water content using double plastic bagging (Casida et al., 1964).

C. Soil Enzymes and Inhibitory Substances

Soil provides an environment that can have major influences on whether a particular enzymatic potential can be expressed or influenced by toxic materials. Important factors include:

1. The presence of soil humic acids, colloids, clays, and organic matter that can adsorb heavy metals and toxic lower-molecular-weight organic compounds, making these constituents less available to react with soil enzymes.
2. The decreased ability of soil enzymes protected by soil clays and colloids to interact with substrates. Where no substrate interaction is possible, modification of enzyme characteristics by energy wastes, even if it might occur, will not have immediately observable effects.
3. The changes in microbial and enzymatic responses that can occur after a first interaction with a toxic material. Microorganisms can respond phenotypically, or genetic changes can result when a material is imposed on an environment. These responses can involve the synthesis of less sensitive enzyme systems, the over-synthesis of inactivating materials such as sulfhydryl groups, or the utilization of pathways that do not involve reactions sensitive to toxic materials, as described by Ernst (1976).

Microbes, with their associated enzymes, also can respond at a community level. Shifts in the composition of complex soil microbial communities can occur after the addition of heavy metals, pesticides, herbicides, and other materials. This important process of microbial adaptation, discussed by Slater and Godwin (1980), suggests that repeated additions or continued exposure of a soil microbial community and its enzymes to particular chemicals will result in a changed sensitivity to subsequent additions of the same material, an illustration of Weber's Law (Adler, 1969).

Enzymatic assessments have not been used routinely in reclamation to the present time, as noted earlier, although this approach has been recommended by Cundell (1977) and more recently by Fresquez and Lindemann (1982). It would appear that enzymatic approaches should be able to play a significant role in reclamation management to provide the following information:

1. Estimates of the relative fertility status of soil and spoil materials.
2. Potential effects of energy wastes on soil enzymatic processes.
3. Information on plant community development with and without effects of an underlying or codisposed waste material.

The major contributions that soil enzymatic approaches can bring to reclamation management are their relatively low cost, analysis speed, and the relative lack of instrumentation required, in comparison with many chemical procedures.

IV. SOIL ENZYMATIC MONITORING: EXPERIENCES WITH RETORTED OIL SHALE

A. Retorted Oil Shale Characteristics

Oil shales result from algal sedimentation and decomposition in lake environments. The resultant resistant organic residues tend to have higher concentrations of trace substances, which can be remobilized in retorting processes. Shendrikar and Faudel (1978), Fox et al. (1979), and Fox (1980a,b) have studied the partitioning of trace metals from shales in retorting and have found that trace metals tend to be retained in the residual spent shales, with only minor amounts of these materials being partitioned to the hydrocarbons, retort process waters, or off-gases. Arsenic, barium, fluoride, strontium, magnesium, vanadium, boron, titanium, and zinc are among the trace elements found in markedly higher concentrations in the residual shale.

Fox (1980a) provided an extensive summary of the partitioning of major, minor, and trace elements during simulated in situ oil shale retorting. Fifty elements were monitored, and most of the 50 elements studied were not significantly mobilized. The elements that were distributed to a greater extent to the by-products (oil, water, gases) included carbon, hydrogen, nitrogen, sulfur, cadmium, and mercury, with lesser quantities of selenium, arsenic, nickel, and cobalt distributed to the waters. Mercury and cadmium were the most mobile elements, with 70% of mercury and 30% of cadmium distributed to the gas phase. Unfortunately, data on fluorine mobility were not provided in these studies.

Information on trace metal mass balances has been complemented by investigations of retort water chemistry, and of waters that result from leaching of various retorted oil shales. Sievers et al. (1982) noted that thiosulfate as well as sulfate was present in these waters, and a wide variety of organic compounds have been identified, including polynuclear aromatic hydrocarbons and pyridines. Trace metals, including arsenic, selenium, cadmium, chromium, molybdenum, and boron also are routinely detected. Stollenwerk (1980) and Stollenwerk and Runnells (1977) provided extensive data sets on the chemical constituents of these waters. This information suggests that there is a wide range of components in retorted oil shales, including trace metals, fluorine, and polynuclear aromatic hydrocarbons, that might influence soil enzyme expression, stability and functioning.

B. Laboratory Assessments: Soil Enzyme-Retorted Shale Interactions

Information is available concerning the effects of retorted shales and shale leachates on pure cultures of bacteria (Adams and Farrier, 1982) and soils (Hersman and Klein, 1979) under more defined laboratory conditions. These studies indicate that direct interactions of retorted shales, and also of soluble components from retorted shales, can result in effects on biological processes. In most cases decreases in activities occur, although with low-level shale additions, some stimulation of plant growth has been observed in a legume-*Rhizobium* system (Hersman, et al., 1981). The major limitation that most of these studies has had is that the interaction times have been relatively short. Under these conditions, the plant-microbe community has not had sufficient time to develop or to modify the environment through organic matter accumulation, and by other immobilization processes that might tend to mitigate effects of the energy waste materials.

Retorted oil shales, in spite of having a high total nitrogen content, are remarkably deficient in plant-available nitrogen. Schmehl and McCaslin (1973) completed a study of properties of spent oil shale in relation to the potential effects of these materials on plant growth. They noted that "deficiencies in both nitrogen and phosphorus were so marked, that satisfactory plant growth was impossible without the combination of these fertilizer elements."

In a study of retorted shale-nitrogen interactive effects on enzymatic processes, samples of shale and shale-soil mixtures were analyzed for total nitrogen and plant-available forms, as noted in Table 1. The low ammonium ion and nitrate nitrogen values of the shale sample are especially striking. When such low nitrogen availability is combined with high salinity and higher pH values, direct effects on microbial functioning and plant development could be expected to occur.

The need to add fertilizer to allow plant growth can also influence dinitrogen fixation, and negative interactive effects of retorted shale

Table 1 Total Nitrogen, Nitrate, and Ammonium Ion in Surface Soil—Paraho Retorted Shale Mixtures[a]

Sample	% total N	ppm NO_3^--N	ppm NH_4^+-N
Surface soil	0.079	10	49
5% shale, 95% soil	0.086	15	41
10% shale, 90% soil	0.078	18	24
25% shale, 75% soil	0.086	16	21
Retorted oil shale	0.145	2	3

[a]Surface soil and Paraho process retorted shales from the experimental area described by Redente and Ruzzo (1979).

and ammonium ion on nitrogen fixation have been observed (Klein et al., 1980). This results from the ability of ammonia nitrogen to repress nitrogen fixation processes (Hardy et al., 1973). By combining shale-soil mixtures with varied ammonium ion levels, the combination of these two potentially repressive components can lead to more pronounced reductions in nitrogen fixation potential (Fig. 2). Even with 10 μg/g ammonium ion nitrogen present with 5% by weight of shale, an accentuated inhibition is observable. These results suggest that fertilizer, when present with retorted shale, may result in more distinct effects on some key soil enzymatic processes.

C. Enzymatic Monitoring: Field Evaluations

A series of retorted shale reclamation plots have been established in the Piceance Basin since the early 1970s, where vegetation responses and leachate characteristics have been monitored.

Plots using Colony TOSCO II materials were established in 1970 at Grand Valley, Colorado, where nine single plant species and one mixture were planted on 60 cm of topsoil over spent shale. In addition, spent shale was amended with sawdust plus phosphorus, and also with talus, jute, or straw. The plots were leached and irrigated with approximately 100 cm of water over a 2-year period (Bloch and Kilburn, 1973), and soil enzymatic measurements were completed in 1981. Only dehydrogenase with glucose showed a significant response to sampling depth on unmulched shale to surface plots, with the maximum activity occurring near the surface in the 0–10-cm depth. No differences in pH, soil moisture, dehydrogenase without glucose, or nitrogen fixation were observed without added surface soil, regardless of amendments.

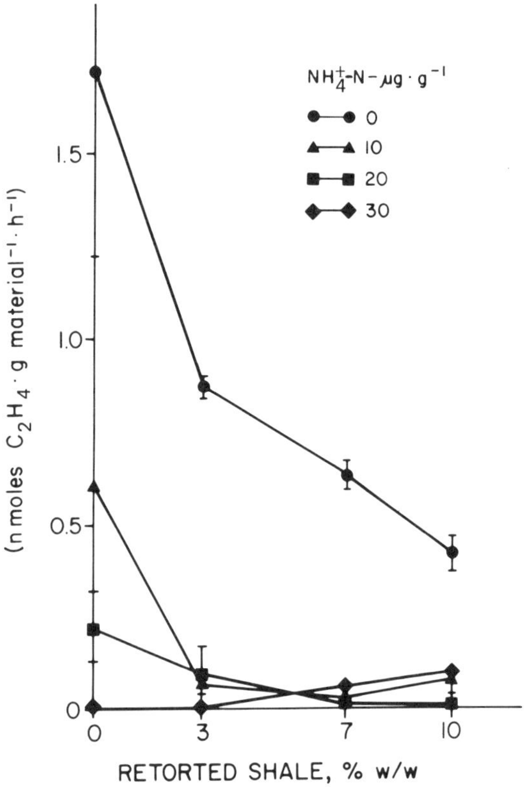

Figure 2 The effects of varied ammonium ion and retorted oil shale additions on the nitrogen fixation potential of surface soil from the intensive study area, Piceance Basin, Colorado. Means and standard deviations are given.

The Anvil Points plots (established in 1973) were designed to evaluate the effects of variations in shale materials (TOSCO and USBM) on plant growth with shale-to-surface samples, and with 15 cm and 30 cm of soil over shale in comparison with a soil control. In this study, south- and north-facing slopes were tested, and the plots were leached with 100 cm of water prior to seeding and fertilization with nitrogen and phosphorus (Harbert and Berg, 1978; Kilkelly et al., 1982). As would be expected, major increases in carbon-processing activity (dehydrogenase) occurred in the surface soil zone, with direct relationships being shown between surface soil presence and activity (Klein et al., 1982). This set of relationships was shown most clearly with the glucose-amended dehydrogenase measurement (Fig. 3), whereas dehydrogenase without glucose gave equivalent responses with all

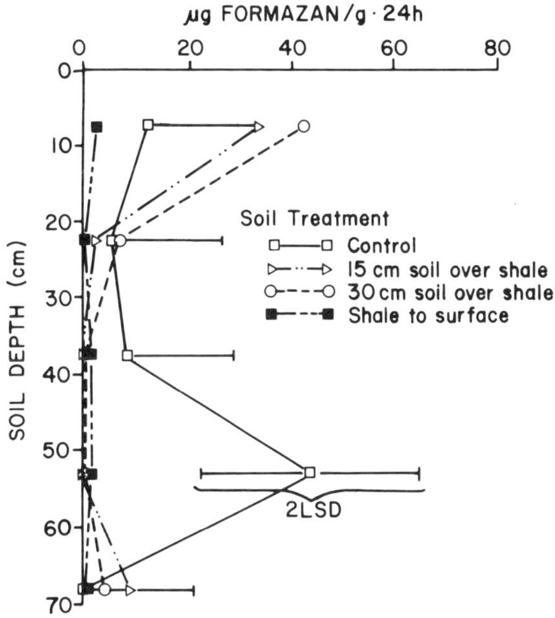

Figure 3 Surface soil depth effect on dehydrogenase with glucose in relation to sampling depth for the TOSCO plots, Anvil Points, Colorado, 1981.

soil depths used over the TOSCO spent shale (Fig. 4). The USBM shales used at this site showed similar dehydrogenase responses across the plots, both with and without glucose amendments, and the shale-to-surface samples had the lowest activities, even after almost 10 years of plant community development.

In contrast to the marked decreases in enzymatic activity with retorted shales to the surface, the plant communities did not show similar reductions (Harbert and Berg, 1978). These results suggest that stable plant communities can be, with extensive leaching and fertilization, established in spite of having a reduced below-ground biogeochemical cycling potential.

A set of higher elevation plots established in the Piceance Basin in 1973 was designed with TOSCO and USBM materials in a manner similar to those used for the lower elevation Anvil Points plots discussed previously (Harbert and Berg, 1978; Kilkelly et al., 1982). Similar analyses of dehydrogenase activity and nitrogen fixation potential were completed. Dehydrogenase with added glucose showed higher activities in the surface zones of plots with 15 cm and 30 cm surface soils over spent shale, and a lesser stimulation was observed

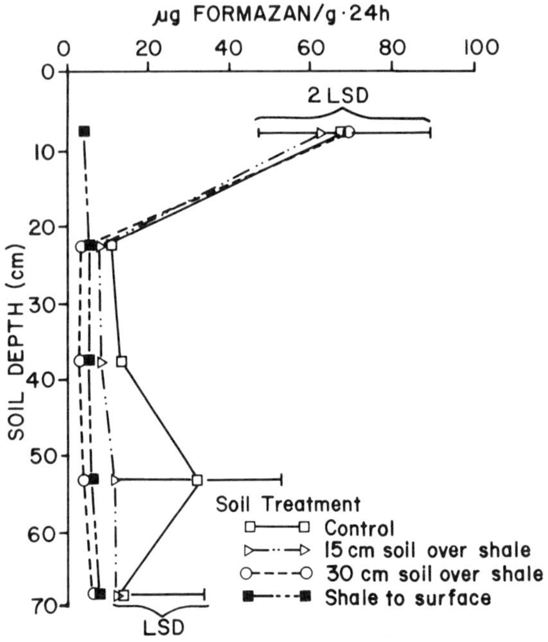

Figure 4 Surface soil depth effects on dehydrogenase without glucose in relation to sampling depth for the TOSCO plots, Anvil Points, Colorado, 1981.

with shale to the surface. The major finding was that with USBM shale to the surface, a higher activity was detected than with the TOSCO materials. In this study, 52–68% vegetation cover developed after seven growing seasons with or without use of surface soil, and the intensive management used at this site also allowed successful plant community establishment on these materials in spite of marked differences in soil enzymatic activities.

These results, using older plots, suggested that stable plant communities can be established on retorted oil shales without soil or organic amendments, even where general soil enzymatic processes are limited. This involves intensive leaching and fertilization, but given sufficient resources, plus natural shale weathering (Richardson et al., 1981) plant communities appear to be able to be established on such retorted shales.

Soil enzymatic responses during revegetation over retorted oil shale have also been monitored in a more extensive experiment initiated in 1977. This experiment has been described by Redente and Ruzzo (1979) and Redente et al. (1982b), and involves varying soil depths

(30, 60, and 90 cm) over shale, 60 cm of soil plus a capillary barrier over shale, and varied fertilizer-planting mixtures. Enzymatic and biochemical activities were assayed as described by Klein et al. (1971), and Sorensen et al. (1981).

The average yearly phosphatase and nitrogen fixation activities across the panels from 1979–1981 are shown in Figs. 5 and 6, respectively, using results from the June sampling. The major trend that was noted in both of these measures of nutrient cycling was the distinct overall decrease in these activities observed during development of the plant community over the 3 years, and the trend towards decreased relative enzymatic activities with less surface soil over shale within each year.

The decrease in acetylene reduction activity that occurred over the 3 years is especially striking (Fig. 6), and may reflect a similar decline in rates of acetylene reduction in old field succession studies

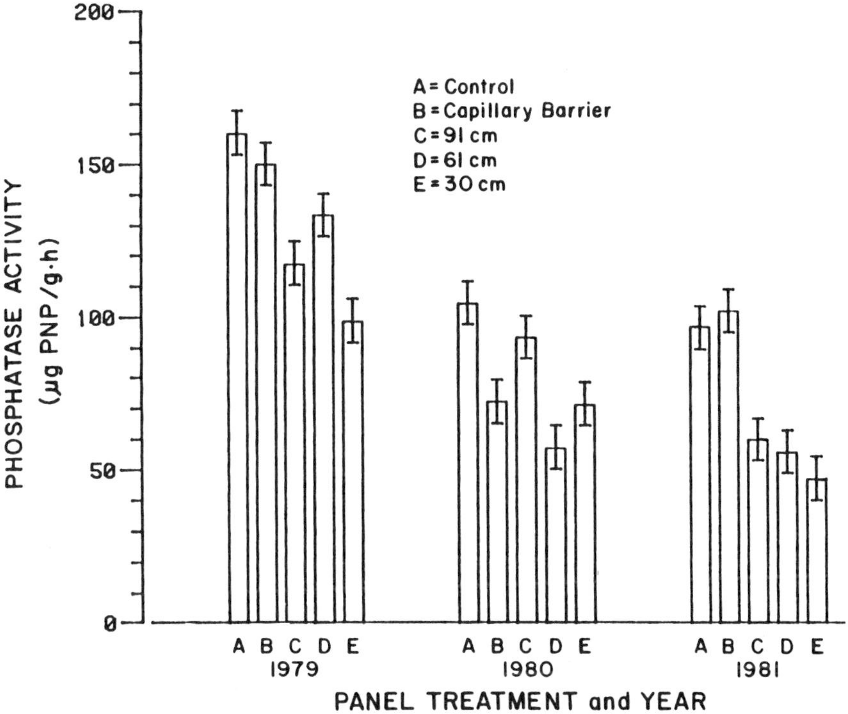

Figure 5 Phosphatase activity of soil of various depths over spent shale and capillary barrier—1979, 1980, and 1981 results. The least significant difference between means is 14.2 μg $PNP \cdot g^{-1} \cdot h^{-1}$, as indicated by the error bars.

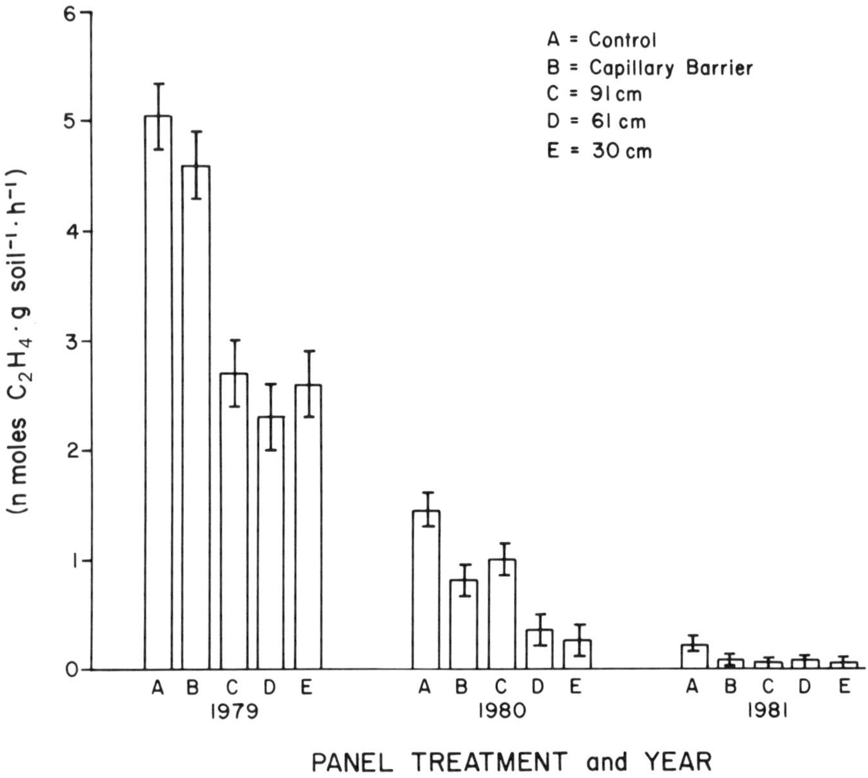

Figure 6 Nitrogen fixation potential for the Retorted Shale Successional Study—1979, 1980, and 1981 responses. The least significant differences are 0.56, 0.31, and 0.075 for 1979, 1980, and 1981, respectively.

with plant community development reported by Kapustka and Rice (1976). Nitrogen fixation sensitivity to soil disturbance and underlying shale materials has resulted in this procedure being recommended by Klein et al. (1979b) for disturbed system monitoring.

In contrast to phosphatase and nitrogen fixation, where decreases in activities were observed, dehydrogenase activities, especially with added glucose, increased over the 3-year period (Fig. 7), with the shallower soils directly over shale showing decreased relative activities within each particular year.

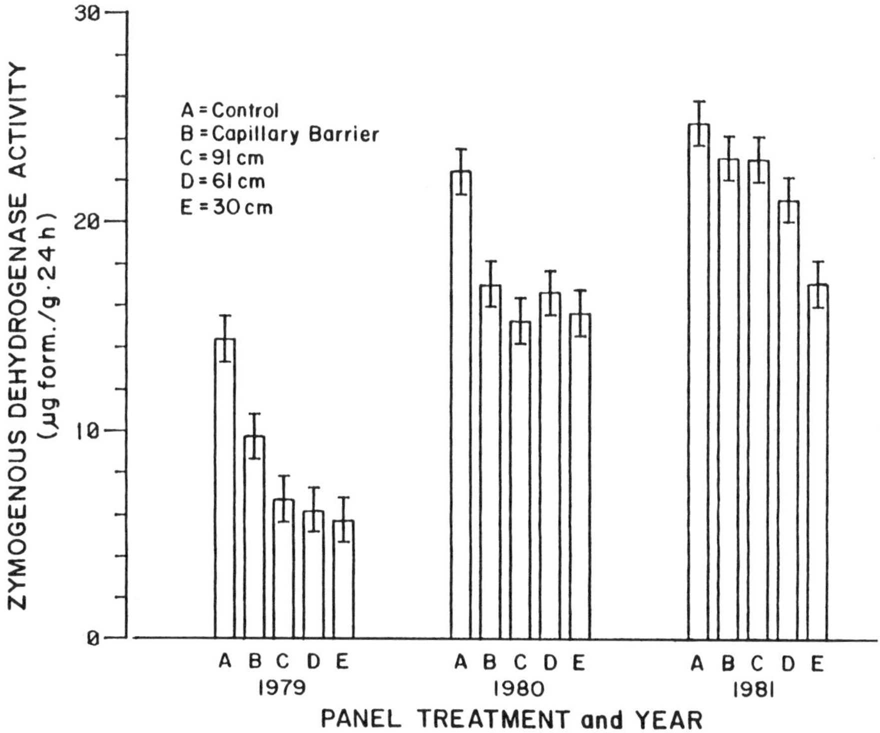

Figure 7 Zymogenous dehydrogenase activity of soil of various depths over spent shale and a capillary barrier—1979, 1980, and 1981 results. The least significant difference between means is 2.2 μg formazan $\cdot g^{-1} \cdot 24\ h^{-1}$, as indicated by the error bars.

D. Interaction Mechanisms-Retorted Shales and Soil Enzymes

Laboratory and field studies of retorted shale interactions with soils and plant-soil systems have shown over-all trends toward decreased soil enzymatic activity, in comparison with control soils, indicating that these materials may provide a unique reclamation challenge. An important part of reclamation management will be understanding the mechanisms by which retorted shales influence surface soils, as indicated by these enzymatic responses. The enzymatic changes that occur probably result from a series of factors. These can result from simple abiotic changes, such as salinity or water variations, or changes that occur naturally as a plant community develops in environments where the effects of different depths of surface soil are being tested.

In addition, these changes may be the result of specific inhibitory components from a waste material. Each of these alternatives should be considered in interpreting and developing recommendations based on soil enzymatic analyses. An essential part of this work is to evaluate possible causes and to make the best possible judgement concerning the reasons for enzymatic responses that might be observed in a particular reclamation situation.

1. *Movement of Soluble Constituents*

One of the concerns regarding successful revegetation of soil overlying spent oil shale is the possible movement of salts (Fransway and Wagenet, 1981) and/or other soluble toxic elements from shales into surface soils. Based on this concern, capillary barriers were used to limit salt movement and to attempt to maintain the functional integrity of the overlying soils. In the experiment where enzymatic responses were monitored, as noted in Figs. 5, 6, and 7, electrical conductivity measurements also were completed. Conductivity for the shallower surface soil panels, where decreased enzymatic activities had been observed, indicate that the 1980 and 1981 samples had lower values than those in 1977, and conductivities were generally very low in all replaced soils. This information suggests that decreases in enzymatic activities had occurred independently of soil conductivity. Soils at the shale-soil interface also were tested in 1981, and some salt movement from the processed shale to the lowest zone of the overlying soil was observable. No evidence was found for increases in conductivity in the surface soils used in this study, making it unlikely that capillary movement of salts and other soluble components would be related to the decreased enzymatic activities observed in the shallower surface soils located over the retorted shale materials.

2. *Soil Moisture Variations*

Soil moisture variations can also be a factor which might have been related to these enzymatic changes. In this study, the average sample moisture content did not vary from a range of 3% by weight over the 3-year period. Correlation coefficients between the biological measurements and the sample moisture contents for each of the three study years were low with the maximum r^2 value being 0.082 (Table 2). This would suggest that less than 10% of the variance in these 1979, 1980, and 1981 biological measurements could be explained on the basis of soil moisture content changes.

Soil moisture data were also evaluated for their possible relationships to the biological responses within each of the experimental years. Analysis of variance procedures showed no significant interactive effects of soil water contents with panel treatments and biological measurements in any of the 3 years.

Table 2 Values of r^2 for Linear Correlations of Soil Moisture Content at the Time of Sampling with Soil Biochemical Activities[a]

Year	Zymogenous dehydrogenase	Autochthonous dehydrogenase	Phosphatase	Nitrogen fixation potential
1979	0.019	NC[b]	0.002	0.001
1980	0.073	0.044	0.082	0.030
1981	0.008	0.015	0.005	0.030

[a]At the site described by Redente and Ruzzo (1979).
[b]NC, Not calculated.

3. *Plant Community Changes*

The enzymatic responses observed in this study may reflect differences in plant community development, as plant growth in mine soils and spoils, (Wilson, 1965; Cresswell, 1973; Visser et al., 1979; and Fresquez and Lindemann, 1982) and in disturbed soils (Ross et al., 1982) can have major short-term effects on organic matter accretion in developing plant-soil systems, and on soil enzymes.

Measurements of percent mean plant cover in the control plots, and corresponding capillary barrier plot with 60 cm of soil, and with 90, 60, or 30 cm of soil directly over shale, indicate no significant decreases in percent mean plant cover when comparing the control plot with only 30 cm of soil, as summarized in Table 3 (Redente et al., 1980, 1981, 1982a,b). An indication of improved plant development with the capillary barrier was also suggested by these results. In comparison, the phosphatase, glucose-amended dehydrogenase and nitrogen fixation potential measurements decreased to a greater extent than the corresponding values for mean percent plant cover on these plots with less surface soil overlying the retorted shale.

4. *Inhibitory Factors*

Inhibitory inorganic or organic compounds, other than soluble salts, could interact with surface soils by a variety of routes and lead to the differences in surface soil biogeochemical processing observed in this study. Such components from retorted shales might be taken up by plants, translocated throughout the root system and aboveground parts, and accumulate in the soil through exudation and litter decomposition, and have direct effects on biogeochemical cycling. In addition, inhibitory compounds, if they play a role in these processes, might diffuse in volatile form through the soil atmosphere to the soil surface. These possibilities have not been investigated at this

Table 3 Relative Percent Mean Plant Cover, Nitrogen Fixation Potential, Phosphatase Activity, and Glucose-Amended Dehydrogenase Activity for Reclamation Plots with Capillary Barrier Plus Topsoil or Varied Topsoil Depths over Retorted Shale. Normalized Averages of 1979, 1980, and 1981 Responses[a,b]

Reclamation treatment	Mean percent canopy cover	Enzymatic assay		
		N_2 fixation	Phosphatase	Dehydrogenase
Control	100	100	100	100
Capillary barrier plus 60 cm soil	137	82	89	84
90 cm soil over shale	123	54	74	76
60 cm soil over shale	113	43	68	72
30 cm soil over shale	104	41	59	64

[a]Control soil responses normalized to 100%.
[b]At the site described by Redente and Ruzzo (1979).

site, and little information is available that would allow evaluation of these possibilities from first principles. However, some conjecture on possible relationships of these types of processes to changes in soil enzymatic activity might be helpful in identifying the possible causes for the lower soil enzymatic activities in soil observed with decreasing surface soil depth over retorted shale. Table 4 provides a summary of information on possible inhibitory factors that might interact with the surface soil in these reclamation systems.

A wide range of trace metals and other constituents of the earth's crust are enriched in surface soils due to translocation by plants and subsequent litter fall, followed by decomposition and incorporation of these materials in surface soil organic matter at higher concentrations (Boyle, 1968). Lindsay et al. (1981) found increased boron and molybdenum in the aboveground portions of plants grown on spent oil shale or on soil overlying spent shale at this site, amplifying results from initial studies carried out by Anderson (1982) immediately after establishment of these plots.

Schwab et al. (1980), found that fluorine, cadmium, lead, and cobalt were at or below detection limits, whereas arsenic and selenium were below toxicity limits. Kilkelly and Lindsay (1982) also studied trace elements in western wheatgrass (*Agropyron smithii*) and fourwing saltbush (*Atriplex canescens*) growing on TOSCO II and USBM materials. Plants were analyzed for boron, molybdenum, arsenic, selenium, and fluorine on a dry-weight basis. Boron in western wheatgrass from the TOSCO plots ranged from 85–101 ppm, in comparison with 4–15 ppm in control plant materials. Fourwing saltbrush samples average 28 ppm of molybdenum, and when combined with low copper, this changed molybdenum-copper ratio might make these plants unacceptable as a forage for ruminants. The microbial role in modifying plant uptake of these elements is not well understood, although Lee and Loutit (1977) found that growth of rhizophere bacteria, with extracellular polysaccharide production, led to decreased plant molybdenum uptake. Arsenic, selenium, and fluorine were below limits set for plant or animal toxicities.

Similar plant translocation processes may occur with organic compounds. Pesticides may be taken up by plants and translocated to aboveground parts (Nash, 1974), and some polynuclear aromatic hydrocarbons can be taken into plants from contaminated soils (Sims and Overcash, 1983). There are no data available to indicate if organic compounds from spent oil shale are being taken up by plants and translocated to the surface, or concerning a microbial role in mediation of these processes. If this is occurring, it may have important consequences for forage quality of the plants as well as possible impacts on the surface soil. Generally, more polar organic compounds are taken up by plant roots and translocated throughout the plant, whereas less polar compounds tend to become adsorbed to the root's outer tissues and are not translocated. It is difficult, however, to predict

Table 4 Possible Mechanisms for Retorted Oil Shale-Surface Soil Interactions

Possible route of transport	Type	Possible specific factors	Probability ranking[a]	Comment	References
Soil solution	Salinity	Major cations and anions	Very low	Low salinity found in all surface soils	Present study
	Inhibitory organics	Ag, As, B, Cd Hg, Mo. Se, F, etc.	Very low	Move as ions with other salinity	Present study
	Inhibitory organics	PAHs[b] (e.g., dibenzothiophene, phenanthrene, fluoranthene, pyrene, thianthrene, aminopyrene, triphenylene, benz[*a*]pyrene, chrysene, benzo[*e*]pyrene, perlene, benzo[*a*]pyrene, etc.	Very low	Must have significant solubility in water. Would not be expected to move further than salinity. Some compounds may stimulate bacterial growth.	Maase (1980), Hass and Applegate (1975), Graf and Nowak (1966), Malaney et al. (1967)
Plant uptake and translocation	Salinity	Major cations and anions	Very low	Low salinity found in all surface soils	Present study

	Toxic metals and elements	As, B, No, Se, etc.	Low to good	Elevated concentrations found only in plants where roots have grown into spent shale.	Lindsay et al. (1981), Kilkelly and Lindsay (1982)
Plant uptake and translocation (continued)	Inhibitory organics	Water extractable PAHs (see list above), etc.	Low to fair	Possible where plant roots have entered shale or contaminated overlying soil. Compounds with higher polarity more likely to be taken up by plants.	Nash (1974), Maase (1980), Sims and Overcash (1983)
Soil atmosphere	Volatile inorganics and organometallics	AsH_3, methylarsines, SeH_3 fluorine	Low to fair	Volatilization depends on physical/chemical conditions and/or biological methylation. Accumulation in surface soil required.	Hassler (1982)
	Volatile toxic organics	Volatile PAHs (e.g., carbazole, acridine, anthracine, phenanthrene), etc.	Low to fair	Low vapor pressure for most organic compounds extracted from spent shale. Buildup in surface soil required.	Maase (1980), Sims and Overcash (1983)

[a] A subjective ranking of the likelihood of the type of factor specified to actually influence the quality of the surface soil environment in a way which would impair microbial activity.
[b] PAH, polynuclear aromatic hydrocarbon.

the plant uptake behavior of individual compounds, and actual measurements are needed to ascertain whether organic compounds from spent shale have been accumulating in plant materials, and what the specific (and possibly manageable) role of soil microorganisms and soil enzymes in these processes might be.

Compounds with relatively high vapor pressure could diffuse through the soil atmosphere in high enough concentrations to inhibit soil microbial activities. There are several immediate candidates which might be considered in relation to this type of interaction.

Arsenic and selenium have been reported to be present in some shales at 59–70 μg/g total arsenic and 2–17 μg/g total selenium, respectively (Shendrikar and Faudel 1978, Fox 1980a, Fox 1980b). Water-soluble arsenic and selenium in spent shales is a small percentage of this amount, ranging from 0.05 μg/g to 0.13 μg/g for arsenic and averaging about 0.05 μg/g for selenium (Fox 1980b). Both arsenic and selenium can be reduced to toxic volatile compounds [AsH_3, CH_3AsH_3, $(CH_3)_2AsH$, $As(CH_3)_3$, SeH_3] by physicochemical and/or biochemical processes. Hassler (1982) found that some spent oil shales can release small amounts of volatile arsenic compounds when organic matter is added to the spent shale to stimulate microbial activity, but volatile selenium production, except in limited experiments where no release were observed, was not monitored. Apparently, the potential for arsenic and selenium volatilization from spent oil shale would be low under field conditions, however, actual field measurements of volatile inorganic or organometallic compound concentrations in the soil atmosphere above spent shales have not been made.

The adsorption of soluble inorganic arsenic to soil and shales has been documented by many workers (Wauchope and Yamamato, 1980; Hassler, 1982; Sievers et al., 1982; Takamatsu et al., 1982). The potential for volatilization in the presence of microorganisms and usable organic matter, if this were to continue over an extended time, might lead to arsenic accumulation in surface soils, with possible changes in soil enzymatic functioning.

Another critical element that could potentially influence surface soil enzymatic activities is fluorine. As summarized by Saether et al. (1980), fluorine is the most abundant halogen in sedimentary rocks, and in oil shales it is found at 10 times the concentration in average coals. Values in the range of 800–1000 ppm of fluorine in retorted oil shales have been reported (Shendrikar and Faudel, 1978). In addition, leaching of shales will yield solutions with 0.8–1.7 ppm of fluorine, exceeding the EPA drinking water standards by 10-fold, and after extensive leaching which results in rapid rinsing of arsenic, selenium, molybdenum, and barium, fluoride concentrations will still remain high, as noted by Stollenwerk and Runnells (1977). In their study, most fluorine was found in association with the illitic clays, and shales with the highest oil yields generally had the highest fluorine concentrations.

As noted by Saether et al. (1980), most fluorine will remain in the residual retorted shale after processing. With the known toxic characteristics of many fluorine compounds (Merck Index, 1976), this element could be an appropriate candidate upon which approaches for understanding possible volatile retorted shale interactions with surface soils could be developed.

In addition to inorganic constituents, potentially toxic organic compounds with relatively high vapor pressures may be present in oil shales after retorting. Polycyclic aromatic hydrocarbons extracted from spent oil shale with vapor pressures comparable to Hg^{0} at 25°C include carbazole, acridine, anthracine, and phenanthrene (Maase 1980; Sims and Overcash 1983). These and/or other organic compounds might be present in the soil atmosphere in sufficiently high concentrations to inhibit certain microbial activities, and continued adsorption by surface soils could lead to accumulation and effects on enzymatic processes. Again, no measurements are available to confirm this possibility either from laboratory experiments or from field measurements, although an excellent review on stress and ecotoxicological assessment methodology for soil microorganisms has been completed by Domsch et al. (1983).

V. SOIL ENZYMES—PROSPECTS AND POTENTIAL APPLICATIONS IN RECLAMATION

Soil enzymes and soil microbiological processes play essential roles in the functioning of stable and resilient plant-soil systems. In reclamation, soil enzymes may or may not be initially present when a particular area is reconstructed, and energy wastes can affect the functioning of soil enzymes.

In considering enzymatic strategies for reclamation management, several characteristics of soil enzymes should be emphasized in relation to their functioning and potential responses:

1. Soil enzymes represent a short-term, "fast response" component of ecosystems. Free or abiontic enzymes that do not normally interact with substrates, or are not associated with microbes or plants, must be considered in any enzymatic analysis strategy for reclamation management.
2. The soil enzymes (at least in a developed mature system) have an excess functional capacity. Especially in a water-limited ecosystem, plant requirements for nutrients can be met with the functioning of only a relatively minor portion of this enzymatic processing potential.
3. With fertilization of arid systems, and the continued cycling and availability of these externally-added nutrients over time, plant dependence on belowground system functioning can be lessened.

This results from fertilizers repressing normal biological nitrogen, phosphorus, and possibly other biogeochemical cycling processes.

In addition, such repression may be enhanced by interactions with components from energy wastes, as observed with retorted shale. Fertilizer use also can lead to decreased rates of soil organic matter accumulation, as has been documented for coal mine reclamation systems (Stroo and Jencks, 1982).

Soil enzymes appear to have a potential for more intensive use in management of reclamation, to assist in providing the following types of information: (1) relative biological quality estimates of different soil materials, and (2) potential effects of particular energy wastes on belowground systems and on plant community functioning. This can include direct interactions, the effects of leachates, and also volatile components. The results of this study suggest that it may be possible to predict future effects of stressed soils on plant community functioning based on comparative enzymatic analyses. (3) To provide information on relative changes in belowground-system functioning over time as a plant community develops in a particular reclamation environment.

In this framework, if soil enzymatic analyses can improve the design, installation, and post-construction management of reclamation systems, this would appear to be useful, especially as these approaches can provide unique information which will complement a series of carefully-selected chemical analyses. The results of this study continue to indicate this potential, and also suggest that enzymatic approaches may be able to provide unique information on mechanisms of energy waste-soil system interactions needed for more effective reclamation management.

ACKNOWLEDGMENTS

This research was supported by the Department of Energy under contract DE-A502-76EV04018, awarded to the Department of Range Science, Colorado State University. Drs. Ralph Franklin and C. W. Cook provided helpful guidance in developing the framework for this chapter.

REFERENCES

Adams, J. C., and Farrier, D. S. 1982. The effect of some oil shale process waters upon the viability of indicator bacteria. *J. Environ. Qual. 11*: 171–174.

Adler, J. 1969. Chemoreceptors in bacteria. *Science 166*: 2588–2597.

Anderson, B. M. 1982. Effects of processed oil shale on the element content of *Atriplex canescens*. Open File Report 82-1051. U.S. Dept. of Interior Geological Survey, 16 pp.

Birch, H. F. 1958. The effect of soil drying on humus decomposition and nitrogen availability. *Plant Soil 10*: 9–31.

Bloch, M. B., and Kilburn, P. D. 1973. Processed shale revegetation studies, 1965–1973. Colony development operation, Atlantic Richfield Co., Denver, Colorado, 209 pp.

Boyle, R. W. 1968. Silver in soils and other products of weathering. Geological Survey of Canada Bulletin 160, Queen's Printers, Ottawa, 264 pp.

Brüning, E., Unger, H., and Dunger, W. 1965. Untersuchungen zur Frage der biologischen Aktivierung alttertiärer Rohbodenkippen des Braunkohlentagebaues in Abhängigkeit von Bodenmelioration und Rekultivierung. *Z. für Landeskultur 6*: 9–38.

Burns, R. G. (Ed.) 1978. *Soil enzymes*. Academic Press, London, 380 pp.

Burns, R. G. 1983. Extracellular enzyme-substrate interactions in soil. In *Mircrobes in Their Natural Environments*, 34th Symposium, Soc. for General Microbiology. J. H. Slater, R. Whittenbury and J. W. T. Wimpenny (Eds.) Cambridge Univ. Press. pp. 249-298.

Casida, Jr., L. E., Klein, D. A., and Santoro, T. 1964. Soil dehydrogenase activity. *Soil Sci. 98*: 371–397.

Clark, F. E., and Paul, E. A. 1970. The microflora of grasslands. *Adv. Agron. 22*: 375–435.

Cresswell, C. F. 1973. Changes in vegetational composition, nutritional status, soils, and microbial populations with the establishment of vegetation on gold-mine dumps on the Witwatersrand. In *Ecology and Reclamation of Devastated Land*, R. J. Hutnick and G. Davis (Ed.) Vol. 2. Gordon and Breach, New York, pp. 335--359.

Cundell, A. M. 1977. The role of microorganisms in the revegetation of strip-mined land in the western United States. *J. Range Managm. 30*: 299–305.

Domsch, K. H., Jagnow, G., and Anderson, T. H. 1983. An ecological concept for the assessment of side-effects of agrochemicals on soil microorganisms. *Residue Rev. 86*: 65–105.

Ernst, W. 1976. Physiological and biochemical aspects of metal tolerance. *Soc. Exp. Biol. Seminar Series. 1*: 115–133.

Fox, J. P. 1980a. The partitioning of major, minor and trace elements during simulated in situ oil shale retorting. LBL-9062. Lawrence Berkeley Laboratory, University of California, Berkeley, California, 441 pp.

Fox, J. P. 1980b. Water related impacts of in situ oil shale processing LBL-6300. Lawrence Berkeley Laboratory, University of California, Berkeley, California, 327 pp.

Fox, J. P., Mason, K. K., and Duvall, J. J. 1979. Partitioning of major, minor and trace elements during simulated in situ oil shale retorting in a controlled-state retort. LBL-9030 Lawrence Berkeley Laboratory, University of California, Berkeley, California.

Fransway, D. F., and Wagenet, R. J. 1981. Salt release and movement in processed oil shale. *J. Environ. Qual. 10*: 107–113.

Fresquez, P. R., and Lindemann, W. C. 1982. Soil and rhizosphere microorganisms in amended coal mine spoils. *Soil Sci. Soc. Am. J. 46*: 751–755.

Graf, W., and Nowak, W. 1966. Growth simulation in lower and higher plants by carcinogenic polycyclic aromatic compounds. *Arch. Hyg. Bakt. 150*: 513–528.

Harbert, III, H. P., and Berg, W. A. 1978. Vegetative stabilization of spent oil shales: vegetation, moisture, salinity, and runoff–1973-1976. U.S. Environ. Prot. Agency Rept. No. EPA-600/7-78-021. Industrial Environ. Res. Lab., Cincinnati, Ohio, 183 pp.

Hardy, R. W. F., Burns, R. C., and Holsten, R. D. 1973. Applications of the acetylene-ethylene assay for measurement of nitrogen fixation. *Soil Biol. Biochem. 5*: 47–81.

Hass, B. S., and Applegate, H. G. 1975. The effects of unsubstituted polycyclic aromatic hydrocarbons on the growth of *Escherichia coli*. *Chem. Biol. 10*: 265--268.

Hassler, R. A. 1982. Microbial contributions to soluble and volatile arsenic dynamics in retorted oil shale. M. S. Thesis. Colorado State University, Fort Collins, Colorado, 73 pp.

Hersman, L. E., and Klein, D. A. 1979. Retorted oil shale effects on soil microbiological characteristics. *J. Environ. Qual. 8*: 520–524.

Hersman, L. E., and Temple, K. L. 1979. Comparison of ATP, phosphatase, pectinolase, and respiration as indicators of microbial activity in reclaimed coal strip mine spoils. *Soil Sci. 127*: 70–73.

Hersman, L. E., Molitoris, E., and Klein, D. A. 1981. Nitrogen fixation by legumes in retorted shale. *Soil Biol. Biochem. 13*: 429–430.

Hoffmann, E., and Seegerer, A. 1951. Uber das Enzymsystem unserer Kulturböden. I. Saccharase. *Biochem. Z. 322*: 174–179.

Jenny, H. 1980. *The Soil Resource. Origin and Behavior.* Ecological Series 37. Springer Verlag, New York, Heidelberg, Berlin, 377 pp.

Kaplan, D. L., and Hartenstein, R. 1979. Problems with toluene and the determination of extracellular enzyme activity in soils. *Soil. Biol. Biochem. 11*: 335–338.

Kapustka, L. A., and E. L. Rice. 1976. Acetylene reduction (N_2-fixation) in soil and old field succession in central Oklahoma. *Soil Biol. Biochem. 8*: 497–503.

Kilkelly, M. K., Berg, W. A., and Harbert, III, H. P. 1982. Project Summary. Field studies on USBM and TOSCO II retorted oil shales: Vegetation, moisture, salinity, and runoff, 1977–1980. U.S. Environmental Protection Agency. Industrial Environmental Research Lab, Cincinnati, Ohio. EPA-600/57-81-139, 9 pp.

Kilkelly, M. K., and Lindsay, W. L. 1982. Selected trace elements in plants grown on retorted oil shales. *J. Environ. Qual.* *11*: 422–427.

Klein, D. A., Hersman, L. E., Nagle, N., Sorensen, D. L., and Wu, S.-Y. 1980. Role of soil microorganisms as indicators and possible controlling factors in plant succession processes on retorted shale and disturbed soils. In *Reclamation Studies on Oil Shale Lands in Northwestern Colorado,* C. W. Cook and E. F. Redente (Ed.). Dept. of Range Science, Colorado State University, Fort Collins, Colorado, pp. 27–34.

Klein, D. A., Hersman, L. E., and Sorensen, D. L. 1979a. Revegetation effects on surface soil microbiological characteristics. In *Symposium on Surface Mining, Hydrology, Sedimentology and Reclamation.* University of Kentucky, Lexington, Kentucky, 4–7 Dec. 1978, pp. 229–233.

Klein, D. A., Hersman, L. E., and Wu, S.-Y. 1979b. Monitoring of retorted oil shale effects on surface soil nitrogen fixation processes: A resource for design and management of land reclamation programs. In *Oil Shale Symposium: Sampling Analysis and Quality Assurance.* C. Gale (Ed.). Denver, Colorado. March 1979. USEPA, Cincinnati, Ohio, EPA 100/9-80-022, pp. 546–554.

Klein, D. A., Loh, T. C., and Goulding, R. L. 1971. A rapid procedure to evaluate the dehydrogenase activity in soils low in organic matter. *Soil Biol. Biochem.* *3*: 385–387.

Klein, D. A., Sorensen, D. L., and Metzger, W. 1982. Soil microorganisms and management of retorted shale reclamation. In *Revegetation Studies on Oil Shale Related Disturbances in Colorado.* E. F. Redente and C. W. Cook (Ed.). Dept. of Range Science, Colorado State University, Fort Collins, Colorado, pp. 27–44.

Lee, T. E., and Loutit, M. W. 1977. Effect of extracellular polysaccharides of rhizosphere bacteria on the concentration of molybdenum in plants. *Soil Biol. Biochem.* *9*: 411–414.

Lindeman, R. L. 1942. The trophic-dynamic aspect of ecology. *Ecology* *23*: 399–418.

Lindsay, W. L., Schwab, A. P., and Smith, P. J. 1981. Concentration of chemical elements in plants growing on retorted oil shales. In *Trace Elements in Oil Shale,* W. R. Chappell (Ed.). DOE-10298-7. Center for Environmental Sciences, University of Colorado, Denver, Colorado, pp. 75–105.

MacMahon, J. A. 1979. Ecosystems over time: Succession and other types of change. In *Forests, Fresh Perspectives from Ecosystem Analysis*, R. H. Waring (Ed.). Proceedings of the 40th annual biology colloquium. Oregon State University Press, Corvallis, Oregon, pp. 27–58.
Maase, D. L. 1980. An evaluation of polycyclic aromatic hydrocarbons from processed oil shales. PhD dissertation, Utah State University, Logan, Utah, 190 pp.
Malaney, G. W., Lutin, P. A., Chibulka, J. J., and Hikerson, L. H. 1967. Resistance of carcinogen organic compounds to oxidation by activated sludge. *J. Water Pollut. Contr. Fed.* *39*: 2022–2029.
Merck Index. 1976. Ninth Edition. Merck and Co., Inc., Rahway, New Jersey, 1313 pp. plus appendices.
Miller, R. M. 1977. Laboratory research studies on the reestablishment of below-ground ecosystems. Land Reclamation Program, Annual Report, Argonne National Laboratory, July 1976–October 1977, pp. 95–119.
Miller, R. M., and Cameron, R. E. 1978. Microbial ecology studies at two coal mine refuse sites in Illinois. Argonne National Laboratory, Argonne, Illinois, Report ANL/LRP-3.
Miller, R. M., and Cameron, R. E. 1977. Some effects on soil microbiota of topsoil storage during surface mining. In *Fourth Symposium Surface Mining and Reclamation.* Louisville, Kentucky, 19–21 Oct. 1976, pp. 131–139.
Nakas, J. P., and Klein, D. A. 1980. Mineralization capacity of bacteria and fungi from the rhizosphere-rhizoplane of a semi-arid grassland. *Appl. Env. Microbiol. 39*: 113–117.
Nash, R. G. 1974. Plant uptake of insecticides, fungicides, and fumigants from soils. In *Pesticides in Soil and Water,* W. D. Guenzi (Ed.). Soil Science Society of America, Madison, Wisconsin, pp. 257–313.
O'Neill, R. V., and Reichle, D. E. 1979. Dimensions of ecosystem theory. In *Forests, Fresh Perspectives from Ecosystem Analysis,* R. H. Waring (Ed.). Proceedings of the 40th annual biology colloquim, Oregon State University Press, Corvallis, Oregon, pp. 11–26.
Odum, E. P. 1969. The strategy of ecosystem development. *Science* *164*: 262–270.
Parkinson, D. 1976. The restoration of soil productivity. In *The Breakdown and Restoration of Ecosystems,* M. W. Holdgate and M. J. Woodman (Ed.). NATO Conference Series I, Ecology. Vol. 3. Plenum Press, New York and London, pp. 213–229.
Redente, E. F., and Ruzzo, W. J. 1979. Topsoiling of retorted oil shale. In *Proc. Symp. Surface Mining, Hydrology, Sedimentation and Reclamation,* S. B. Carpenter (Ed.). University of Kentucky, Lexington, Kentucky, 4–7 Dec. 1979, Office of

Eng. Serv., College of Eng., Univ. of Kentucky, Lexington, Kentucky, pp. 285–291.

Redente, E. F., Ruzzo, W. J., Grygiel, C. E., Jepson, R. F., and Mount, C. B. 1980. Effect of plant species, soil material, and cultural practices upon plant establishment and succession. In *Reclamation Studies on Oil Shale Lands in Northwestern Colorado*, C. W. Cook and E. F. Redente (Ed.). Dept. of Range Science, Colorado State University, Fort Collins, Colorado, pp. 1–17.

Redente, E. F., Doerr, T. B., Mount, C. B., Grygiel, C. E., Sievers, T. E., and Biondini, M. 1981. Effects of plant species, soil material, and cultural practices upon plant establishment and succession. In *Revegetation Research on Oil Shale Lands in the Piceance Basin*, E. F. Redente and C. W. Cook (Ed.). Department of Range Science, Colorado State University, Fort Collins, Colorado, pp. 1–31.

Redente, E. F., Doerr, T. B. Grygiel, C. E., Allerdings, L., Start, J. J., and Biondini, M. E. 1982a. Effects of plant species, soil material, and cultural practices upon plant establishment and succession. In *Revegetation Studies on Oil Shale Related Disturbances in Colorado*, E. F. Redente and C. W. Cook (Ed.). Dept. of Range Science, Colorado State University, Fort Collins, Colorado, pp. 1–25.

Redente, E. F., Mount, C. B., and Ruzzo, W. J. 1982b. Vegetation composition and production as affected by soil thickness over retorted oil shale. *Reclam. Reveg. Res. 1*: 109–122.

Reichle, D. E., O'Neill, R. V., and Harris, W. F. 1975. Principles of energy and material exchange in ecosystems. In *Unifying Concepts in Ecology*, W. H. van Dobben and R. H. Lowe-McConnel (Ed.). Dr. W. Junk B. V. Publishers, The Hague, pp. 27–43.

Richardson, S. G., McKell, C. M., George, M. N., and Gray, G. 1981. Weathering effects on some chemical and physical properties of retorted oil shale. *J. Environ. Qual. 10*: 221–224.

Ross. D. J., Speir, T. W., Tate, K. R., Cairns, A., Meyrick, K. F., and Pansier, E. A. 1982. Restoration of pasture after topsoil removal: Effects on soil carbon and nitrogen mineralization, microbial biomass and enzyme activities. *Soil. Biol. Biochem. 14*: 475–481.

Schaller, F. W., and Sutton, P. (Ed.) 1978. *Reclamation of Drastically Disturbed Lands.* American Society of Agronomy, Madison, Wisconsin, 742 pp.

Saether, O. M., Runnells, D. D., Ristinen, R. A., and Smythe, W. R. 1980. Fluorine: Its mineralogical residence in the mahogany zone of the green river formation, Piceance Creek Basin, Colorado. In *Trace Elements in Oil Shale*, W. R. Chappell (Ed.). Center for Environmental Sciences, University of Colorado, Denver, Colorado, pp. 85–112.

Schmehl, W. R., and McCaslin, B. D. 1973. Some properties of spent oil shale significant to plant growth. In *Ecology and Reclamation of Devastated Land*, Vol. 1, R. Hutnik and G. Davis (Ed.). Gordon & Breach, New York, New York, pp. 27–43.

Schwab, A. P., Lindsay, W. L., and Marx, G. P. 1980. Uptake of chemical elements by plants growing on spent shale. In *Trace Elements in Oil Shale*, W. R. Chappell (Ed.). Center for Environmental Sciences, University of Colorado, Denver, Colorado, pp. 85–112.

Schuman, G. E., Berg, W. A., and Power, J. F. 1976. *Management of Mine Wastes in the Western United States*. Land Application of Waste Materials, Soil Conservation Society of America, pp. 182–194.

Shendrikar, A. D., and Faudel, G. B. 1978. Distribution of trace metals during oil shale retorting. *Environ. Sci. Technol. 12*: 332–334.

Sievers, R. E., Stanley, J. S., Conditt, M. K., Hawthorne, S. B., Caolo, M. A., and Barkley, R. M. 1982. Characterization and treatment of organic compounds in wastewater from oil shale processing. In *Trace Elements in Oil Shale*, W. R. Chappell (Ed.). Center for Environmental Sciences, University of Colorado, Denver, Colorado, pp. XI-1-22.

Sims, R. L., and Overcash, M. R. 1983. Fate of polynuclear aromatic compounds in soil plant systems. *Residue Rev. 88*: 1–68.

Skujins, J. 1967. Enzymes in soil. *Soil Biochem. 1*: 371–414.

Skujins, J. 1973. Dehydrogenase: An indicator of biological activities in arid soils. *Bull. Ecol. Res. Comm. Stockholm, 17*: 235–239.

Skujins, J. 1976. Extracellular enzymes in soil. *CRC Crit. Rev. Microbiol. 4*: 383–421.

Skujins, J. 1978. History of abiontic soil enzyme research. In *Soil Enzymes*, R. G. Burns (Ed.). Academic Press, London, pp. 1–49.

Skujins, J., Braal, L., and McLaren, A. D. 1962. Characterization of phosphatase in a terrestrial soil sterilized with an electron beam. *Enzymologia 25*: 125–133.

Skujins, J., Pukite, A., and McLaren, A. D. 1974. Adsorption and activity of chitinase on kaolinite. *Soil Biol. Biochem. 6*: 179–182.

Slater, J. H., and Godwin, D. 1980. Microbial adaptation and selection. In *Contemporary Microbial Ecology*, D. E. Ellwood et al. (Ed.). Academic Press, New York, pp. 137–160.

Sorensen, D. L., Klein, D. A., Ruzzo, W. J., and Hersman, L. E. 1981. Enzyme activities in revegetated surface soil overlying spent Paraho process oil shale. *J. Environ. Qual. 10*: 369–371.

Sorenson, L. H. 1969. Fixation of enzyme protein in soil by the clay mineral montmorillonite. *Experientia 25*: 20–21.
Speir, T. W. and Ross, D. J. 1981. A comparison of the effects of air-drying and acetone dehydration on soil enzyme activities. *Soil. Biol. Biochem. 13*: 225–229.
Stollenwerk, K. G. 1980. Geochemistry of leachate from retorted and unretorted Colorado oil shale. PhD thesis, University of Colorado, Boulder, Colorado, 218 pp.
Stollenwerk, K. G., and Runnells, D. D. 1977. Leachability of arsenic, selenium, molybdenum, boron, and fluoride from retorted oil shale. Proc. 2nd Pacific Chemical Engineering Congress (PACHEM '77), pp. 1023–1030.
Stroo, H. F. and Jencks, E. M. 1982. Enzyme activity and respiration in mine soils. *Soil. Sci. Sco. Am. J. 46*: 548–553.
Takamatsu, T., Aoki, H., and Yoshida, T. 1982. Determination of arsenate, arsenite, monomethylarsonate, and dimethylarsinate in soil polluted with arsenic. *Soil Sci. 133*: 239–246.
Visser, S., Zak, J., and Parkinson, D. 1979. Effects of surface mining on microbial communities and processes. In *Ecology and Coal Resource Development*, Vol. 2, M. K. Wali (Ed.). Pergamon Press, New York, pp. 643–651.
Wauchope, R. D., and Yamamoto, M. 1980. Extraction, speciation, and analysis of arsenic and arsenical herbicides in runoff: Evaluation of simple methods at the ppb level. *J. Environ. Qual. 9*: 597–601.
Wilson, H. A. 1965. The microbiology of strip mine spoils. West Virginia Agri. Exp. Stn. Bull. 506T, 44 pp.
Wilson, H. A. 1961. Rhizosphere bacteria of some strip mine vegetation. *Proc. W. Virginia Acad. Sci. 33*: 15–20.
Wilson, H. A., and Hedrick, H. G. 1957a. Carbon dioxide evolution from some strip mine spoils. *Appl. Microbiol.* 5: 17–21.
Wilson, H. A., and Hedrick, H. G. 1957b. Some qualitative observations of the microflora in a strip mine spoil. *Proc. W. Virginia Acad. Sci. 29-30*: 35–38.
Wilson, H. A., and G. Stewart. 1956. The number of bacteria, fungi, and actinomycetes in some strip mine spoil. West Virginia Agri. Exp. Stn. Bull. 388T, 15 pp.
Wilson, H. A., and G. Stewart. 1955. Ammonification and nitrification in a strip mine spoil. West Virginia Agri. Exp. Stn. Bull. 379T, 16 pp.

6

Mycorrhizae and Reclamation of Stressed Terrestrial Environments

ROBERT M. DANIELSON / *The University of Calgary, Calgary, Alberta, Canada*

I. INTRODUCTION

Studies of mycorrhizal relationships on mine spoils and tailings have spanned nearly two decades. In most instances, the mycorrhizal status of plants growing on mine spoils has been of concern, whereas only relatively few evaluations of the actual effects of the plant-fungus symbioses on plant growth and survival have been made. The essential nature of mycorrhizae for the successful colonization of certain mine wastes was established by the landmark work of J. R. Schramm in the early 1960s. Since that time, research has been expanded to include vesicular-arbuscular (VA) and ericoid mycorrhizae and the development of techniques for inoculating hosts with fungi specifically adapted to spoils. In spite of these advances, mycorrhizal aspects of revegetation programs remain largely in an experimental phase. Reclamation of mined land requires innovative techniques that reduce costs and increase the chance of success of plant establishment and survival. Understanding the role of mycorrhizal associations in relation to establishment and development of plant communities may offer solutions to some of the problems encountered in operational reclamation programs.

Previous reviews (Meyer, 1973; Marx, 1975, 1980a,b; Jurgensen, 1978; Parkinson, 1979) have stressed the difficulties of establishing plants on spoils and tailings due to poor physical and chemical conditions, the lack of organic matter, and often the presence of extremes in temperature and pH. Although these factors are evident to individuals working in revegetation programs, the biological properties, especially mycorrhizal potential, are much less appreciated. The purpose of this review is to update the previous reviews and to determine if mycorrhizae are indeed successful at overcoming the variety of stresses encountered by plants being established or functioning on mine wastes.

II. MECHANISMS BY WHICH MYCORRHIZAE MAY ALLEVIATE STRESS ON MINE SITES

Schramm (1966) has convincingly shown that two factors were essential for vegetation to become established on barren, black coal wastes in Pennsylvania. First, the young seedlings had to be protected from the extremely high surface temperatures to prevent heat girdling of stems, and second, that the plants had to become mycorrhizal to increase their ability to exploit the spoil for nitrogen. Given these circumstances, woody plants did survive and were able to colonize the spoil habitat successfully. Since his work, it has become apparent that other spoil types exert different demands on colonizing plants, and that mycorrhizae can have other benefits beyond scavenging impoverished substrates for their meager supplies of available nitrogen. Some of these mechanisms are well established and others are more speculative, awaiting carefully controlled and analyzed experimentation to be properly evaluated.

It is now recognized that nutrients other than nitrogen may be limiting on mine spoils and that mycorrhizal associations provide an elegant adaptation alleviating deficiencies encountered by potentially mycorrhizal plants. Foremost among these is the enhanced ability of mycorrhizal plants, especially VA mycorrhizal species. to obtain available phosphorus from soils (Malajczuk and Lamont, 1981). Although suggestions have been made to the contrary, evidence now indicates that mycorrhizae are not capable of utilizing sources of phosphorus unavailable to nonmycorrhizal roots but are simply exploiting a larger pool of labile phosphorus (Malajczuk and Lamont, 1981; Smith, 1980; Thomas et al., 1982).

The uptake of nutrients other than phosphorus may also be enhanced by the mycorrhizal condition but evidence is scanty. This enhanced nutrient uptake would be most likely to occur with poorly mobile ions such as zinc and copper and to a lesser extent, ammonium (Lambert et al., 1979; Bowen and Smith, 1981). Bowen and Smith (1981) have pointed out that with less mobile ions, transfer through soil limits nutrient uptake by plant roots, and rooting density and geometry determines total uptake. It would be expected that mycorrhizae would most enhance the uptake of these ions. The degree of development of external mycelium may be of prime importance in the stimulation of host growth. Graham et al. (1982) have shown that for five *Glomus* species, differences in host growth enchancement were related to the amount of extramatrical mycelium produced. This may also be true of ectomycorrhizae, with mycelial strands and rhizomorphs having special importance (Bowen, 1973; Marx, 1980b).

Related to nutrient uptake are the effects of mycorrhizae on the accumulation or exclusion of toxic elements (Smith, 1980). It has often been stated that mycorrhizal plants may be more tolerant to heavy metals than their nonmycorrhizal counterparts but until recently, evidence to support this contention has been lacking. Bradley et al. (1981) pointed out that whereas enhanced uptake of ions such as phosphorus by mycorrhizae is beneficial, increased uptake of heavy metals may be very deleterious to plant growth and survival in soils containing high concentrations of metals. In field soils, it has been shown that mycorrhizae substantially increase the uptake of copper and zinc, both in terms of total uptake and concentration in the tissues (Lambert et al., 1979). This suggests that the mycorrhizal condition of plants on areas contaminated with copper, zince, and other heavy metals, as well as on acid soils where aluminum toxicity may be a problem (Brady, 1974), could influence plant growth and perhaps the types of plants present.

Bradley et al. (1981, 1982) studied the tolerance of ericaceous plants and their endophytes from contaminated and normal soils to various concentrations of copper and zinc in sand cultures. In all cases, the growth of mycorrhizal plants was superior to the growth of nonmycorrhizal plants when exposed to either copper or zinc. Tolerance to copper and zinc occurred regardless of the origin of the host or the endophyte. It appeared that protection was due to the restriction of translocation of the metals from roots to shoots with the metals accumulating in the roots. Bradley et al. (1981) suggest that protection is afforded by the adsorption of ions on the cell wall of the hyphae in the root and, further, that plants with VA mycorrhizae would be less tolerant due to less biomass of internal hyphae and the phycomycetous nature of the endophyte. This suggestion has been substantiated by a preliminary report by Killham and Firestone (1982). They reported that mycorrhizal grass plants were more sensitive to copper and nickel and had higher metal concentrations in the shoots than nonmycorrhizal plants. Although additional studies are required, these results imply that VA mycorrhizal plants are poorly adapted to sites with high levels of toxic metals. It would seem likely that ectomycorrhizal plants would behave similarly to ericaceous plants, in that mycorrhizae have a large fungal biomass component and thus large adsorptive surfaces. An organism such as the E strain, which produces intracellular infections in pine (Mikola, 1965), might be particularly effective in ameliorating the effects of heavy metals.

Another factor that has been repeatedly cited as a critical factor of revegetation of mine spoils, and one that might be alleviated by mycorrhizae, is drought tolerance. However, the opinion expressed by Powell (1982) was that conclusive evidence showing that mycorrhizal infection of plants results in increased water uptake is lacking. Most frequently, the work by Safir et al. (1972) has been cited as demonstrating the beneficial effect of mycorrhizae on the water relations of plants. They reported reduced resistance of water transport in mycorrhizal soybean roots as compared with nonmycorrhizal roots. In a separate experiment, the addition of a complete nutrient solution eliminated the differences in resistance between inoculated and noninoculated plants (the roots were not examined for infection) suggesting the mycorrhizal effect was simply due to more favorable nutrition. However, it should be kept in mind the technique used by Safir et al. (1972) to determine resistances involved saturating the soil (Boyer, 1971) and thus did not account for soil or rhizosphere resistance to water movement. As VA mycorrhizae are known to enhance phosphorus uptake by more effective exploitation of the soil than possible with nonmycorrhizal plants, possibly the most likely function of mycorrhizae in water relations (increased utilization of water distant from the rhizosphere) may have been overlooked. In a study of *Pinus radiata*, Sands and Theodorou (1978) concluded that soil resistance was a more important factor in water relations than plant resistance. This would suggest that root distribution and the nature of the extramatrical mycelium might be as important for water uptake as for uptake of poorly mobile ions. Allen (1982) showed that mycorrhizal infection of the grass *Bouteloua gracilis* reduced soil resistance as well as within-plant resistances and proposed that the improved water relations might be an important factor in plant survival in arid habitats. In contrast, Sands and Theodorou (1978) found that mycorrhizal infection of *Pinus radiata* with *Rhizopogon luteolus* resulted in increased within-plant resistance and increased water stress. These apparently conflicting results indicate that caution should be exercised in attributing mycorrhizae with the property of increasing drought tolerance until field or greenhouse studies actually demonstrate that infected plants survive and reproduce better than noninfected plants when under water stress.

Goss (1960) conducted one of the few tests to determine if mycorrhizae would influence drought tolerance. *Pinus ponderosa* is used in shelter belts in the Great Plains where rainfall is low, and Goss felt that the presence of mycorrhizae might be critical for survival of planting stock. Pot experiments were conducted with mycorrhizal and nonmycorrhizal seedlings and water was withheld until the seedlings exhibited drought symptoms followed by water addition and monitoring of recovery patterns. The water stress in all cases was severe and in three tests all the seedlings died. In a fourth test, 4 of 30 mycorrhizal seedlings died and 12 of 29 nonmycorrhizal seedlings died. Goss attributed the differences in survival to the better poststress lateral root development of the mycorrhizal seedlings, which in turn may have been due to the greater size and vigor of the mycorrhizal plants. Thus, in a test of a sudden and severe water deficiency, Goss' work indicates that mycorrhizal plants may survive better than nonmycorrhizal plants, but this may be due to seedling vigor rather than a direct mycorrhizal effect. A preliminary report by Duddridge et al. (1980) states that *Pinus sylvestris* seedlings infected with *Suillus bovinus* were able to survive a dry period whereas uninfected seedlings died, the mycorrhizal seedlings were able to obtain water from distant wet soil via transport through a system of rhizomorphs. This strongly indicates that rhizomorphic ectomycorrhizal species may increase drought tolerance by allowing exploitation of larger soil volumes than nonmycorrhizal seedlings or mycorrhizal plants lacking rhizomorphs. It remains to be shown that the mycorrhizal status per se improves plant water relations and drought tolerance under field conditions.

Mine spoils may also be extremely acidic (Jurgensen, 1978) and have high surface temperatures. Schramm (1966) found the surface temperatures on black coal wastes frequently reached 40–60°C and occasionally reached 75°C. It is expected that temperatures in the root zone of mine sites would substantially exceed root zone temperatures in undisturbed areas where the vegetation would reduce insolation. Thus, indigenous fungi may not be adapted to postdisturbance temperature regimes. However, few studies have compared mycorrhizal associations (as opposed to pure culture studies) over a range of temperatures. Marx et al. (1970) and Marx and Bryan (1971) demonstrated in monoxenic culture that not only could *Pisolithus tinctorius* infect roots at higher temperatures than *Thelephora terrestris*, but that the tolerance of the pine seedlings to high temperatures was increased considerably when infected with *Pisolithus*. Similar experiments have not been performed with other fungi or in open systems where the humidity is low and concurrent demands are placed on water uptake.

Nonlethal temperatures also affect mycorrhizal development with substantial differences existing among species and strains (Theodorou and Bowen, 1971; Schenck and Smith, 1982). Colonization and infection of roots by certain species of VA fungi has been shown to be slow at 18°C (Schenck and Smith, 1982) and at 16°C for ectomycorrhizal fungi (Theodorou and Bowen, 1971), temperatures that may be well above those normal in temperate regions (Theodorou and Bowen, 1971). Thus, if mycorrhizal associations are to be manipulated on disturbed sites, temperature relations may be a critical factor in symbiont selection (Trappe, 1977).

The effects of soil pH on the development of mycorrhizae have been reviewed several times, e.g., Slankis (1974) and Mosse et al. (1981). From these reviews it appears that individual species of endophytes can form mycorrhizae over a wide pH range and that mycorrhizae develop with appropriate host plants at any soil pH were the host occurs. However, the symbionts of both VA and ectomycorrhizae probably differ in extremely acid or alkaline soils from soils with moderate pH values, with few species being involved at the extremes (Trappe, 1977). The high solubility of potentially toxic heavy metals (Jurgensen, 1978) is related to low pH levels the possibility of micronutrient deficiences is related to high pH values.

III. MYCORRHIZAL SYMBIONTS INDIGENOUS TO MINE SPOILS

From the restricted number of published reports, the number of fungal symbionts reported to occur on mine spoils is low (Table 1). The limited number of observations, as well as the risk of relating fruit body occurrence with mycorrhizal abundance, makes a conservative interpretation of the data essential. Reports of gasteromycetes and hymenomycetes with persistent fruit bodies have been numerous but, with the exception of *Pisolithus tinctorius*, there is little evidence that these fungi are dominant symbionts on mine spoils. Due to the bright yellow color of the mycelium of some strains of *P. tinctorius*, it can be readily recognized, unlike most other ectomycorrhizal fungi. Schramm (1966) observed the natural infection of a variety of woody plants by *Pisolithus* and described in detail the penetration of the spoil and cracks of shale by the yellow rhizomorphs. Schramm showed conclusively that *Pisolithus* naturally colonized roots in barren spoils and developed an extensive system of mycelium and rhizomorphs. No other fungus has been assessed quantitatively for actual mycorrhizal abundance on mine spoils. The collection data suggest however that few agarics fruit on spoils and only a few species may be involved in mycorrhizal associations on barren spoils. *Inocybe* and *Laccaria* appear to be two genera containing species capable of forming mycorrhizae on barren spoils whereas Schramm (1966) found *Amanita*, *Cenococcum*, and *Boletus* only associated with older trees where a litter layer had developed. However, Benson et al. (1980) reported five genera in the Agaricales fruiting on largely uncolonized arsenic mine wastes.

Ascomycetes appear to be rare as ectomycorrhizal symbonts with the possible exception of E-strain fungi. Since a teleomorph is unknown for the E strain, its presence must be determined by direct examination of the mycorrhizae. E-strain fungi may be the dominant symbionts in nurseries in north temperate regions (Laiho, 1965; Mikalo, 1965) as well as in tropical regions (Mikola, 1980), a situation similar to that found with *Thelephora terrestris*. It would appear likely that E-strain fungi could be readily introduced with planting stock into minespoils, although observations indicate that E-strain fungi do not persist after outplanting (Mikola, 1965; Thomas and Jackson, 1979). However, mine spoils may offer little competition from indigenous fungi and E-strain fungi have been observed to be the dominant symbionts of white spruce in a coal mine spoil for a period exceeding 4 years (Danielson, unpublished data).

The inclusion of *Helvella corium* and *Sepultaria arenose* [*Geopora arenicola* (Lev.) Kers] is provisional because Schumacher (1979) reports both species to fruit in open areas. However, other species of *Geopora* are almost certainly mycorrhizal (Trappe, 1971) and the possibility of these ascomycetes forming mycorrhizal associations deserves further attention.

The ascomycete *Pezizella ericae* and perhaps close relatives are the only recognized fungal associates of ericoid mycorrhizae (Read, 1974; Mosse et al., 1981). As ericaeous plants are the dominant colonizers of spoils containing excesses of heavy metals (Bradley et al., 1981), it is certain that *Pezizella* is an important endophyte on mine spoils.

VA mycorrhizal endophytes differ from ectomycorrhizal fungi in many ways but prominent among the general differences are taxonomic, lack of host specificity, slowness of dispersal, and the small number of species of the VA fungi. The species spectrum found on mine spoils is not unlike that reported for undisturbed habitats (Gerdemann and Trappe, 1974), indicating a lack of selection or adaptation to spoils at the species level. In general, Mosse et al. (1981) observed that most VA fungal species were worldwide in distribution but that within a small distance spore populations may be highly variable and dependent upon the host species. Thus, one might not expect distinct groups of species on mine spoils owing to the poor dispersal mechanisms, the relatively few species available for colonization, and the lack of host specificity. However, intraspecific adaptation to the spoil environment may occur as illustrated by the much greater tolerance to zinc and cadmium of a strain of *Glomus mosseae* from a site polluted with zinc and cadmium as compared with a strain from a field soil (Gildon and Tinker, 1981). Strain adaptation to other extreme features of mine spoils, e.g., temperature, probably also occurs and may be as important as differences among species.

IV. MYCORRHIZAL DEVELOPMENT ON MINE SPOILS AND TAILINGS

A majority of the mycorrhizal studies on mine spoils have been conducted either to determine the presence or absence of mycorrhizae or to monitor the development of mycorrhizal root systems with time. These studies imply that there are beneficial effects of infection, but unaided by supplemental studies, they do not prove it. Schramm (1966) found nearly all the successful colonizers of black coal wastes in Pennsylvania were mycorrhizal. These included species of *Pinus*, *Betula*, *Populus*, and *Salix* and all were ectomycorrhizal. The wastes were 30–36 years old and on some there had been no spontaneous revegetation for 30 years. However, when pine was seeded on these areas and protected from high surface temperatures, the seedlings quickly become mycorrhizal. Schramm concluded that inoculum was abundant and generally not a decisive factor in plant establishment. With the exception of the legume, *Robinia*, which is capable of fixing N_2 when nodulated with *Rhizobium*, VA host plants were generally unsuccessful colonizers. *Acer rubrum* seedlings were occasionally found and were mycorrhizal, but did not live past the second growing season. Schramm's observations suggest that VA plant failure was due to their inability to obtain

Table 1 Mycorrhizal Fungi Reported to Occur on Mine Spots

Mycorrhizal species	Spoil type	Reference
Amanita muscaria (L. ex Fr.) Hooker	Arsenic spoil	Benson et al. (1980)
Amanita rubescens (Fr.) S.F. Gray	Coal spoil	Schramm (1966)
Amphinema byssoides (Fr.) J. Erikss.	Coal spoil	Danielson (unpublished data)
Astraeus hygrometricus (Pers.) Morgan	Coal spoil Slag heaps	Schramm (1966) Gaie and Heinemann (1980)
Boletus sp.	Uranium tailings Coal spoil	Kalin and Stokes (1981) Schramm (1966)
Cenococcum geophilum Fr.	Coal spoil Lignite spoil	Schramm (1966) Mejstrik (1971)
E strain	Coal spoil	Zak et al. (1982)
Helvella corium (Weberb.) Massee	Uranium tailings	Kalin and Stokes (1981)
Inocybe dulcamara (A. & S. ex Pers.) Kummer	Mine waste	Malloch (1982)
Inocybe immigrans Malloch	Uranium tailings Mine wastes	Kalin and Stokes (1981) Malloch (1982)
Inocybe lacera Fr.	Uranium tailings Coal spoil	Kalin and Stokes (1981) Schramm (1966)
Inocybe sp.	Coal spoil	Schramm (1966)
Leccinum scabrum (Bull. ex Fr.) S.F. Gray	Arsenic spoil	Benson et al. (1980)
Laccaria laccata (Scop. ex Fr.) Cooke	Arsenic spoil	Pyatt (1973) Benson et al. (1980)

Laccaria proxima	Uranium tailings	Kalin and Stokes (1981)
Lactarius rufus (Scop. ex Fr.) Fr.	Arsenic spoil	Benson et al. (1980)
Paxillus involutus (Batsch ex Fr.) Fr.	Arsenic spoil	Benson et al. (1980)
	Uranium tailings	Kalin and Stokes (1981)
Pisolithus tinctorius (Pers.) Coker & Couch	Coal spoil	Schramm (1966)
	Coal spoil	Medve and Gill (1982)
	Many spoils (Review)	Marx (1977)
	Nickel tailings	Malloch and Kuja (1979)
Rhizopogon vulgaris	Minespoils	Miller (1982)
Scleroderma aurantium Pers.	Coal spoil	Medve and Gill (1982)
	Coal spoil	Schramm (1966)
	Coal spoil	Meyer (1973)
Scleroderma citrinum Pers.	Arsenic spoil	Benson et al. (1980)
Sepultaria arenosa (Fuckel) Boud.	Uranium tailings	Kalin and Stokes (1981)
Russula sp.	Coal spoil	Medve et al. (1977)
Suillus luteus	Coal spoil	Medve and Gill (1982)
	Coal spoil	Medve et al. (1977)
Suillus tomentosus	Oil sand tailings	Parkinson et al. (1980)
Thelephora caryophyllea Schaeffer ex Fr.	Coal spoil	Schramm (1966)
Thelephora terrestris Ehrh. ex Fr.	Arsenic spoil	Benson et al. (1980)
	Uranium tailings	Kalin and Stokes (1981)
	Coal spoil	Schramm (1966)
	Coal spoil	Medve and Gill (1982)
	Oil sand tailings	Parkinson et al. (1980)
	Coal spoil	Medve et al. (1977)

Table 1 (continued)

Mycorrhizal species	Spoil type	Reference
Tomentella sp.	Coal spoils + peat	Danielson (unpublished data)
Pezizella ericae Read	Metal-contaminated spoil	Bradley et al. (1981, 1982)
Entrophospora infrequens (Hall) Ames & Schneider	Coal spoil	Zak et al. (1982)
"fine endophyte"	Coal spoil	Daft et al. (1975)
	Coal spoil	Lambert and Cole (1980)
	Coal spoil	Daft and Nicolson (1974)
	Coal spoil + peat	Zak et al. (1982)
	Oil sand tailings + peat	Zak et al. (1982)
Gigaspora calaspora	Coal spoil	Daft et al. (1975)
	Coal spoil	Daft and Nicolson (1974)
Gigaspora gigantea	Coal spoil	Daft et al. (1975)
	Coal spoil	Daft and Hacskaylo (1976)
	Spoil + topsoil	Langkamp and Dalling (1982)

Gigaspora gilmoreii Trappe & Gerdemann	Spoil + topsoil	Langkamp and Dalling (1982)
Gigaspora sp.	Spoil + topsoil	Langkamp and Dalling (1982)
Glomus aggregatum Schenck & Smith	Coal spoil	Zak et al. (1982)
	Oil sand tailings	Zak et al. (1982)
Glomus fasciculatum	Stockpile	Rives et al. (1980)
	Coal spoil	Daft et al. (1975)
	Coal spoil	Daft and Hacskaylo (1976)
	Coal spoil	Daft and Nicolson (1974)
	Surface spoil	Bohn and Liberta (1982)
Glomus macrocarpus var. *geospora*	Coal spoil	Daft et al. (1975)
Glomus macrocarpus var. *macrocarpus*	Coal wastes	Khan (1978)
Glomus mosseae (Nicol. & Gerd.) Gerd. & Trappe	Zn-Cd spoil	Gildon and Tinker(1981)
	Coal spoils	Zak et al. (1982)
	Oil sand tailings	Zak et al. (1982)
Sclerocystis rubiformis Gerdemann & Trappe	Coal Wastes	Khan (1978)

sufficient soil nitrogen, as neither water nor phosphorus were limiting and nitrogen-fixing species of *Alnus* and *Robinia* were successful. Furthermore, grasses colonizing the sites responded to ammonium but not phosphate amendments.

The ectomycorrhizae that Schramm described on the spoil were characterized either by extensive rhizomorph development (*Astraeus, Pisolithus, Boletus, Thelephora*) or hyphae which fused soil particles tightly to the mycorrhizal surface (*Inocybe, Boletus*). These mycorrhizal features may represent especially efficient methods of extracting and transferring meager nitrogen supplies to the roots. The adhesion of soil particles to roots in dry soil has been suggested as a mechanism to increase water conductivity (Coutts, 1982) and it may also function to enhance the movement of mobile ions such as nitrate. Hyphae of VA fungi may extend 8 cm from the root surface (Rhodes and Gerdemann, 1975), but rhizomorphs of *Pisolithus* may extend 4 m into the soil (Schramm, 1966), a result that suggests ectomycorrhizae are better adapted to long-distance transport than VA mycorrhizae. Schramm observed that *Astraeus* formed a dense network of rhizomorphs and mycelia near the spoil surface where intermittent drying occurred. He suggested that these fungal structures may survive drying and revive quickly to use nitrogen from precipitation percolating into the spoil.

Studies since 1966 on additional mining wastes have shown colonization patterns of plants and mycorrhizal types differing from those observed by Schramm. Ten-year-old untreated coal tips in Scotland were colonized by both grasses and dicotyledonous plants and all but one species (which was nonmycorrhizal) were infected with VA fungi (Daft and Nicolson, 1974). Extractable phosphorus levels were low (0.3–2.3 μg/g) and the authors considered that successful plant growth was dependent on VA infection and the resulting increase in phosphorus uptake. Similar results were found in coal wastes in Pennsylvania in which 23 of 24 colonizing plant species were VA mycorrhizal (Daft and Hacskaylo, 1976). Some of the same woody nitrogen-fixing plants that Schramm (1966) observed to be successful colonizers (*Robinia hispida, R. pseudoacacia* and *Alnus glutinosa*) were determined by Daft and Hacskaylo (1976) to have VA mycorrhizae. On lignite spoil banks, *Alnus glutinosa* was observed to be ectomycorrhizal (Mejstrik, 1971). Khan (1978) examined plants colonizing three bituminous coal tips in Australia that were composed of mine dirt and washer discard and determined that 22 of 26 species were infected with VA fungi and four were nonmycorrhizal. Spores were extracted from spoil collected from the root zone of all the 26 species.

Plants grown in unamended fresh coal spoils have been observed to become VA mycorrhizal by Ponder (1979), Mott and Zuberer (1982), and Zak and Parkinson (1982). Unlike the examples cited above, some mine spoils have been observed to be noninfective, i.e., devoid of mycorrhizal inoculum. Aldon (1978) planted *Atriplex canescens* in 4-year-old spoil in New Mexico and did not detect any infection in noninoculated seedlings after the second growing season. A 10-year-old Pennsylvania spoil bank used in a field study by Lambert and Cole (1980) was completely noninfective for white clover. The addition of amendments containing inoculum can substantially decrease the delay in infection, and this may be a desirable procedure on some mine spoils or tailings (Zak and Parkinson, 1982). However, Allen and Allen (1980) found that even when stockpiled grassland topsoil was mixed with spoil, infection levels of the grass *Agropyron smithii* were nil or much lower than on undisturbed sites. Even after 3 years, infection levels on reclaimed sites were less than half those on the undisturbed grassland. The low levels of infection may have been due to stockpiling and the incorporation of surface materials from as deep as 2 m. Schwab and Reeves (1981) showed that very little VA inoculum was present below a depth of 0.5 m in an arid sage community. In contrast, however, Lambert and Cole (1980) obtained high levels of infection of white clover after one growing season on spoil amended with topsoil from a grassy field. Three legumes responded well to the addition of soil but grasses did not show a response that could be attributed to mycorrhizal status. Lambert and Cole (1980) did not examine the roots of the grasses for infection, but it may be that the grasses were infected less than the legumes and were more sensitive to low inoculum levels.

The harsh procedures used to remove bitumen from sand mined in northern Alberta results in mine tailings that are completely devoid of mycorrhizal inoculum (Parkinson, 1979). To obtain early establishment of mycorrhizae, inoculum must be introduced to tailings of this type. Zak and Parkinson (1982) found that less than 1% of slender wheatgrass roots grown in untreated sand were infected after one growing season, whereas tailings amended with peat from a forested site resulted in 23% infection. After 2 and 4 years, infection of slender wheatgrass roots in untreated sand reached 4% and 36%, respectively, illustrating the extremely slow introduction of effective inoculum via wind, water, or animal transport even though the experimental plots were surrounded by pastures (Zak and Parkinson, 1983). The addition of a mineral fertilizer to the tailings sand resulted in much larger root systems but did not result in a percent increase in infection although the infected root length of fertilized plants was greater after 2 years than plants in unamended tailings. The fertilizer level used was not high enough to suppress infection because mycorrhizae formed readily on slender wheatgrass in adjacent coal mine spoil test plots (Zak and Parkinson, 1982).

The source or quality of ectomycorrhizal inoculum can influence the rates of mycorrhizal development on roots growing in tailings sand. Containerized jack pine seedlings that were infected with *Thelephora terrestris* were also grown on the experimental plots described by Zak and Parkinson. Although *Thelephora* is thought to be highly aggressive, insignificant root colonization occurred in tailings sand after one growing season (Parkinson et al., 1980). In peat-amended sand, 25% of the short roots were infected by E-strain fungi indigenous to the peat (Danielson, 1982). After two seasons' growth, infection by *Thelephora terrestris* in untreated and fertilized tailings sand did not exceed 33% whereas in the peat-amended tailings infection by E-strain reached 72%. These studies on slender wheatgrass and jack pine illustrate the slow establishment of mycorrhizae in spoils devoid of inoculum. The introduction of effective inoculum may be beneficial, as Mosse et al. (1981) state that the seedling stage may be the most mycorrhizal dependent stage both for tree species and VA annuals. Furthermore, Abbott and Robson (1981) conclude that the rate of mycorrhizal formation is particularly important with annuals grown in areas with short growing seasons.

The pioneer plants colonizing disturbed areas and spoils in the arid southwestern U.S. have been observed to be largely nonmycorrhizal annuals (Reeves et al., 1979; Miller, 1979; Allen and Allen, 1980). Reeves et al. (1979) studied the mycorrhizal status of plants colonizing a road bed abandoned 3 years previously and an adjacent sage community. Their results showed that the road bed was dominated by nonmycorrhizal plants (99% plant cover by *Salsola kali*) whereas nonmycorrhizal plants were rare in the undisturbed community. The authors suggested that the presence of nonmycorrhizal species may be detrimental to the establishment of mycorrhizal-dependent species by reducing inoculum levels in disturbed soils. The disturbed soil was far less infective than the undisturbed soil (Moorman and Reeves, 1979), but it was not shown that inoculum levels were so low as to prevent establishment of mycorrhizal species. Miller (1979) observed a similar phenomenon on spoil covered with 30 cm of topsoil as compared with undisturbed plant communities. No mycorrhizae were detected on the disturbed site although viable inoculum was present and some species on the disturbed site were mycorrhizal in the native sites. Allen and Allen (1980) also observed the dominance of nonmycorrhizal species on some mining sites treated with stockpiled topsoil. In a disked area dominated by the nonmycorrhizal species *Chenopodium pratericola* and *Salsola kali*, the infection of roots of *Agropyron smithii* exceeded 62% 1 and 2 years following disturbance.

It remains to be determined if indeed the presence of weedy nonmycorrhizal annuals inhibits the establishment of mycorrhizal species by reducing soil infectivity or if other factors are responsible for the success of these plants. In any case, many of these species are autoallelopathic, e.g., *Kochia scoparia* and *Salsola kali*, and thus usually disappear rapidly (Lodhi, 1979a,b). Mosse et al. (1981) have also stated that such species are opportunistic and thus inherently less competitive than mycorrhizal species that follow in the successional sequence.

Although stockpiling and replacement of topsoil is often required by legal regulations and is desirable for plant establishment, the question has been raised whether the practice has unfavorable effects on mycorrhizal inoculum and hinders plant establishment. Two methods have been used to determine the effects of stockpiling on VA mycorrhizal inoculum, spore counts and a greenhouse bioassay utilizing a rapidly growing test plant. Spore counts have been shown to provide an unreliable index of soil infectivity (Abbott and Robson, 1981; Gould and Liberta, 1981) and the corn bioassay is considered to provide a better indication of the presence of total infective propagules (Moorman and Reeves, 1979; Gould and Liberta, 1981). In a similar manner, the infective potential of forest soils subjected to disturbances (Schoenberger and Perry, 1982) and of oil sand overburden materials (Danielson et al., 1983) have been evaluated.

Stockpiling of topsoil for up to 3 years has been shown to have no effect on rates of infection after 1 month when undiluted soil was used (Rives et al., 1980). After 4.5 years, infection had decreased from 93% to 80% (Gould and Liberta, 1981). Using a series of different aged stockpiles, Liberta (1981) estimated that infectivity of undiluted soil decreased from 85% to 12% after 3 years. When the soil was diluted with an inoculum-free medium, the effect of stockpiling on reducing viable mycorrhizal inoculum was much more evident. There is little doubt that stockpiling will result in decreased infectivity of topsoil but the studies conducted so far have failed to demonstrate that the decreases will significantly affect plant growth and survival.

In the reclamation of extracted oil sands, amendation is absolutely necessary (Parkinson et al., 1980). The amendments will be the sole sources of plant nutrients, organic matter, and mycorrhizal inoculum except for slow introduction slowly by wind and precipitation. The principal amendment used is stockpiled peat from black spruce-tamarack muskeg sites. Due to the critical nature of the amendment characteristics used on mine tailings, such as oil sands and retorted oil shale, the amendments should be carefully evaluated for mycorrhizal inoculum. Danielson et al. (1983) examined several potential oil sand amendments for mycorrhizal infectivity and found that peat contained very small quantities of VA mycorrhizal inoculum, as indicated by infection levels of slender wheatgrass plants grown in the greenhouse. However, substantial quantities of VA inoculum were present at 1-m depths in mixed forest stands, and this material could conceivably be used as a source of VA inoculum in operational procedures. Ectomycorrhizal inoculum compatible with jack pine was present in the peat as well as in soils from the mixed stands with the dominant symbiont being identified as the E strain. Fungi forming mycelial strands were rare and this could indicate that these materials may not provide the most effective symbionts (Marx, 1980b).

V. BENEFITS OF MYCORRHIZAL ASSOCIATIONS ON MINE WASTES

Relatively few studies have shown unequivocal benefits of mycorrhizae on mine spoils. Data from some studies are difficult to interpret due to the lack of thorough quantification of mycorrhizal status of the root system, the failure to determine the persistance of introduced symbionts, or the lack of a sound statistical experimental design. The benefits of mycorrhizae on mine spoils need to be clearly shown to be substantial to alter policies and mining practices that are primarily concerned with moving huge masses of material at high expense. Regardless, several studies have demonstrated that the mycorrhizal symbiosis may provide a valuable and practical tool for achieving the desired end point of reclamation practices on mining wastes.

Schramm's (1966) detailed studies of volunteer and planted woody plants on black mine wastes in Pennsylvania were the first to indicate that mycorrhizae were not only beneficial but essential for plant growth on certain mine wastes. He selected two types of materials, black shale mixed with coal waste and fine grained sludge, both of which supported very sparse vegetation or were entirely barren. The spoils were 30–36 years old, acid, and with surface temperatures exceeding 60°C during the summer on level locations. Schramm determined that it was the high surface temperatures and not acidity or lack of moisture, that prevented the initial establishment of plants. Once established, nonmycorrhizal plants (mostly pines) became chlorotic and eventually died whereas all of hundreds of healthy individual plants he examined were ectomycorrhizal. The predominant mycorrhizal symbiont was *Pisolithus tinctorius*, which could be identified by the bright yellow color of the mycelium. Through irrigation and fertilizer experiments, Schramm established that nitrogen was the decisive factor limiting seedling growth. Phosphorus additions did not relieve deficiency symptoms and combining nitrogen and phosphorus was no better than nitrogen alone. Nonmycorrhizal seedlings responded to nitrogen additions, but after fertilization was terminated growth stopped and deficiency symptoms reoccurred. Reclamation of these black coal wastes with nonnitrogen-fixing plants was only possible by altering the microenvironment around seedling stems and by the establishment of ectomycorrhizal fungi on the seedlings' root systems.

The major advance made since 1966 for the practical application of Schramm's observations has been the development of procedures for producing seedlings inoculated with specific fungi by Marx and coworkers. They recognized that *Pisolithus tinctorius* was particularly well adapted to the harsh environments encountered on mine wastes in the southeastern United States (Marx, 1975) and have shown the practical advantages of using planting stock inoculated with *Pisolithus* (Marx, 1980a,b). Berry and Marx (1978) outplanted *Pinus taeda* and *Pinus virginiana* seedlings inoculated with *Pisolithus* or naturally infested with *Thelephora terrestris* on acid, barren, metallic mine wastes in which previous reforestation attempts had largely failed. After 2 years, seedlings inoculated with *Pisolithus* were significantly larger than the naturally infested seedlings on one of the two sites, with high rates of survival in all treatments. Both *Pisolithus* and *Thelephora* persisted and colonized new roots. On acid coal mine spoils, Marx and Artman (1979) found similar results, with *Pisolithus*-infected seedlings exhibiting both better survival and growth after 3–4 years than seedlings infected with *Thelephora*. In this study, based on visual identifications, *Thelephora* did not appear to persist as well as *Pisolithus*. In another study on coal spoils in the southeastern United States, Berry (1982) demonstrated that container-grown pines planted on spoils untreated in any way other than contouring grew successfully and that inoculation with *Pisolithus* stimulated growth. All of these studies by Marx, Berry, and Artman show a clear benefit in growth performances of inoculation with a specific symbiont. Although the studies did not compare mycorrhizal with nonmycorrhizal plants, they represent a comparison of experimental inoculation with operational forestry practices in the southeastern United States were seedlings became infected with the weedy species *Thelephora terrestris*. Comparisons in northern climates where other fungi [e.g., E strain (Laiho, 1965; Thomas and Jackson, 1979) *Inocybe lacera* or *Laccaria laccata* (Trappe and Strand, 1969; Sinclair, 1974)] may be the dominant nursery fungi are required to determine the fungi of choice in these regions.

The study by Grossnickle and Reid (1982) illustrates some of the difficulties of testing symbiont efficiency with slow-growing hosts in a cold climate with a short growing season, a situation quite unlike that encountered by Marx and his colleagues. An initial problem confounded interpretation of field data in that the intended mycorrhizal-free control and a portion of the inoculation treatments became infected with a greenhouse contaminant. This precluded a determination of the essentiality of the mycorrhizal condition but may have better represented operational conditions. Seedlings of *Pinus contorta*, *Pinus flexilis* and *Picea engelmannii* were inoculated with *Pisolithus tinctorius*, *Suillus granulatus*, and *Cenococcum geophilum* and outplanted on a molybdenum tailings pond covered with alkaline waste rock. Phosphorus, phosphorus plus nitrogen, and sewage sludge plus wood chips were added to some treatments to ameliorate the poor physical and nutritional conditions of the spoil. Regardless of treatment, growth of all three conifer species was minimal for the first 2 years in the field. Height response to sewage sludge as compared with the other amendments was observed in the third year with *P. flexilis* and after 4 years by all three species. The maximum annual height increase was less than 1 cm for the first 2 years and less than 5 cm after 4 years. Height growth appeared to be enhanced by inoculation with *Suillus granulatus*, but again differences in height growth were very small. From the results of Grossnickle and

Reid (1982) it appears that detection of a growth response at such sites will be much more difficult than in warm climates and will require increased times.

In warm climates with fast-growing host species height and stem diameter measurements appear to be adequate for determining response to mycorrhizal inoculum, but it appears unlikely that these measurements are sensitive enough to be used with slow growing species. In a continuation of the amendation experiments on coal spoils described by Zak et al. (1982), no differences in the height growth of white spruce could be detected for one to 4 years among spoils treated with peat, sewage sludge, fertilizer, in comparison with controls (Danielson et al., 1979 and unpublished data). It was not until after the fourth season's growth that differences in shoot growth could be detected in terms of dry weight. In contrast, with a rapidly growing *Salix* species planted under the same conditions, amendation significantly affected height growth over the first year, with differences among amendments being even more distinct when shoot weights were measured. With *Pinus banksiana* grown on oil sand tailings amended with the same materials (Parkinson et al., 1980), seedling heights in the four amendation treatments did not differ after the second growing season whereas up to 16-fold differences existed in shoot weights (unpublished data). In that there were very pronounced nutritional and microbial differences introduced to the spoils by the amendments described by Zak et al. (1982) and Grossnickle and Reid (1982) and the difficulties in detecting treatment effects by height growth with *Picea* and *Pinus* species, it is unlikely that a host response to inoculation would be detected by similar means. It would appear that destructive sampling will be necessary to evaluate inoculation trials in cold climates.

Due to slow host growth in mountainous or cool temperature climates, prolonged persistence of introduced symbionts is essential and this must be monitored to determine real effects of inoculation treatments. Grossnickle and Reid (1982) reported that neither *Pisolithus* or *Cenococcum* persisted on test plant root systems at a high elevation site in Colorado. Indeed, they estimated that infection of short roots in the waste material exceeded 50% in only one of the 12 mycorrhizal treatments after 4 years and in many cases was less than 20%. In that Marx et al. (1982) have estimated that infection by *Pisolithus* should exceed 50% to expect a growth response to inoculation, it is clear that long-term persistence and colonization of new roots represent key criteria for successful symbiont performance in sites with short growing seasons.

The effect of VA mycorrhizae on the growth of three legume species was evaluated on field plots established on a 10-year-old spoil bank by Lambert and Cole (1980). The acid spoil was amended with limestone and NH_4NO_3 and a minimal amount of topsoil (1 cm depth) to serve as VA inoculum. This inoculation treatment was highly effective as mycorrhizal infection was zero where no soil was added and exceeded 90% with soil additions. Both the growth and survival of all three species were markedly increased in the mycorrhizal treatments, e.g., survival after 3 years was improved 6-, 8-, and 15-fold for birdsfoot trefoil, crownvetch, and flat pea, respectively, with similar dry weight increases observed. In a separate experiment, phosphorus was shown to be limiting plant growth and that a combination of phosphorus fertilizer and topsoil resulted in better growth than when either one was applied separately. In the greenhouse, mycorrhizae did not result in improved growth of two grass species indicating that benefits may not be generalized for all species. Further, Lambert and Cole (1980) used five different vegetated mine spoils and a forest soil as inoculum for white clover in the greenhouse with their spoil adjusted to either pH 4.5 or 6.5. Although infection rates were similar among the inoculation treatments, yield response varied four-fold at pH 4.5 and five-fold at pH 6.5 with the different inocula. They considered these differences to indicate better adaptation to conditions in the acid spoil by some symbionts than by others. They concluded that the introduction of VA mycorrhizae to mine wastes would be useful if (1) inoculum is lacking or if the indigenous symbionts are ill-adapted, (2) if available phosphorus is low, and (3) if the plant species of choice respond to VA infection. The striking growth response of legumes demonstrated by Lambert and Cole (1980) appears to justify the time and expense of verifying if these conditions exist on other mine spoils.

Shrubs have received little attention in terms of mycorrhizal growth enhancement on mine spoils, and additional research appears to be warranted. Williams et al. (1974) determined that the growth of *Atriplex canescens*, a shrub native to arid regions in the southwestern United States, grew better in untreated soil than in autoclaved soil and suggested this was due to the VA mycorrhizal status of plants in the native soil. Lindsey et al. (1977) were unable to obtain infections of *Atriplex canescens* with extracted chlamydospores in coal spoil but demonstrated that another shrub, *Chrysothamnus nauseosus*, grew 13-fold better in autoclaved spoil inoculated with spores of *Glomus fasciculatus* than without inoculation and three-fold better in unautoclaved spoil when comparing inoculated and uninoculated materials. These greenhouse studies suggested that the growth and survival of woody shrubs might be enhanced on mine spoils in arid regions. Aldon (1978) demonstrated this by outplanting seedlings of *Atriplex canescens*, which had been inoculated with soil containing *Glomus mosseae*, onto a 4-year-old coal mine spoil. Uninoculated shrubs remained nonmycorrhizal, inoculated plants became infected, and shoot growth and survival of the latter were superior to nonmycorrhizal plants.

VA mycorrhizal tree species used on mine spoils may also benefit from inoculation. Inoculation of red maple in pots containing anthracite wastes stimulated growth when nutrients other than phosphate were supplied (Daft and Hacskaylo, 1977). Because VA mycorrhizal trees may grow well on mine spoils (Shuffstall and Medve, 1979), mycorrhizal dependencies and responses should be determined to provide information for management planning.

It has frequently been noted that some mine spoils may contain high levels of potentially toxic metals and this has led to repeated speculative statements that mycorrhizae might reduce toxicity and increase plant tolerance to contaminated soils or spoils. As stated previously, the data of Bradley et al. (1981, 1982) strongly suggest that mycorrhizal relationships are responsible for influencing patterns of plant colonization of sites characterized by high levels of available heavy metals. The colonization of metal contaminated and highly acid soils is characteristic of some ericaceous plants and Bradley et al. (1982) have argued convincingly that this colonization was related to complexing of the metals by the fungal element of ericoid mycorrhizae. Endophytes obtained from metal contaminated sites were no more beneficial to the host than an endophyte from natural heathland, perhaps as a consequence of high acidity and high availability of metals in undisturbed soils where ericaceous plants thrive (Bradley et al., 1981).

In contrast, VA mycorrhizal plants are usually found in less acid environments (e.g., agricultural soils) and not exposed to high levels of toxic metals. Bradley et al. (1982) state that colonization by grasses is rare on the same areas where *Calluna vulgaris* is common. However, VA mycorrhizal symbionts can become adapted to metal contaminated soils as Gildon and Tinker (1981) have shown. Copper and zinc were added to pots containing a clay loam soil and seeded with clover inoculated with either an isolate of *Glomus mosseae* from metal-contaminated soil or an isolate from an agricultural soil. Infection at all metal levels was greater with the isolate from the contaminated soil without associated increases in copper or zinc concentrations being observed in the plant tissue. Gildon and Tinker (1981) concluded that endophyte tolerance was important, implying that the mycorrhizal symbiont must be considered as a component of whole plant tolerance to potentially toxic metals. In a preliminary report, Killham and Firestone (1982) reported that under very acid conditions, a VA mycorrhizal grass accumulated higher levels of copper and nickel in the shoots than nonmycorrhizal plants, and that yield was reduced by mycorrhizal infection. Thus, it appears that infection by a VA fungus not adapted to heavy metals may have adverse effects on plant growth.

Whether ectomycorrhizae may impart tolerance to toxic metals has not been determined, but needles of pines infected with *Pisolithus* have been reported to contain lower levels of manganese and zinc (Berry, 1982) and manganese, iron, and aluminum (Marx and Artman, 1979) than plants infected with *Thelephora*. Heavy metal toxicity may have been the cause of the lack of mycorrhizae on *Salix* and *Populus* planted in copper mine tailings reported by Harris and Jurgensen (1977), although additional study indicated that the physiological condition of the host rather than copper toxicity prevented mycorrhizal formation (Jurgensen, 1978).

VI. CONCLUSIONS

Relatively few species of ectomycorrhizal symbionts have been identified as occurring on mine wastes, and of those, very few have been properly quantified with respect to their actual importance. To determine the degree of fungal symbiont adaptation to conditions on mine wastes, infection levels of each species must be quantified. It has been established that *Pisolithus tinctorius* is admirably adapted to spoil materials in warm climates, but in boreal and mountainous habitats it seems likely that other symbionts will be more effective and persistent. Research still remains largely concentrated on one symbiont although Parkinson (1979) warned of the dangers of directing all efforts on a single organism despite the enormous variety of mine waste habitats. Of the ectomycorrhizal symbionts listed as occurring spontaneously on spoils and tailings in Table 1, only *Pisolithus tinctorius*, *Thelephora terrestris*, and *Cenococcum geophilum* have been field tested successfully (Marx, 1980a). It is not possible to make decisions on species desirability for use in cold climatic regions until quantitative data on abundance and persistence of the various spontaneous or introduced symbiont colonizers are obtained and host performance in the field is evaluated. In other regions, *Pisolithus* is the obvious current choice and the standard with which other symbionts must be compared.

It appears that considerable strain differences exist among the VA fungi, and that it would be profitable to screen isolates for adaptability to mine spoils. It is likely that indigenous symbionts or those introduced with topsoil will not be those best adapted to mine sites (Lambert et al., 1980; Mosse et al., 1981). Old mine spoils with established vegetation may prove to be valuable sources of inoculum of adapted strains.

The general beneficial effect of the mycorrhizal condition on establishing and enhancing plant growth on mine wastes is evident. The principal mechanism appears to be improved mineral nutrition, although in special cases increased tolerance to toxic metals may be crucial. Plant inoculation is possible but in many cases it may not be justifiable, as in practice economic factors must be considered. If the main benefit is enhanced nitrogen or phosphorus uptake, inoculation must produce a "sustained improvement in plant performance so that it is superior to and more persistent than that obtained by phosphorus fertilization" (Lambert et al., 1980). Inoculation can more easily be justified in situations where inoculum is totally lacking or where nonnutrient stresses are decisive. Prior to any inoculation efforts, mine spoils, tailings, and any amending materials should be evaluated for mycorrhizal inoculum, as well as properties that affect plant growth and the potential benefits of embarking on a time consuming program of mycorrhizal manipulation.

Schramm's detailed examinations of the factors limiting plant establishment and success on mine spoils set the stage for employing one of the most common and essential links in disturbed ecosystems to repair the impacts of mineral and energy resources extraction. Although the potential value of mycorrhizae for improvement of disturbed site rehabilitation is appreciable, many problems must be considered and dealt with in an effective manner. Well-directed efforts must be guided by past observations and experiments to the end purpose of establishing both above-ground and below-ground processes that are both conducive to rehabilitating drastic disturbances to visually pleasing landscapes as well as establishing a recovery process that will persist after intensive management is completed.

ACKNOWLEDGMENTS

In many ways, both large and small, my colleagues at the University of Calgary, Dennis Parkinson, Suzanne Visser, and John Zak, have contributed to the form and substance of this chapter. Funding for this review was provided by a Natural Science and Engineering Research Council of Canada grant to D. Parkinson and the Research Secretariat, Alberta Environment.

REFERENCES

Abbott, L. K. and Robson, A. D. (1981). Infectivity and effectiveness of vesicular arbuscular mycorrhizal fungi: Effect of inoculum type. *Aust. J. Agr. Res. 32*:631–639.

Aldon, E. F. (1978). Endomycorrhizae enhance shrub growth and survival on mine spoils. In *The Reclamation of Disturbed Arid Lands*, R. A. Wright (Ed.), University of New Mexico Press, Albuquerque, pp. 174–179.

Allen, M. F. (1982). Influence of vesicular-arbuscular mycorrhizae on water movement through *Bouteloua gracillis* (H.B.K.) Lag ex Streud. *New Phytol. 91*: 191–196.

Allen, E. B., and Allen, M. F. (1980). Natural re-establishment of vesicular-arbuscular mycorrhizae following stripmine reclamation in Wyoming. *J. Appl. Ecol. 17*: 139–147.

Benson, L. M., Evans, R. L., and Peterson, P. J. (1980). Occurrence of basidiomycetes on arsenic-toxic mine waste. *Trans. Br. Mycol. Soc. 74*: 199–201.

Berry, C. R. (1982). Survival and growth of pine hybrid seedlings with *Pisolithus* ectomycorrhizae on coal spoils in Alabama and Tennessee. *J. Environ. Qual. 11*: 709–715.

Berry, C. R., and Marx, D. H. (1978). Effects of *Pisolithus tinctorius* ectomycorrhizae on growth of loblolly and Virginia Pines in the Tennessee Copper Basin. U.S. For. Serv. Res. Note SE-264.

Bohn, K. S., and Liberta, A. E. (1982). A laboratory bioassay for assessing the effects of sludge supernatant on plant growth and vesicular-arbuscular mycorrhizae formation. In *1982 Symposium on Surface Mining, Hydrology, Sedimentology and Reclamation*, D. H. Graves (Ed.), OES Publ., College of Engineering, University of Kentucky, Lexington, pp. 699–703.

Bowen, E. D. (1973). Mineral nutrition of ectomycorrhizae. In *Ectomycorrhizae: Their Ecology and Physiology*, E. C. Marks and T. T. Kozlowski (Ed.), Academic Press, New York, pp. 151–205.

Bowen, G. D., and Smith, S. E. (1981). The effects of mycorrhizas on nitrogen uptake by plants. In *Terrestrial Nitrogen Cycles. Processes, Ecosystem Strategies and Management Impacts. Ecol. Bull. 33*, F. E. Clark and T. Rosswall (Ed.), Swedish National Science Research Council, Stockholm, pp. 237–247.

Boyer, J. S. (1971). Resistances to water transport in soybean, bean and sunflower. *Crop Sci. 11:* 403–407.

Bradley, R., Burt, A. J., and Read, D. J. (1981). Mycorrhizal infection and resistance to heavy metal toxicity in *Calluna vulgaris*. *Nature 292:* 335–337.

Bradley, R., Burt, A. J., and Read, D. J. (1982). The biology of mycorrhiza in the Ericaceae. VIII. The role of mycorrhizal infection in heavy metal resistance. *New Phytol. 91:* 197–209.

Brady, N. C. (1974). *The Nature and Properties of Soils*. Macmillan, New York.

Coutts, M. P. (1982). Growth of sitka spruce seedlings with roots divided between soils of unequal matric potential. *New Phytol. 92:* 49-961.

Daft, M. J., and Hacskaylo, E. (1976). Arbuscular mycorrhizas in the anthracite and bituminous coal wastes of Pennsylvania. *J. Appl. Ecol. 13:* 523–531.

Daft, M. J., and Hacskaylo, E. (1977). Growth of endomycorrhizal and nonmycorrhizal red maple seedlings in sand and anthracite spoil. *For. Sci. 23:* 207–216.

Daft, M. J., Hacskaylo, E. and Nicolson, T. H. (1975). Arbuscular mycorrhizas in plants colonizing coal spoils in Scotland and Pennsylvania. In *Endomycorrhizas*, F. E. Saunders, B. Mosse, and P. B. Tinker (Ed.), Academic Press, New York, pp. 561–580.

Daft, M. J., and Nicolson, T. H. (1974). Arbuscular mycorrhizas in plants colonising coal wastes in Scotland. *New Phytol. 73:* 1129–1138.

Danielson, R. M. (1982). Taxonomic affinities and criteria for identification of the common ectendomycorrhizal symbiont of pines. *Can. J. Bot. 60* : 7–18.

Danielson, R. M., Visser, S., and Parkinson, D. (1983). Microbial activity and mycorrhizal potential of four overburden types used in the reclamation of extracted oil sands. *Can. J. Soil Sci. 63* : 363–375.

Danielson, R. M., Zak, J. and Parkinson, D. (1979). Plant growth and mycorrhizal development in amended coal spoil material. In *Ecology and Coal Resource Development*, M. K. Wali (Ed.), Pergamon Press, New York, *Vol. 2*, pp. 912–919.

Duddridge, J. A., Malibari, A., and Read, D. J. (1980). Structure and function of mycorrhizal rhizomorphs with special reference to their role in water transport. *Nature 287* : 834–836.

Gaie, W., and Heinemann, P. (1980). Synthese et morphologie de l'ectomyocorhize *Betula pendula/Astraeus hygrometricus*. *Bull. Soc. Roy. Bot. Belg. 113* : 161–165.

Gerdemann, J. W., and Trappe, J. M. (1974). The Endogonaceae in the Pacific Northwest. Mycologia Memoir No. 5, New York Botanical Gardens, New York.

Gildon, A., and Tinker, P. B. (1981). A heavy metal tolerant strain of a mycorrhizal fungus. *Trans. Br. Mycol. Soc. 77* : 648–649.

Goss, R. W. (1960). Mycorrhizae of ponderosa pine in Nebraska grassland soils. Res. Bull. 192, Nebr. Agr. Expt. Sta.

Gould, A. B., and Liberta, A. E. (1981). Effects of topsoil storage during surface mining on the viability of vesicular-arbuscular mycorrhiza. *Mycologia 73* : 914–922.

Graham, J. H., Linderman, R. G., and Menge, J. A., (1982). Development of external hyphae by different isolates of mycorrhizal *Glomus* spp. in relation to root colonization and growth of troyer citrage. *New Phytol. 9* : 183–189.

Grossnickle, S. C., and Reid, C. P. P. (1982). The use of ectomycorrhizal conifer seedlings in the revegation of a high-elevation mine site. *Can. J. For. Res. 12* : 354–361.

Harris, M. M., and Jurgensen, M. F. (1977). Development of *Salix* and *Populus* mycorrhizae in metallic mine tailings. *Plant Soil 47* : 509–518.

Jurgensen, M. F. (1978). Microorganisms and the reclamation of mine wastes. In *Forest Soils and Land Use*, C. T. Youngberg (Ed.), Dept. of For. and Wood Sci., Colorado State Univ., Ft. Collins, Colorado, pp. 251–286.

Kalin, M., and Stokes, P. M. (1981). Macrofungi on uranium mill tailings-associations and metal content. *Sci. Total Environ. 19* : 83–94.

Khan, A. G. (1978). Vesicular-arbuscular mycorrhizae in plants colonizing black wastes from bituminous coal mining in the Illawarra region of New South Wales. *New Phytol. 81* : 53–63.

Killham, K., and Firestone, M. K. (1982). V. A. Mycorrhizal mediation of grass response to heavy metal and acidic depositions. Agron. Abstr., Amer. Soc. Agron., Madison, p. 190.

Laiho, O. (1965). Further studies on the ectendotrophic mycorrhiza. *Acta Forest. Fenn.* *79* : 1–35.

Lambert, D. H., Cole Jr., H., and Baker, D. E. (1980). Adaptation of vesicular-arbuscular mycorrhizae to adaphic factors. *New Phytol.* *85* : 513–520.

Lambert, D. H., Baker, D. E., and Cole, H., Jr. (1979). The role of mycorrhizae in the interactions of phosphorus with zinc, copper and other elements. *Soil Sci. Soc. Am. J.* *43* : 976–980.

Lambert, D. H., and Cole Jr., H. (1980). Effects of mycorrhizae on establishment and performance of forage species in mine spoil. *Agron. J.* *72* : 257–260.

Langkamp, P. J., and Dalling, M. J. (1982). Nutrient cycling in a stand of *Acacia holosericea* A. Cunn. ex G. Don. II. Phosphorus and endomycorrhizal associations. *Austr. J. Bot.* *30* : 107–119.

Liberta, A. E. (1981). Effects of topsoil-storage duration on inoculum potential of vesicular-arbuscular mycorrhizae. In *Symposium on Surface Mining Hydrology, Sedimentology and Reclamation*, D. H. Graves (Ed.), OES Publ., College of Engineering, Univ. of Kentucky, Lexington, pp. 45-48.

Lindsey, D. L., Cress, W. A., and Aldon, E. F. (1977). The effects of endomycorrhizae on growth of rabbitbrush, fourwing saltbush and corn in coal mine spoil material. USDA For. Serv. Res. Note RM-343, Rocky Mtn. For. Range Expt. Sta., 6 pp.

Lodhi, M. A. K. (1979a). Germination and decreased growth of *Kochia scoparia* in relation to its autoallelopathy. *Can. J. Bot.* *57* : 1083–1088.

Lodhi, M. A. K. (1979b). Allelopathic potential of *Salsola kali* L. and its possible role in rapid disappearance of weedy stage during revegetation. *J. Chem. Ecol.* *5* : 429–437.

Malajczuk, N., and Lamont, B. B. (1981). Specialized roots of symbiotic origin in heathlands. In *Heathland and Related Shrublands of the World, B, Analytical Studies*, R. L. Specht (Ed.), Elsevier Publ. Co., Amsterdam, pp. 165–182.

Malloch, D. (1982). An undescribed species of *Inocybe* from mine wastes in Ontario. *Can. J. Bot.* *60* : 40–45.

Malloch, D., and Kuga, A. L. (1979). Occurrence of the ectomycorrhizal fungus *Pisolithus tinctorius* in Ontario. *Can. J. Bot.* *57* : 1848–1849.

Marx, D. H. (1975). Mycorrhizae and establishment of trees on strip-mined land. *Ohio J. Sci.* *75* : 288–297.

Marx, D. H. (1977). Tree host range and world distribution of the ectomycorrhizal fungus *Pisolithus tinctorius*. *Can. J. Microbiol.* *23* : 217–223.

Marx, D. H. (1980a). Role of mycorrhizae in forestation of surface mines. Proc. Trees for Reclamation, sponsored by Intrastate Mining Compact Comm. and USDA Forest Service, Lexington, Kentucky.

Marx, D. H. (1980b). Ectomycorrhizal fungus inoculations : a tool for improving forestation practices. In *Tropical Mycorrhizal Research*, P. Mikola (Ed.), Clarendon Press, Oxford, pp. 13–71.

Marx, D. H., and Artman, J. D. (1979). *Pisolithus tinctorius* ectomycorrhizae improve survival and growth of pine seedlings on acid coal spoils in Kentucky and Virginia. *Reclam. Rev.* 2:23–31.

Marx, D. H., and Bryan, W. C. (1971). Influence of ectomycorrhizae on survival and growth of aseptic seedlings of loblolly pine at high temperature. *For. Sci.* 17 : 37–41.

Marx, D. H., Bryan, W. C., and Davey, C. B. (1970). Influence of temperature on aseptic synthesis of ectomycorrhizae by *Thelephora terrestris* and *Pisolithus tinctorius* on loblolly pine. *For. Sci.* 16 : 424–431.

Marx, D. H., Ruehle, J. L., Kenney, D. S., Cordell, C. E., Riffle, J. W., Molina, R. J., Pawuks, W. H., Navratil, S., Tinus, R. W., and Goodwin, O. C. (1982). Commercial vegetative inoculum of *Pisolithus tinctorius* and inoculation techniques for development of ectomycorrhizae on container-grown tree seedlings. *For. Sci.* 28 : 373–400.

Medve, R. J., and Gill, S. M. (1982). Distribution and ecology of *Pisolithus tinctorius* on bituminous and stripmine spoils in western Pennsylvania. *Bull. Torrey Botan. Club* 109 : 35–38.

Medve, R. J., Hoffman, F. M., and Gaither, T. W. (1977). The effects of mycorrhizal-forming amendments on the revegetation of bituminous stripmine spoils. *Bull. Torrey Botan. Club* 104: 218–225.

Mejstrik, V. (1971). Ecology of mycorrhizae of tree species applied in reclamation of lignite spoil banks. *Nova Hedw.* 22 : 675–698.

Meyer, F. H. (1973). Distribution of ectomycorrhizae in native and man-made forests. In *Ectomycorrhiza: Their Ecology and Physiology*, G. C. Marks and T. T. Kozlowski (Ed.), Academic Press, New York, pp. 79–105.

Mikola, P. (1965). Studies on the ectendotrophic mycorrhizae of pine. *Acta. Forest. Fenn.* 79 : 1–56.

Mikola, P. (1980). Mycorrhizae across the frontiers. In *Tropical Mycorrhizal Research*, P. Mikola (Ed.). Clarendon Press Oxford, pp. 3–10.

Miller, R. M. (1979). Some occurrences of vesicular-arbuscular mycorrhizae in natural and disturbed ecosystems of the Red Desert. *Can. J. Bot.* 57 : 619–623.

Miller, S. L. (1982). Mycorrhizal hypogeous fungi in the southeastern United States. *Mycol. Soc. Amer. Newsletter* 33 : 39.

Moorman, T., and Reeves, F. G. (1979). The role of endomycorrhizae in revegation practices in the semi-arid west. II. A bioassay to determine the effect of land disturbance on endomycorrhizal populations. *Am. J. Bot. 66* : 14–18.

Mosse, B., Stribley, D. P., and LeTucon, F. (1981). Ecology of mycorrhizae and mycorrhizal fungi. *Adv. Microb. Ecology 5* : 137–210.

Mott, J. B., and Zuberer, D. A. (1982). Abundance of *Rhizobium trifolii* and endomycorrhizal fungi in surface mine spoils in Texas. Agron. Abstr., Am. Soc. Agron., Madison, p. 190.

Parkinson, D. (1979). Microbes, mycorrhizae and mine spoil. In *Ecology and Coal Resource Development*, M. K. Wali (Ed.), Pergamon Press, *Vol. 2*, New York, pp. 634–642.

Parkinson, D., Visser, S., Danielson, R. M., and Zak, J. (1980). Restoration of fungal activity in tailing sand (oil sands). In *Soil Biology as Related to Land Use Practices*, D. L. Dindal (Ed.), Proc. VII Intern. Coll. Soil Zoology, U. S. Environ. Prot. Agency, pp. 362–370.

Ponder, F. Jr. (1979). Presence of endomycorrhizal fungi in recently graded coal mine spoil. *J. Soil Water Conser. 34* : 186–187.

Powell, C. L. (1982). Mycorrhizae. In *Experimental Microbial Ecology*, R. G. Burns and J. H. Slater (Ed.), Blackwell Scientific Publ., London, pp. 447--471.

Pyatt, F. B. (1973). Sporophores of *Laccaria laccata* on metal spoil. *Trans. Br. Mycol. Soc. 61* : 189–190.

Read, D. J. (1974). *Pezizella ericeae* sp. nov. the perfect state of a typical mycorrhizal endophyte of Ericaceae. *Trans. Br. Mycol. Soc. 63* : 381--383.

Reeves, F. B., Wagner, D., Moorman, T., and Kiel, J. (1979). The role of endomycorrhizae in revegation practices in the semi-arid west. I. A comparison of incidence of mycorrhizae in severely disturbed vs natural environments. *Am. J. Bot. 66* : 6–13.

Rhodes, L. H., and Gerdemann, J. W. (1975). Phosphate uptake zones of mycorrhizal and nonmycorrhizal onions. *New Phytol. 75* : 555--561.

Rives, C. S., Bajwa, M. I., Liberta, A. E., and Miller, R. M. (1980). Effects of topsoil storage during surface mining on the viability of VA mycorrhiza. *Soil Sci. 129* : 253–257.

Safir, G. R., Boyer, J. S., and Gerdemann, J. W. (1972). Nutrient status and mycorrhizal enhancement of water transport in soybean. *Plant Physiol. 49* : 700–703.

Sands, R., and Theodorou, C. (1978). Water uptake by mycorrhizal roots of *Radiata* pine seedlings. *Austr. J. Plant Physiol. 5* : 301–309.

Schenck, N. C., and Smith, G. C. (1982). Responses of six species of vesicular-arbuscular mycorrhizal fungi and their effects on soybean at four soil temperatures. *New Phytol.* *92* : 193–201.

Schoenberger, M. M., and Perry, D. A. (1982). The effect of soil disturbance on growth and ectomycorrhizae of Douglas-fir and western hemlock seedlings : A greenhouse bioassay. *Can. J. For. Res.* *12* : 343–353.

Schramm, J. R. (1966). Plant colonization studies on black wastes from anthracite mining in Pennsylvania. *Trans. Am. Phil. Soc.* *56* : 1–194.

Schumacher, T. (1979). Notes on taxonomy, ecology and distribution of operculate discomycetes (Pezizales) from river banks in Norway, *Norw. J. Bot.* *26* : 53–83.

Schwab, S., and Reeves, F. B. (1981). The role of endomycorrhizae in revegetation practices in the semi-arid west. 3. Vertical distribution of vesicular-arbuscular (VA) mycorrhiza inoculum potential. *Am. J. Bot.* *68* : 1293–1297.

Shuffstall, W. C., and Medve, R. J. (1979). Growth performances and mycorrhizae of native and exotic hardwoods on bituminous strip-mine spoils. *Ohio J. Sci.* *79* : 274–279.

Sinclair, W. A. (1974). Development of ectomycorrhizae in a Douglas-fir nursery. I. Seasonal characteristics. *For. Sci.* *20* : 51–56.

Slankis, V. (1974). Soil factors influencing formation of mycorrhizae. *Ann. Rev. Phytopath.* *12* : 437–457.

Smith, S. E. (1980). Mycorrhizas in autotrophic higher plants. *Biol. Rev.* *55* : 475–510.

Theodorou, C., and Bowen, G. D. (1971). Influence of temperature on the mycorrhizal associations of *Pinus radiata* D. Don. *Austr. J. Bot.* *19* : 13–20.

Thomas, G. W., Clarke, C. A., Mosse, B., and Jackson, R. M. (1982). Source of phosphate taken up from two soils by mycorrhizal (*Thelephora terrestris*) and non-mycorrhizal *Picea sitchensis* seedlings. *Soil Biol. Biochem.* *14* : 73–75.

Thomas, G. W., and Jackson, R. M. (1979). Sheathing mycorrhizas of nursery grown *Picea sitchensis*. *Trans. Br. Mycol. Soc.* *73* : 117–125.

Trappe, J. M. (1971). Mycorrhiza-forming ascomycetes. In *Mycorrhizae.* Misc. Publ. 1189, USDA For. Serv., pp. 19–37.

Trappe, J. M. (1977). Selection of fungi for ectomycorrhizal inoculation in nurseries. *Ann. Rev. Phytopathol.* *15* : 203--222.

Trappe, J. M., and Strand, R. F. (1969). Mycorrhizal deficiency in a Douglas-fir region nursery. *For. Sci.* *15* : 381–389.

Williams, S. E., Wollum II, A. G., and Aldon, E. F. (1974). Growth of *Atriplex canescens* (Pursh) Nutt. improved by formation of vesicular-arbuscular mycorrhizae. *Soil Sci. Soc. Am. Proc.* *38* : 962–965.

Zak, J. C., Danielson, R. M., and Parkinson, D. (1982). Mycorrhizal fungal spore numbers and species occurrence in two amended mine spoils in Alberta, Canada. *Mycologia 74*: 785–792.
Zak, J. C., and Parkinson, D. (1982). Initial VA mycorrhizal development of slender wheatgrass on two amended mine spoils. *Can. J. Bot. 60*: 2241–2248.
Zak, J. C., and Parkinson, D. (1983). Effects of surface amendation of two mine spoils in Alberta, Canada on vesicular-arbuscular mycorrhizal development of slender wheatgrass: a four-year study. *Can. J. Bot. 61*: 798–803.

7

Management of Microbial Processes in Surface Mined Land Reclamation in Western Canada

SUZANNE VISSER / *The University of Calgary, Calgary, Alberta, Canada*

I. INTRODUCTION

In Western Canada, a large proportion of the energy resources that can be extracted by surface mining (i.e., coal and oil) is located in Alberta. These resources are situated in a wide variety of habitats— anthracitic or bituminous coal, which is used in the metallurgical industry and mainly exported, is mined in the mountainous regions; subbituminous coal, which is one of the main sources of thermoelectric power, is mined in the plains region; whereas heavy oil is currently being extracted from large oil sands deposits located in the boreal region of northern Alberta. Smith (1977) emphasized the importance of considering environmental diversity when planning a reclamation strategy. Geographic (including climatological) and geological factors have resulted in a large number of "ecological regions," which include shortgrass and midgrass prairie, aspen parkland, subalpine montane, and boreal forest. The method of reclamation of land disturbed by surface mining depends on the region in which it is located, e.g., the mountainous regions with their steep terrain, extreme climate, thin, poor soil, short growing season, and coniferous forests or tundra-like vegetation, require a completely different reclamation program than would be necessary for a mine site on the plains where the topography is gentler and soils generally better but where precipitation is low.

A. Effect of Surface Mining on Selected Chemical and Physical Properties of Three Alberta Mine Spoils

Three study sites located in different climatic and vegetational zones in Alberta have been chosen to illustrate the effects of mining on selected soil characteristics. These sites include a short grass prairie, a subalpine forest and a northern mixed wood or boreal forest (Table 1). The prairie area was surface mined for coal during the 1940s and was abandoned until 1974 when revegetation attempts met with failure. Mining at the subalpine sites is currently in progress and requires massive excavation to expose the often deeply buried and folded coal seams. The oil sands are being actively surface mined for heavy oil—a process that involves removing the overburden and muskeg (which is stockpiled for reclamation) lying above the bitumen-bearing sand, excavation of the oil-bearing sand, and separation of the oil using hot water, caustic soda, and steam. This extraction process results in two types of mine tailings: a liquid slurry consisting of extractant, residual bitumen, sand, silt and clay which is disposed of in tailings ponds, and almost pure oil-free sand. The sand is used to dyke the tailings ponds and is then revegetated to prevent erosion.

Aside from topographical features such as slope gradients and aspect, the main factors that require consideration to ensure sucessful establishment and plant growth on these mine spoils are:

1. The climate of the mined area. Droughty conditions in the semi-arid plains (mean growing season moisture deficit of 370 mm can occur) and possibly in the boreal forest regions could potentially result in plant failure, whereas low temperatures and short growing seasons (Table 1) would restrict plant growth in the subalpine areas. Extreme wind conditions in all areas could cause tremendous erosion problems unless mine sites are rapidly revegetated.
2. Soil nutrient status. Mine spoils are notoriously low in available plant nutrients (Table 2). All three mine wastes require nutrient input (particularly nitrogen and phosphorus) to encourage plant productivity.
3. Spoil texture. Each mine spoil under consideration has specific soil textural problems which, if possible, should be alleviated to ensure rapid revegetation. The prairie coal mine spoil has a high clay and exchangeable sodium content that renders the spoil surface impermeable, thereby restricting root penetration and plant establishment, and increasing erodibility (Power et al., 1978). The rocky nature of the subalpine spoil may result in plant establishment difficulties particularly if the outplanting of trees is required to fulfill a forestry land use goal. The coarse texture of the highly rocky spoil (77% of particles exceeded 2 mm in diameter) may also accelerate the loss of nutrients through leaching. The sandy mine tailings (96% fine- to medium-grained quartz sand) are also highly

Table 1 General Data for Three Mine Sites in Alberta

Soil	Location	Dominant vegetation	Soil type	Annual mean precipitation (mm)	Mean Monthly temperature range January–July	Frost-free days
Shortgrass prairie site at Bow City, Alberta	50°25'N, 112°12'W Altitude: 755 m	Wheatgrasses (*Agropyron* spp.) Blue grama (*Bouteloua gracilis*) Sagebrush (*Artemisia cana*)	Solonetzic or chernozemic	335	−13.9 to 18.5°C	100
Subalpine forest at Luscar, Alberta	53°N 117°24'W Altitude: 1680 m	Spruce (*Picea engelmanii*) Fir (*Abies lasiocarpa*)	Brunisolic	500	−14.2 to 16.1°C	< 60
Northern boreal forest near Mildred Lake, Alberta	57°03'N 111°35'W Altitude: 370 m	Jack pine (*Pinus banksiana*) Lichen (*Cladina mitis*)	Brunisolic	470	−21.8 to 16.4°C	60–75

Table 2 Selected Chemical and Physical Characteristics of Test Mine Spoils[a]

Location of mine	Mining activity	pH	SAR[b]	C(%) Organic C	C(%) CO_3–C	N(%)	NO_3–N (μg g^{-1})	PO_4–P (μg g^{-1})
Shortgrass prairie	Subbituminous coal mining (surface mine)	7.2	27.8	6.4[c]	0.1	.11	6.4	1.3
Subalpine forest	Bituminous coal mining (pit mine)	7.1	0.5	3.5[c]	0	.09	4.0	3.4
Boreal forest	Heavy oil extraction from sand (surface mine)	7.3	0.5	0.2	NM[d]	0	0.1	1.2

[a]Method of analysis according to McKeague (1976).
[b]SAR, sodium adsorption ratio.
[c]This value includes carbon present in coal particles contaminating the soil samples.
[d]NM, not measured.

leachable, and because of their instability are easily eroded. Wind erosion of the sand tailings is a major factor in the sand blasting and burial of young seedlings.

4. Organic matter content. The removal of the organic surface soil during the mining process substantially reduces the fertility and water and nutrient retention properties of the soil. Segregation of the organic soil layers during the mining process and subsequent replacement of this soil after mining (as is currently being done in the prairie and oil sands regions) aids in alleviating not only nutritional deficiencies but also adverse textural characteristics.

Major shifts in pH were recorded as a result of mining the subalpine and boreal forest (oil sands) sites, i.e., in the subalpine site the pH increased from 4.5 in the undisturbed organic mat (F_2/H layer) to 7.1 in the mine spoil, whereas the extraction of oil from the oil sands resulted in tailings with a pH of 7.3 compared with 5.1–6.0 in the surface (0–15 cm) soil in an adjacent undisturbed jack pine woodland. The acidity problems often associated with mine wastes in the eastern United States, where high levels of iron pyrite or iron sulfide in waste materials can stimulate sulfur-oxidizing bacteria (*Thiobacillus* spp.) to generate large quantities of sulfuric acid, are generally not encountered in coal mine spoils in western Canada. The coal in this region usually has a low pyritic sulfur content, hence the potential for acid production by iron and sulfur-oxidizing bacteria would be expected to be low. It has been demonstrated that iron and sulfur-oxidizing bacteria are present and active in oxidizing some forms of pyrites that occur in Southeastern Montana coal spoils, but that the acid produced is quickly neutralized by carbonates present in the overburden and groundwater (Olson et al., 1979; 1981). This phenomenon may also occur in coal mine wastes in Alberta, particularly those in the prairie regions where soil carbonate levels can be high (Table 2).

B. Role of Microorganisms in the Reclamation Process

Although the chemical and physical properties of mine spoils are important in determining initial plant establishment and growth, biological factors are essential for the successful maintenance of soil fertility and the evolution of a stable soil/plant system. Research concerning the restoration of land destroyed by mining has emphasized the physical/chemical aspects of mine wastes and methods of alleviating adverse conditions to ensure a productive plant cover. In comparison, the soil biological characteristics, both before and after mining, have received very little attention, as discussed in Chapter 5. It has become accepted that in undisturbed ecosystems, primary production, decomposition, and nutrient cycling are the main processes upon which the stability of an ecosystem depends. There is no doubt that the rapid establishment of the primary producers on disturbed land should be the main aim of a reclamation program, but it should also be realized that once the vegetation is established the rate at which essential nutrients are released from dead plant residues and returned to the plant system and the rate at which stable organic matter accumulates on these disturbed systems will be largely determined by the biological characteristics of the mine waste. The soil microbial biomass is both a source and sink for essential plant nutrients and is fundamental to the transformation and flow of carbon, nitrogen, sulfur, and phosphorus (Ausmus et al., 1976; Paul and Voroney, 1980). The transformation of the organic nitrogenous and sulfur compounds in dead plant tissues to forms where they can be readily utilized by plants is highly dependent on the microbial oxidation of these compounds. Owing to their potential for high metabolic and reproductive rates, the soil microorganisms also have the capability to rapidly immobilize nutrients that would otherwise be lost to leaching. As well, soil microbes exhibit great metabolic versatility which allows them to adapt to the low nutrient levels and adverse chemical and physical conditions characteristic of many minewastes (Parkinson, 1979). Since mine spoils often lack such key nutrients as nitrogen and phosphorus, it is essential to determine the roles of symbiotic microorganisms (i.e., nitrogen-fixing bacteria and mycorrhizal fungi) in improving the nutrient status of plants used to colonize mine wastes. The importance of microorganisms in restoring soil fertility and productivity on derelict land has been stressed by Cundell (1977), Jurgensen (1978), and Parkinson (1979).

The microbial status of a mine spoil is highly dependent on factors such as temperature, moisture availability, nutrient levels, presence of toxic elements, and pH, but unlike the primary producers, the soil microorganisms, being mainly heterotrophic, require external sources of organic carbon. Therefore, it can be assumed that for the majority of mining activities, it is the degree of destruction of the surface organic mat, and whether this organic material is replaced after mining that determine the microbial status of a spoil immediately after mining. The microbiology of the spoil material itself depends to a large extent on the procedures used to mine and extract the ore. For example, the harsh physical and chemical procedures used to extract oil from bitumen-bearing sand results in mine tailings that are microbially sterile, whereas the overburden replaced after the removal of coal contains some microbial material, depending on the depth from which the overburden originated. The survival and development of microorganisms that are either introduced or indigenous to a mine spoil depend in large part on the chemical and physical properties of the mine waste. Müller (1973) provided evidence that suggested that numbers of microorganisms in coal mine dumps in Poland decreased with increasing acidity, while Lawrey (1977a) attributed the reduced microbial activity measured in coal mine spoils in Ohio to toxic levels of heavy metals made available by the acidity of the overburden. It must be noted that the direct effects of acidity on microbial populations are difficult to determine since pH is a major factor in determining nutrient availability (Brady, 1974) which in turn may influence the microbial status of a mine spoil.

The pH status of Alberta mine spoils (Table 2) would not be expected to be inhibitory to the majority of microorganisms. The properties of these mine spoils that could cause the greatest limitations to microbial growth include:

1. Low moisture status. The coarse texture and lack of organic matter in the subalpine and oil sands mine spoils would reduce the moisture retention capacity of these wastes, while the high sodium adsorption ratio of the prairie mine spoil would decrease water infiltration into this particular spoil.
2. Soil temperature. Climatic conditions at all three study sites vary from extreme highs during the summer to below freezing during the winter (Table 1). Widely fluctuating temperatures, particularly at the surface of bare mine spoils, would inhibit microbial growth and adversely affect survival of more sensitive organisms.
3. Soil chemical/physical factors. The sodic nature of the prairie mine spoil may adversely affect selected microbial components.
4. Low levels of organic matter. Raising the organic matter status of mine wastes either through amendation or plant growth would improve the energy available for microbial growth, thereby increasing the rate of microbial development.

II. MICROBIOLOGICAL CHARACTERISTICS OF SURFACE MINED SOILS AND MINE TAILINGS IN ALBERTA

The microbial characterization of mine spoils or mine tailings has generally involved three main approaches:

1. The quantitative and qualitative determination of bacteria, actinomycetes, and fungi using isolation and identification techniques (Wilson and Stewart, 1956; Müller, 1973; Cresswell, 1973; Lawrey, 1977b; Williams et al., 1977; Fresquez and Lindemann, 1982).
2. The measurement of soil biological activity including soil respiration (Hedrick and Wilson, 1956; Lawrey, 1977b; Hersman and Klein, 1979; Stroo and Jencks, 1982), adenosine 5'-triphosphate levels (Hersman and Temple, 1978; Schafer et al., 1979; Hersman and Klein, 1979), and dehydrogenase, amylase, phosphatase, urease activities (Hersman and Temple, 1979; Fresquez and Lindemann, 1982; Stroo and Jencks, 1982).
3. The assessment of leaf litter decay processes (Lawrey, 1977a; Carrel et al., 1979; Lanning and Williams, 1979) and rates at which essential plant nutrients, particularly nitrogen, are mineralized (Wilson, 1965; Williams and Cooper, 1976; Lodhi, 1979; Lanning and

Williams, 1980a,b; Hons and Hossner, 1980). The rates of nitrogen fixation by both symbiotic and free-living bacteria has also been considered an important parameter since this process can accelerate nitrogen accumulation on mine spoils (Hersman and Klein, 1979; Langkamp et al., 1979; Skeffington and Bradshaw, 1980; Hersman et al., 1981).

The data obtained by the first approach yields a descriptive and static microscale view of the changes in microbial components as a result of mining. The information collected by the second and third approaches yields a much broader and more dynamic picture of the effects of mining disturbance. An holistic approach to describing the microbial status and processes in a particular mine waste would encompass not only the measurement of biological activity, organic matter decomposition, and nutrient transformation rates, but also the identification of the microorganisms involved in these processes.

To assess the effects of severe soil disturbance on the soil microbiology at the study sites described in Table 1, a combination of three approaches was used. The parameters evaluated in replicated soil samples removed from both undisturbed and disturbed areas included:

1. Numbers of bacteria and actinomycetes using dilution plates.
2. Nature of the colonizing bacteria using methods outlined by Nelson and Visser (1978).
3. Mycelial lengths using a modified Jones and Mollison (1948) agar film technique.
4. Types of fungi using soil wasing and plating techniques.
5. Soil respiration measured as CO_2 evolved at 22–23°C.
6. Microbial biomass carbon based on the measurement of glucose-stimulated CO_2 efflux where 1 ml CO_2 hr^{-1} measured at 22°C is equivalent to 40 mg microbial biomass carbon (Anderson and Domsch, 1978).
7. ATP estimations following the methods outlined by Paul and Johnson (1977).

The cellulose and lignin decay potentials of disturbed versus undisturbed soils were determined by following the weight loss of cellulose filter paper and wood (Douglas-fir) dowel segments in the field over a period of up to 45 months. A more detailed explanation of the methods employed is presented in Visser et al. (1983).

A. Microbial Populations

In the grassland site, bacterial numbers in the surface layers of the disturbed soil were greater than in the undisturbed topsoil (Table 3).

Table 3 Numbers of Bacteria and Actinomycetes and Fungal Hyphal Lengths in Undisturbed and Disturbed (Surface Mined) Soils

Microbial parameter	Site	Undisturbed Surface (0–15)	Undisturbed Subsurface (15–25)	Disturbed Surface (0–5)	Disturbed Subsurface (5–15)
Bacteria ($\times 10^6$ g^{-1} dwt soil)	Prairie[a]	8^{ab}	3^{a}	96^{c}	26^{bc}
	Subalpine[a]	210^{c}	3^{a}	79^{b}	55^{b}
	Oil Sands[b]	34^{b}	3^{a}	8^{a}	9^{a}
Actinomycetes ($\times 10^5$ g^{-1} dwt soil)	Prairie[b]	100^{b}	55^{ab}	6^{a}	114^{ab}
	Subalpine[a]	5^{a}	0^{a}	1.5^{a}	1.8^{a}
	Oil sands	0	0	0	0
Total fungal mycelium (m g^{-1} dwt soil)	Prairie[a]	$1,200^{b}$	$1,200^{b}$	260^{a}	220^{a}
	Subalpine[a]	$27,100^{c}$	$2,700^{b}$	77^{a}	78^{a}
	Oil sands	ND[c]	ND	95	100
Mycelium with cell contents (m g^{-1} dwt soil)	Prairie[a]	130^{b}	170^{b}	22^{a}	18^{a}
	Subalpine[a]	$5,000^{c}$	270^{b}	9^{a}	6^{a}
	Oil sands	ND	ND	77	81

[a]Data analyzed by one-way ANOVA and differences detected by Scheffé multiple contrasts for pairwise comparisons. Values in each row followed by the same italicized letter(s) do not differ significantly ($p \leqslant 0.05$).
[b]Data analyzed by Kruskal-Wallis test and differences detected by post-hoc comparisons. Values in each row followed by same italicized letter(s) do not differ significantly ($p \leqslant 0.05$).
[c]ND, Not determined.

Fresquez and Lindemann (1982), studying a similar spoil, recorded significantly fewer bacteria in the barren spoil than in the undisturbed soil. A possible reason for the discrepancy in results between the present study and that of Fresquez and Lindemann (1982) may be the differences in the ages of the spoils in the two studies, the Alberta spoil being much older. Also reclamation attempts during the 1970s resulted in the luxurious growth of *Kochia scoparia*, and stimulation of bacterial growth may have occurred in the rhizosphere of these plants Wilson (1965) observed that numbers of bacteria in the rhizosphere and on the surface of the plant roots in a vegetated spoil were greater than bacterial numbers in an unvegetated spoil.

In both the subalpine and oil sand sites, mining resulted in a general reduction in bacterial and actinomycete counts and hyphal lengths (Table 3). Decreases in the individual microbial components were especially evident when results from the surface undisturbed soil were compared with those from the disturbed soil. Differences in the microbial numbers and hyphal lengths were not as obvious when the undisturbed subsurface (mineral) soil was compared with the mined soil. The loss of organic matter during the mining process was presumably responsible for the reductions in the microbial populations. In the two forest sites, mining resulted in the almost complete removal of biologically available organic carbon. Wilson (1965) and Müller (1973), working with acidic coal mine spoils, also observed that bacterial and actinomycete numbers were lower in barren mine spoils than in vegetated or undisturbed soils. Both researchers came to the conclusion that the presence of vegetation was one of the most important factors in reestablishing the soil microflora. Input of organic matter, i.e., dead leaves, roots, and root exudates, was probably responsible for stimulating microbial growth in the revegetated mine spoils they studied.

Neither Wilson (1965) nor Müller (1973) measured a significant effect of mining and revegetation on the fungi. Their data, however, are difficult to interpret since soil dilution plates were used. Warcup (1955) and more recently Wainwright (1982) presented data that confirmed that dilution plate counts of fungi are largely counts of fungal spores and provide no indication of the amounts of fungal hyphae and types of fungi active in the soil at the time of sampling. In the present study, mining effects on fungi were determined by a direct observation technique, and the results show (Table 3) that mining significantly decreased fungal hyphal lengths (both total and those with cell contents) at two of the study sites.

It is difficult to compare the quantitative importance of the bacterial, actinomycete, and fungal groups in the undisturbed and disturbed soils due to the different methods used to assess each component. However, the data do suggest that the bacteria were less sensitive to the nutrient deficiencies and environmental extremes typical of mine spoils than the fungi were, or bacterial resting stages survived better in nonactive states than the fungal hyphae did.

It was observed that in the grassland and subalpine sites, mining caused a significant increase in the proportion of the soil bacterial community occupied by the coryneform-type bacteria (Table 4). Prior to disturbance of the grassland, *Bacillus* and the unpigmented Gram-negative rods were common in the soil, whereas the bacterial community in the undisturbed subalpine site was dominated by *Cytophaga, Flavobacterium* (organic soil only), and the unpigmented Gram-negative bacteria. The unpigmented Gram-negative bacteria were also the most frequently isolated bacteria from the unmined oil sands site, together with the coryneforms. The predominance of *Arthrobacter* in the grassland mine spoil may be related to its adaptation to low-nutrient regimes. Nelson (1978) and Nelson and Parkinson (1978) demonstrated that *Arthrobacter* had the capability to survive temperature and nutrient stress whereas *Bacillus* seemed less adaptable to these conditions. The disappearance of *Cytophaga* as a result of mining at the subalpine site may have been due to lack of moisture and organic matter in the mine spoil. *Cytophaga* spp. are commonly isolated from very wet soils and are significant for their ability to degrade relatively decay-resistant substrates such as cellulose and chitin (Christensen, 1977). If *Cytophaga* spp. in the organic mat of the undisturbed spruce-fir forest at the subalpine site were active in decomposing the chitinous walls of the large fungal standing crop measured in that soil, the loss of this mycelial substrate through mining would undoubtedly decrease the frequency of *Cytophaga*.

The fungal taxa in barren mine spoils have been found to be less diverse than those in undisturbed soil (Lawrey, 1977b; Fresquez and Lindemann, 1982). However, these investigations have also demonstrated that the incorporation of organic amendments (hay, sewage sludge) or the development of a vegetative cover on the mine spoil will increase the fungal diversity, i.e., it is apparently the lack of organic matter that limits the number of fungal genera in mine spoils. Each of the three Alberta mine spoils was different in terms of the fungal genera present (Table 5). At the grassland site, the spoil contained a large number of fungal taxa with the dominants being *Cladosporium* spp., *Alternaria* spp., *Penicillium* spp., sterile dark forms, and yeasts. *Cladosporium, Alternaria*, and yeasts are widely distributed and commonly colonize leaf litter and soil (Domsch et al., 1980). They are frequently isolated from the atmosphere and hence would be the first group of organisms to have the opportunity to colonize the surface of recently mined soil. *Penicillium* spp. were the fungi most frequently isolated from the subsurface mine spoil. Members of this genus can utilize a wide variety of substrates and survive extremes in moisture and temperature (Domsch et al., 1980). The number of plated soil particles that yielded fungi was approximately the same for the undisturbed and disturbed sites, indicating that the grassland mine spoil was well colonized.

The diversity of fungal taxa isolated from the subalpine spoil was not as great as that observed in the undisturbed soil. Also, with the exception of *Candida* sp. and *Chrysosporium* spp., many of the species

Table 4 Frequencies of Most commonly Isolated Bacterial Groups from Undistrurbed and Disturbed (Surface Mined) Soils[a]

		Percent of total viable isolates			
		Undisturbed		Disturbed	
Bacterial group	Site	Surface (0-15)	Subsurface (g5-25)	Surface (0-5)	Subsurface (5-15)
Arthrobacter	Prairie	14^{a}	3^{a}	5^{a}	10^{a}
	Subalpine	5^{a}	3^{a}	18^{a}	19^{a}
Bacillus	Prairie	30^{ab}	33^{b}	7^{ab}	3^{a}
	Subalpine	3^{a}	9^{a}	12^{a}	8^{a}
Coryneforms (excluding *Arthrobacter* and *Nocardia*)	Prairie	26^{a}	40^{ab}	72^{b}	68^{b}
	Subalpine	6^{a}	14^{ab}	46^{c}	28^{abc}

Cytophaga	Prairie	4	0	0	0
	Subalpine	16	23	0	6
Nonpigmented Gram-negative rods	Prairie	21^{b}	20^{b}	3^{ab}	0^{a}
	Subalpine	48^{b}	45^{ab}	4^{a}	15^{ab}
Total Gram-negative	Prairie	28^{b}	20^{ab}	8^{a}	3^{a}
	Subalpine	84^{c}	71^{bc}	9^{a}	37^{ab}
Total Gram-positive	Prairie	72^{a}	80^{ab}	92^{b}	97^{b}
	Subalpine	16^{a}	29^{ab}	91^{c}	62^{c}
% viability after storage	Prairie	87^{a}	98^{ab}	100^{b}	100^{b}
	Subalpine	58^{a}	58^{a}	92^{b}	90^{b}

[a]Where possible, data were analyzed by one-way ANOVA and differences detected by Scheffé multiple contrasts for pairwise comparisons. Values in each row followed by same italicized letter(s) do not differ significantly ($p \leqslant 0.05$).

Table 5 Percent frequency of Occurrence of Most Commonly Isolated Fungal Groups from Undisturbed and Disturbed (Surface-Mined) Soils[a]

		Soils and depth (cm)			
		Undisturbed		Disturbed	
Site	Fungal group	Surface (0-15)	Subsurface (15-25)	Surface (0-5)	Subsurface (5-15)
Prairie	*Alternaria*	0^{a}	0^{a}	24^{a}	10^{a}
	Chrysosporium	20^{ab}	67^{b}	3^{a}	7^{a}
	Cladosporium	0.3^{a}	0^{a}	14^{b}	2^{ab}
	Penicillium	1^{a}	5^{ab}	6^{ab}	23^{b}
	Sterile dark forms	24^{a}	4^{a}	17^{a}	13^{a}
	Yeasts[b]	4^{ab}	1^{a}	35^{b}	13^{ab}
	% particles plated which yielded fungi	76	87	70	88
Subalpine	*Candida*[c]	0	0	20^{a}	9^{a}
	Chrysosporium	3^{a}	9^{ab}	18^{b}	22^{b}
	Mortierella	84^{c}	35^{b}	0.3^{a}	3^{a}

	Mucor	14^{ab}	27^{b}	1^{a}	2^{a}
	Oidiodendron	20^{b}	4^{ab}	1^{a}	0.3^{a}
	Penicillium	55^{b}	9^{a}	1^{a}	1^{a}
	Trichoderma polysporum	27	15	0	0
	% particles plated which yielded fungi	100^{c}	99^{c}	45^{ab}	40^{a}
Oil sands	*Cladosporium*	0	0	10	0
	Mortierella	69^{b}	10^{a}	2^{a}	0
	Mucor	13	11	0	0
	Oidiodendron	33	7	0	0
	Penicillium	87^{b}	7^{a}	1^{a}	0
	Trichoderma viride	44	15	0	0
	% particles plated which yielded fungi	100^{c}	96^{c}	27^{b}	0.4^{a}

[a]Where possible data were analyzed by one-way ANOVA and differences detected by Scheffé multiple contrasts for pairwise comparisons. Values in each row, followed by same italicized letter(s) do not differ significantly ($p \leqslant 0.05$).
[b]Data analyzed by Kruskal-Wallis test.
[c]Data analyzed by Student's *t*-test.

in the mine spoil occurred with very low frequencies. Although mining at the grassland site resulted in a significant decrease in the occurrence of *Chrysosporium*, mining at the subalpine site, significantly increased *Chrysosporium* (mainly *C. pannorum*) frequency. The increased frequency of *C. pannorum* in the subalpine mine spoil may be the result of the rise in soil pH after mining (pH 4.2–4.5 in undisturbed soil and 7.0–7.3 in mined spoil), since this organism appears to prefer neutral soils (Domsch et al., 1980). Also, laboratory studies of substrate utilization by this fungus (Visser, unpublished data) have demonstrated that this organism is capable of utilizing a wide variety of substrates (starch, pectin, cellulose, chitin, tannins). This capability may give *C. pannorum* a competitive advantage over other fungi, hence its predominance in this nutritionally impoverished spoil. The pH optimum for *Chrysosporium* is about 7 and this could explain why it has not been isolated from acidic mine spoils. In contrast to the grassland spoil, less than 50% of the plated subalpine mine spoil particles yielded fungi. The older age of the grasslands spoil may have allowed the fungi to colonize that spoil to a greater extent than the subalpine spoil, as the latter was spread immediately prior to sampling.

The harsh methods used to remove oil from the oil sands results in a fine-grained material devoid of organic matter and nutrients, and mycologically sterile. Fungal isolations made from the extracted sand demonstrated that only the surface sand was colonized to any extent and that the fungi consisted mainly of *Cladosporium* spp. and yeasts, organisms that are common in the atmosphere. The poor fungal colonization of the extracted sand was further substantiated by the low number of plated particles that yielded fungi. These results suggest that the tailing sand would require a highly organic amendment to establish a fungal community approaching the diversity observed in the undisturbed soils.

In general, the fungal isolation data imply that mining procedures and spoil age may have significant effects on fungal communities and fungal growth in spoils. It appears that in mine wastes where indigenous fungi are lacking, fungi characteristic of the atmosphere (*Cladosporium, Alternaria*, and yeasts) are the first colonizers, particularly in the spoil surface.

B. Microbial Activity and Biomass

Although it may be inferred from the microbial population data that mining results in reduced microbial activity, plate counts and hyphal length measurements give very little indication of the activity of the organisms isolated. Bacterial and actinomycete plate counts can include both active vegetative cells and inactive spores, whereas the presence of protoplasm in fungal hyphae at the time of sampling does not necessarily mean the mycelium is active, but rather indicates a potential for being active. Also, the type of media used to isolate bacteria or actinomycetes may select only a certain proportion of the community, whereas major difficulties are encountered in the separation of mycelium from the soil (particularly in highly organic or clayey soils). As a result, plate counts and hyphal length measurements should be coupled with information on soil respiration or ATP in an effort to estimate more accurately the level of activity exhibited by both isolated and nonisolated organisms.

At the prairie and subalpine sites, significantly lower CO_2 effluxes were recorded in the disturbed than undisturbed soils while microbial biomass carbon and ATP were substantially less in the disturbed than in undisturbed soils at all three sites (Table 6). Presumably the decrease in microbial activity in the disturbed surface soil resulted mainly from the loss of the organic mat, although high sodium levels in the grassland spoil may also have contributed to some inhibition of microbial activity. Malik and Azam (1980) reported a decrease in soil microbial biomass with increasing salinity. The reduced respiration in strip-mined spoils studied by Hedrick and Wilson (1956) was believed to be due to insufficient nitrogen, whereas Lawrey (1977b) attributed lower respiratory activity in mined soils to high levels of trace metals made available by the low soil pH. Recently, Stroo and Jencks (1982), in a study on enzyme activity and respiration in mine spoils at various stages of reclamation, concluded that the accumulation of organic matter and nitrogen in a mine spoil would lead to a slow recovery of microbial activity.

ATP measurements have been considered by Hersman and Temple (1979) as one of the best and most rapid methods for determining biological activity in mine spoils. Their results (Hersman and Temple, 1978), like those obtained in this study, demonstrated that ATP levels in undisturbed soils were higher than those in mine spoils (Table 6). This method may be rapid, but in the present study, ATP data were found to be highly variable. Also, ATP in the extracted oil sand was so low that it was not measureable, although some microbial biomass carbon was measured in these samples. Unsuitable ATP extraction techniques and the analysis of small quantities of soil (1 g in this study) probably contributed to the observed variation.

C. Decomposition Processes

The evolution of a self-sustaining vegetative cover on mine spoils is ultimately dependent on the establishment of decomposition and mineralization processes, since these processes ensure a continual nutrient supply to the primary producers. Decay rates for relatively easily degradable (cellulose) and recalcitrant (wood) carbon were evaluated in both undisturbed and disturbed systems to provide comparative data on the effects of mining on decomposition processes.

Table 6 CO_2 Efflux, Microbial Biomass Carbon, and ATP in Undisturbed Soils and Soils Disturbed by Surface Mining[a]

		Soil type and depth (cm)	
Microbial	Site	Undisturbed surface (0–15)	Disturbed surface (0–5)
CO_2 efflux (μl 100 g^{-1} hr)	Prairie	430[a]	160[b]
	Subalpine	5300[a]	140[b]
	Oil sands	100[a]	50[a]
Microbial biomass carbon (mg 100 g^{-1} dwt soil)	Prairie	85[a]	15[b]
	Subalpine	970[a]	24[b]
	Oil sands	21[a]	4[b]
ATP (μg g^{-1} dwt soil)	Prairie[b]	32[a]	1[b]
	Subalpine	20[a]	1[b]
	Oil sands	3	0

[a]Data analyzed by Student's *t*-test. Values in each row followed by the same letter do not differ significantly ($p \leqslant 0.05$).
[b]Data analyzed by Wilcoxan test.

The use of a woody substrate proved to be a poor indicator of the effects of mining on decay, since very little weight loss occurred at either the grassland or subalpine sites over the course of the study (98–100% dry weight remained after 49 months in the field). Woody substrates are comprised mainly of cellulose and lignin, which are characterized by their recalcitrant carbon and low nitrogen levels (the wood in this study contained 0.03% nitrogen). Therefore, this substrate provided very few readily available nutrients for microbial growth resulting in slow decomposition. Amendation of the wood dowel with ammonium nitrate to reduce the carbon:nitrogen ratio did not accelerate decay. Possibly the decay resistant quality of the carbon in this lignified substrate controlled the decomposition rate to a greater degree than the levels of other nutrients such as nitrogen. Also, the very dry conditions and the extremes in temperature occurring at the grassland site were not conducive to rapid microbial growth and turnover. Low soil temperatures, characteristic of soils such as those in the subalpine site, have also been regarded as a major factor in the inhibition of the decay process (Moore, 1981).

In contrast to the wood degradation measurements, decomposition of the cellulose filter paper in litter bags proceeded at a more rapid rate. Short-term differences in cellulose decomposition were not detected at any of the three sites, but over the long term (i.e., 24–33 months), faster decay rates were observed in the disturbed than in the undisturbed plots for the grassland and subalpine study areas. The percent dry weight remaining of filters placed in the subalpine site for 33 months was 86% and 53% for undisturbed and disturbed areas, respectively, whereas those placed in the grassland undisturbed and disturbed plots had 87% and 26% dry weight remaining after 33 months. The grassland site results may have been due to heavy spring rains that embedded and sometimes buried the filters in the spoil resulting in ameliorated microenvironmental conditions around the filters. Once buried, the filters decayed rapidly so that 44 months after the initiation of the study, less than 1% of the filter mass remained.

Filters placed in the undisturbed grassland, however, were not readily incorporated into the soil and the vegetation would occasionally elevate the filters above the soil surface. Therefore, the filters were more subject to drying and high temperatures which inhibited decomposition. Similar observations were made at the subalpine study area. Filters placed on this particular mine spoil were eventually buried, whereas those in the undisturbed forest remained on the surface for the full term of the study. The incorporation of the filters into the forest floor appeared to be largely dependent on the growth of the feather moss (*Hylocomium splendens*) over the bag containing the filter. Once overgrown by the moss, the ameliorated moisture and temperature conditions would probably accelerate decay. The results obtained in this study imply that mine spoil instability can influence substrate decomposition by increasing burial, which protects the decomposing substrate from temperature and moisture extremes—variables that are major controlling factors affecting decay rates (Bunnell et al., 1977).

Although respiration and microbial biomass were significantly lower in the disturbed grassland and subalpine site than in undisturbed soils (Table 6), cellulose degradation proceeded rapidly in the spoil. The very low carbon status of the mine spoils affected the microbial biomass, but once an easily available carbon source such as cellulose was added, rapid utilization occurred. Consequently, it is possible that due to the lack of available carbon in disturbed soils, decomposition may be more rapid than in undisturbed soils where other forms of carbon are available. Although both the filter paper and the mine spoils generally lacked nitrogen, decay was not accelerated when the filters were amended with ammonium nitrate.

The relationship between cellulose filter paper and natural litter decomposition is questionable. Filters placed in the undisturbed grassland site decayed slowly in comparison with grass litter, while at the undisturbed subalpine site filters decayed at the same rate as fir needles (Visser, unpublished data). At the undisturbed oil sands plot, the filters decomposed more rapidly than jack pine needles (Visser, unpublished data). Therefore, for undisturbed systems it is difficult to generalize about relationships between filter and natural litter decomposition, since the relationship appears to be site specific and is probably dependent on the litter type, climate, and the speed at which decomposing substrates are incorporated into the soil system. Similarly, Ross et al. (1978) observed that the weight loss of cellulose squares placed on the soil surface was generally correlated positively, but not significantly with weight loss of grass litter. The authors also found high variability among replicates of the cellulose squares and concluded that microenvironment was important in determining decomposition rates. A higher degree of variability among weight losses of filter paper versus weight losses of native herbage was also measured by Ratliff (1980) in meadow sites. Lack of essential nutrients in the filters could possibly explain some of the variation, since microbial utilization of the cellulose would be largely dependent on the nutrient status of the soil under the filter. Park (1976) emphasized the importance of mineral nutrients (particularly nitrogen) as one of the variables affecting the decay rate of such substrates as cellulose. Natural plant litter would have a better nutritional status than pure substrates, making the litter more amenable as a food source for the soil microbes.

Few sites disturbed by mining have been examined for their decomposition potential. Lawrey (1977a) investigated the relative decomposition potential of strip-mined and undisturbed areas in eastern Ohio using leaf litter of *Pinus, Robinia,* and *Acer.* His results showed that decomposition was initially retarded in soils affected by mining, but after 6 months there were no significant differences in weight loss of litter between mined and unmined areas. In contrast, decay rates for litter placed on strip mines in Missouri were reported to be 2–10 times less than those for litter placed on vegetated soils (Carrel et al., 1979). The filter paper decay data collected for mine spoils in this study are difficult to compare with the decomposition rates of natural litter measured by other investigators. Although filters degraded more rapidly over the long term in the mined areas at both the grassland and subalpine sites, it would not be warranted to assume that similar results would have been obtained with natural litter.

D. Microbial Activity and Organic Matter Accumulation

Since microorganisms require greater quantities of carbon than other nutrients for growth, one would expect that as revegetation of a mine spoil progresses and organic matter accumulates, microbial activity and biomass carbon would increase accordingly. But, with the exception of Schafer et al. (1979) and Stroo and Jencks (1982), very few studies have been conducted on the relationship between microbial activity in mine spoils and soil genesis. Stroo and Jencks (1982) observed that microbial activity in mine spoils in West Virginia increased with increasing age of the revegetated mine spoil and concluded that recovery was related to organic matter and nitrogen levels. Respiratory activity was highly correlated with biologically available carbon and mineralizable nitrogen, and microbial respiration in unamended sites revegetated for 20 years approached that in the native soils. However, in an investigation of microbiological properties of 1–50-year-old mine spoils and undisturbed soils in Montana, Schafer et al. (1979) found no differences in ATP activity in natural soils and those disturbed by mining.

In this study, microbial respiration, biomass carbon, and organic matter levels in the subalpine mine site all increased in older mine spoils (Table 7). CO_2 evolution, biomass carbon, and organic matter levels were not significantly different in the undisturbed mineral soil and stockpiled regolith (material salvaged for reclamation of the minesite once it is backfilled), which suggests that vegetation and organic matter removal and not physical disturbance of the soil controls microbial activity in the soil after mining. Although microbial activity and biomass carbon in the 7-year-old site were significantly greater than that measured in other sites, it was still substantially less than that in the undisturbed organic mat. As in the study by Stroo and Jencks (1982), a highly significant relationship was detected between microbial biomass carbon and percent organic matter. A regression analysis performed on biomass carbon and organic matter measurements from only the disturbed soils demonstrated a strong relationship ($r^2 = 0.601$) between the two variables but not as strong as that detected when measurements from the undisturbed soils were included ($r^2 = 0.867$).

Wilson (1965) concluded that organic matter on vegetated acidic coal mine spoils and iron-ore spoils accumulated at a very slow rate. In contrast, Schafer et al. (1980) recorded a relatively rapid accumulation of carbon on Montana coal mine spoils, particularly those that had been heavily fertilized, seeded to grasses and legumes, and left ungrazed. Within 30 years after mining, the organic matter levels within the surface 10 cm of these mine spoils were similar to those in natural soils. Anderson (1977), working on mine spoils in a semiarid region of Saskatchewan, calculated that 250–350 years of organic matter accumulation was required to reach levels found in native soils. It would, therefore, appear that the rate of organic matter accumulation on mine spoils can be highly variable, depending on climate, revegetation programs, and maintenance practices once the vegetation is established. In the subalpine regions of Alberta, the rate of organic matter accumulation resulting from plant litter input would be expected to be slow, due to the harsh climate and short growing season.

Table 7 Microbial Respiration and Biomass in Soil from an Undisturbed Subalpine Forest (Spruce/Fir) and from Adjacent Areas in Various Stages of Reclamation[a]

Soil description	pH (0.01 M $CaCl_2$)	Organic matter[b]	Basal respiration (µl CO_2 100 g^{-1} dwt hr^{-1})	Microbial biomass carbon (mg C 100 g^{-1} dwt hr^{-1})
Undisturbed organic (0–10 cm)	4.5 *a*	86 *c*	6900 *d*	1150 *d*
Undisturbed mineral (10–20 cm)	4.7 *a*	6 *a*	86 *a*	22 *ab*
Mineral (0–10 cm) (revegetated for 7 years)	7.3 *d*	19 *b*	1000 *c*	110 *c*
Mineral (0–10 cm) (revegetated for 6 years)	7.5 *d*	7 *a*	440 *b*	46 *bc*
Mineral (0–10 cm) (revegetated for 2 years)	6.8 *c*	2 *a*	120 *d*	26 *ab*
Stockpiled regolith (0–10 cm)	5.9 *b*	4 *a*	50 *a*	13 *a*

[a]Data analyzed by one-way ANOVA and Scheffé multiple contrasts. Values in each column followed by same italicized letter(s) do not differ significantly ($p \leqslant 0.05$).
[b]Determined by modified Walkley and Black (1934) digestion method as described in McKeague (1976).

III. METHODS OF RESTORING BIOLOGICAL ACTIVITY TO SOIL DISTURBED BY MINING

As would be expected, the destruction of the organic mat at the three Alberta mine sites resulted in reduced microbial activity and major shifts in bacterial and fungal community composition. Also, root growth tended to be concentrated in the organic horizons of the soil profile, hence the disruption of this horizon eliminated microbial activity associated with the roots (i.e., rhizoplane, rhizosphere, mycorrhizae). Therefore, if the reestablishment of microbial activity is to be considered essential to the successful reclamation of mine wastes, the presence of organic matter (given that the spoil has no extreme chemical/physical problems) should be regarded as the most important aspect in a reclamation strategy. The poor organic matter status characteristic of most mine spoils can be improved by (1) establishing a vegetative cover that will ultimately result in the accumulation of stable organic matter through plant residue decomposition, and (2) ameliorating the mine waste with an organic amendment followed by subsequent revegetation.

The development of microbial activity in mine spoils following amendation and revegetation has received only limited attention. Consequently, large quantities of mine spoil from the subalpine and oil sands mine sites were amended, planted, and monitored regularly for plant growth and the development of microbial populations, microbial activity, and biomass and decomposition processes. The amendments applied to each of the mine wastes included fertilizer (112 kg N and P_2O_5 ha^{-1}, 90 kg K_2O ha^{-1}), anaerobically digested sewage sludge (46 mT ha^{-1}), and feather moss peat (15 cm). Details of the experimental design can be found in Parkinson et al. (1980). Slender wheatgrass, alsike clover, laurel leaf willow, and white spruce were planted in the subalpine spoil while slender wheatgrass, sainfoin, jack pine, and bearberry were tested on the oil sands. The methods used to determine the microbiological status of the variously amended and planted mine spoils over the term of the study were identical to those discussed earlier. Plant shoot material produced over the third growing season was utilized in the decomposition studies. Shoots from the peat-treated plots only (to avoid resource quality related effects on decay rates) were collected at the conclusion of the growing season, bagged in 1-mm mesh, and returned to the plots within 1 month of collection. Treatment and plant growth effects on the chemical and microbiological characteristics were limited mainly to the top 5 cm of the mine spoils.

A. Selected Soil Chemical Characteristics After Amendment

Table 8 exemplifies the effects of the various treatments on chemical characteristics of the extracted oil sands. A similar pattern was observed when the subalpine spoil was treated with the same amendments.

Table 8 Amendment Effects on Selected Soil Characteristics of the Extracted Oil Sand (Depth; 0–5 cm)[a]

	Treatment			
Parameter	Unamended control	Fertilizer	Sewage sludge	Peat
Organic matter (%)	0.4^{a}	0.3^{a}	1.5^{b}	26.9^{c}
Total nitrogen (%)	0	0.01^{a}	0.02^{a}	0.82^{b}
NO_3-N ($\mu g\ g^{-1}$)	0.1^{a}	40^{b}	16^{b}	250^{c}
PO_4-P ($\mu g\ g^{-1}$)	1.2^{ab}	53^{de}	26^{cd}	2.7^{b}
pH	7.3^{b}	6.4^{a}	6.8^{ab}	7.1^{b}

Fertilizer applied 3 days prior to sampling; sewage sludge and peat applied 2–3 weeks prior to sampling. Sludge and peat were mixed into the sand and samples were removed before planting.
[a]Values in each row followed by the same italicized letter(s) do not differ significantly ($p \leqslant 0.05$) as determined by one-way ANOVA and Scheffé confidence intervals for pairwise comparisons.

The organic matter status of the spoil materials was most enhanced by the addition of peat and to a lesser degree by sewage sludge. The sewage sludge was allowed to dry on the surface prior to being mixed into the mine wastes resulting in a very heterogeneous dispersion of nutrient-rich sludge aggregates within the spoil matrix. Very high NO_3-N levels were recorded in the peat, but extractable phorphorus was so low that it was potentially limiting to plant growth. Inorganic fertilizer and sewage sludge significantly raised the nitrogen and phosphorus status of the subalpine and oil sands mine wastes, particularly in the sand tailings where nutrient levels were exceedingly low. With the exception of the fertilizer, which tended to acidify the mine soils, amendation did not significantly alter the neutral pH of either the subalpine or oil sands spoils.

B. Microbial Populations After Amendment

Prior to their application to the mine spoils, sewage sludge and peat exhibited the greatest microbial activity (6300 and 1100 $\mu l\ O_2\ 100\ g^{-1}$ dwt hr^{-1} for sludge and peat, respectively) with the highest bacterial numbers (42 and 7 $\times 10^8\ g^{-1}$ dwt for each of the two amendments, respectively) and largest amounts of fungal hyphae (2500 and 2300 $m\ g^{-1}$ dwt sludge and peat, respectively). The peat also contained a high level of actinomycetes (1500 $\times 10^4\ g^{-1}$ dwt peat). Numbers of bacteria in the fertilizer were very low ($1.3 \times 10^5\ g^{-1}$), as was the length of mycelium (32 $m\ g^{-1}$).

Bacterial, actinomycete, and fungal mycelial values in the sand tailings, initially and 15 months after amendation and planting of slender wheatgrass (Table 9), followed a pattern similar to that observed for the subalpine spoil. Bacterial counts were highest in the sludge-and peat-treated mine spoils, whereas actinomycetes and fungal mycelium were most prevalent in the peat treatment. Fertilization had no significant immediate effects on the microbial populations. The addition of peat to the microbially impoverished tailing sand also resulted in a substantial increase in the number of fungal taxa (Parkinson et al., 1980). Sewage sludge or fertilizer treatment did not alter the fungal community from that observed in the untreated tailing sand, suggesting that very few fungi were introduced via these amendments. It is not surprising that the mine spoils treated with sewage sludge and peat exhibited the highest levels of microbes, since these two amendments introduced not only large numbers of microorganisms but also carbon and essential nutrients such as nitrogen and phosphorus required by the microbes for growth and maintenance.

Growth of slender wheatgrass for two growing seasons (15 months) did not significantly alter the soil microbial populations, although bacterial numbers in the fertilized treatment appeared to be increasing, whereas those in the sludge amended mine spoils were decreasing. In

Table 9 Bacterial and Actinomycete Numbers and Fungal Hyphal Lengths in 0–5-cm-Deep Oil Sands Tailings Before Planting (0.5 Month) and After Planting Slender Wheatgrass (15 Months)[a]

Microbial parameter	Sampling time (months)	Treatment			
		Control	Fertilizer	Sewage sludge	Peat
Bacteria	0.5	7.5^{a}	8^{ab}	1200^{d}	650^{d}
($\times 10^6$ g^{-1} dwt soil)	15.0	8^{ab}	67^{bc}	220^{cd}	600^{cd}
Actinomycetes	0.5	0.6^{a}	1^{a}	2^{ab}	3300^{c}
($\times 10^4$ g^{-1} dwt soil)	15.0	4^{a}	1^{a}	20^{ab}	1200^{c}
Total mycelium	0.5	95^{a}	37^{a}	86^{a}	1300^{b}
(m g^{-1} dwt soil)	15.0	41^{a}	57^{a}	66^{a}	680^{b}

[a]Data for each microbial parameter were analyzed by a two-way ANOVA and Scheffé multiple contrasts for pairwise comparisons. Values within each category followed by the same italicized letter(s) do not differ significantly ($p \leqslant 0.05$).

the fertilized mine spoils the readily available carbon provided to the soil microorganisms through sloughed root material and root exudates may have stimulated bacterial growth. It is widely recognized that actively growing plant roots release a wide variety of substances (Russell, 1977). Barber and Martin (1976) attempted to quantify root exudation and estimated that 12–18% of the photosynthetically fixed carbon of wheat and cereal plants was released by the roots. In the nutrient-poor conditions characteristic of mine spoils, this amount of carbon could represent a major food resource for the soil microflora, particularly if amendation had not introduced other forms of carbon. The importance of root activity in stimulating microbial growth and activity in mine spoils has been substantiated by the studies of Wilson (1965) and Fresquez and Lindemann (1982).

Although microbial populations demonstrated little response to plant growth, fungal community composition appeared to be very much influenced by plant growth (Parkinson et al., 1980). The number of fungal taxa in the sewage sludge-treated plots exhibited the greatest increase after two growing seasons, presumably because of the nutrient-rich composition of the sludge and enhanced nutrient input by the plants. Fresquez and Lindemann (1982) noted a substantial increase in the number of fungal genera isolated from the rhizosphere of fourwing saltbush grown in sludge-treated mine spoil.

C. Microbial Activity and Biomass After Amendment

Both microbial activity and biomass carbon had similar responses to amendation and planting, hence only data for microbial biomass are presented here (Table 10). Although microbial respiration and biomass were generally greater in the untreated, unvegetated subalpine spoil than in the untreated, unvegetated tailing sand, the effects of the various amendments followed the same pattern in the two mine spoils—a pattern that was still evident three to four growing seasons after amendation (Table 10). Peat was the most effective in increasing the microbial biomass of the mine spoils. This amendment was characterized by high levels and a wide variety of microbes and was applied at a heavier rate than the other two amendments. Sewage sludge with its high microbial inoculum load and nutrient-rich composition was also highly effective in raising the microbial activity and biomass. Eiland (1981) and Macgregor and Naylor (1982) recorded similar observations when they applied sewage sludge to cropland. Wilson and Hedrick (1957) recorded a stimulation of CO_2 production from unvegetated mine spoils when treated with NPK fertilizer, but this was not observed to be the case for the fertilized Alberta mine spoils where carbon availability appeared to be limiting microbial activity.

Table 10 Microbial Biomass (mg C 100 g^{-1} dwt soil) in Unvegetated and Vegetated Mine Spoils[a]

Minespoil	Time after planting (months)	Vegetation type	Amendment			
			Control	Fertilizer	Sewage sludge	Peat
Subalpine	27	Unvegetated	36^{ab}	44^{abc}	100^{de}	290^{fgh}
		Slender wheatgrass	85^{cd}	130^{def}	230^{efg}	620^{h}
		White spruce	32^{a}	42^{abc}	80^{bcd}	450^{gh}
Oil sand tailings	39	Unvegetated	4^{ab}	4^{a}	24^{cde}	120^{fgh}
		Slender wheatgrass	4^{a}	16^{bcd}	80^{efg}	490^{h}
		Jack pine	9^{abc}	10^{abcd}	40^{def}	220^{gh}

[a]Data for each mine spoil analyzed by two-way ANOVA and Scheffé, multiple contrasts for pairwise comparisons. Values for each mine spoil followed by same letter(s) do not differ significantly ($p \leqslant 0.05$).

In previous investigations it has been suggested that a vegetative cover is essential for microbial activity in mine spoils (Stroo and Jencks, 1982) and that the plant species may determine the rate of microbial activity development (Müller, 1973). With the exception of the unamended sand tailings where nutrient deficiencies resulted in plant failure and limited microbial development, the patterns of plant (slender wheatgrass, jack pine, white spruce) growth, litter input (Visser, unpublished data) and microbial development in the subalpine and oil sands mine spoils were very similar (Table 10). Jack pine and white spruce had negligible effects on the soil microbial biomass while the growth of slender wheatgrass generally stimulated microbial growth and productivity. Peak primary production by slender wheatgrass occurred during the second growing season, resulting in litter accumulation on the spoil surface. Also, root weights in the grass plots were substantially greater than in the white spruce or jack pine plots. Sewage sludge amendation was particularly effective in enhancing grass production, hence the greatest litter input occurred in this treatment, whereas root production was highest in the peat-treated mine spoils. Since microbial growth would be greatly stimulated by the addition of carbon sources in litter, roots, and root exudates, their activity was presumably enhanced by the input of these plant materials. Therefore, it appears that microbial productivity may be strongly linked with primary production through the regulation of microbial metabolic activity by plant-derived energy sources. Consequently, it would be expected that the development of microbial biomass would be more rapid in a mine spoil planted with a fast-growing, highly productive grass species than a slow-growing tree species, particularly if productivity was enhanced by the addition of a nutrient-rich amendment. Müller (1973) also reported a more active development of microflora in coal mine spoils planted with a grass-legume mixture than in tree-vegetated spoils.

Because microbial development in mine spoils appears to be closely associated with the type of vegetative cover, the choice and longevity of a plant species are important factors to consider when formulating a reclamation strategy. To increase or maintain the microbial biomass stimulated by the grass over the long term, it would be essential that this vegetative cover continue to survive and grow. A failure in the vegetative cover could return the soil microbial biomass to levels found prior to planting, as reported by Stroo and Jencks (1982) for a mine spoil in West Virginia. When using short-lived species, such as slender wheatgrass, to provide initial erosion control on a mine spoil, it may be advantageous to plant longer-lived woody shrubs or conifers once the grass cover has failed. The choice of plant species and maintenance (i.e., fertilization) regimes would, however, be heavily dependent on final land use goals.

D. Decomposition Processes as Affected by Amendment

Although the subalpine and oil sands mine spoils were markedly different in terms of their physical, chemical, and biological characteristics, the short-term (12 month) decay of slender wheatgrass leaves and stems was surprisingly similar in the two mine spoils, regardless of treatment (Table 11). A two-way analysis of variance, applied to the grass leaf and stem data for each of the mine spoils, demonstrated that stems decomposed more slowly than leaves ($p \leqslant 0.05$) and that both litter components decayed equally fast in all treatments ($p \leqslant 0.05$). Weight losses of the leaves placed in the slender wheatgrass plots three growing seasons after the initiation of the study, were about 20% of their original weight after 6 months and 50% after 1 year in the field. Leaf litter is a composite of a variety of substrates including sugars, polysaccharides, hemicellulose, cellulose, and lignin, all of which have been shown to decompose at varying rates (Minderman, 1968) with the soluble sugars and polysaccharides disappearing most rapidly while the more resistant components (cellulose, lignin) accumulate. The rapid initial weight loss (6 months) is presumed to have been due to the loss of the soluble fraction of the litter (resulting from both leaching and microbial utilization), which for *Agropyron* was found to be approximately 16%. The lack of treatment effects on the decay of the grass litter is interesting in that it implies that over the short term (i.e., 1 year) the chemical and physical properties of a mine spoil and the soil microbial biomass in the spoil immediately beneath the decaying litter do not regulate the decay rate of that litter. When correlation coefficients were calculated between the remaining percent dry leaf weight after 12 months decomposition and soil microbial activity and biomass immediately below the slender wheatgrass litter bags, no significant relationships were detected. These data suggest that controls for the initial decomposition of leaf litter accumulating on mine spoils in the early stages of revegetation are independent of factors controlling decay in the soil beneath the litter. Lawrey (1977a) showed that once various litter types had been in the field for 6 months, their decay rates on bare strip-mined, vegetated strip-mined, and unstripped areas were not significantly different and concluded that leaf litter decomposition processes were relatively resistant to mining disturbance. Data from the Alberta mine spoils support his conclusion. It should be emphasized that the decay of slender wheatgrass was followed over a short term and that once the decomposing litter is incorporated into the soil, mineralization of the remaining nutrients could be more influenced by soil amendation and spoil type.

The equal rate of grass litter decay regardless of treatment may have long-term consequences in that plant growth stimulated by nutrient additions may result in the accumulation of a thick litter mat which acts as a nutrient sink, thereby reducing plant growth and perhaps plant establishment. Other investigators (Marrs et al., 1980; Schaefer et al.,

1980; Stroo and Jencks, 1982) have commented on the development of undecomposed litter mats on mine spoils and have suggested that management techniques such as controlled grazing, burning, or harvesting may be required to accelerate decomposition and nutrient (particularly nitrogen) cycling. Sheep grazing has been shown to be an effective means of reducing standing dead plant material and increasing the rate of nitrogen cycling on reclaimed china clay wastes in England (Marrs et al., 1980). The blockages in nutrient cycling which could result from slow decomposition rates emphasize the importance of taking into account decay rates (both long and short term) when planning a reclamation strategy.

In the slender wheatgrass plots, slender wheatgrass leaves decayed most rapidly, followed by the stems and lastly the filter paper (Table 11). As mentioned previously, natural plant litter is nutritionally superior to cellulose, hence microbial growth and turnover would be expected to be more active in litter than in pure cellulosic materials. Comparison of the three study materials demonstrated that the decay rate of filter paper could not be extrapolated to predict decay rates of natural litter. However, decomposition of buried cellulose filters may be a good indicator of the levels of microbially-available nutrients in amended mine spoils, since the degradation of pure cellulose is probably highly dependent on nutrient availability.

A three-way analysis of variance, applied to the slender wheatgrass and alsike clover weight remaining values (Table 11), showed that alsike clover litter (leaves plus stems) was mineralized more rapidly than grass litter. Differences in decay rates between the two litter types may have been due to differences in resource quality as clover litter contained more nitrogen (3.5%) than grass litter (2.3%). Litter nitrogen content is believed to be a major factor in regulating decay rates on mine spoils (Carrel et al., 1979). In contrast to the clover litter, the sainfoin litter (2.9% nitrogen) placed on the tailings sand plots decayed more slowly than the grass litter, suggesting that factors other than nitrogen content were important in controlling the decay of this particular legume litter. Of the litter types tested on the Alberta mine spoils, alsike clover was observed to decay the fastest, followed by slender wheatgrass (particularly leaves), and lastly sainfoin. The short-term weight loss of the alsike clover and slender wheatgrass litter demonstrated that if a mixture of these two plant species was used to revegetate a mine spoil, nutrient release from the above-ground litter would be expected to occur at three different rates (alsike clover leaves would mineralize most rapidly, followed by grass leaves and clover stems and lastly grass stems). This pattern of nutrient release may be an advantage on mine wastes in the early stages of recovery, where rapidly decaying litter could provide a continuous supply of nutrients required to maintain primary production, whereas the slowly decaying litter would contribute to the accumulation of stable organic matter. Since legume shoots contain more nitrogen than grass shoots,

Table 11 Short-Term (12 month) Decay of Various Litter Types and Cellulose Filter Paper Placed in Subalpine and Oil Sands Tailings Plots 28 Months After Amendment and Planting

		Decomposition (% dwt remaining)			
Mine spoil	Plant and litter type	Control	Fertilizer	Sewage sludge	Peat
Subalpine	Grass	44 (5)	47 (6)	48 (5)	47 (3)
	leaves stems	77 (2)	78 (3)	73 (6)	78 (3)
	Clover	29 (2)	28 (2)	33 (2)	33 (3)
	leaves stems	47 (5)	45 (3)	48 (3)	54 (6)
	Filter paper	97 (3)	92 (10)	84 (11)	87 (15)
Oil sand tailings	Grass	46 (9)	50 (25)	46 (8)	44 (18)
	leaves stems	81 (11)	75 (10)	66 (6)	72 (7)
	Sainfoin	58 (15)	59 (11)	60 (4)	51 (7)
	leaves stems	61 (3)	67 (9)	57 (2)	62 (7)
	Filter paper	84 (14)	38 (20)	70 (23)	75 (26)

Each litter type was placed in plots of its respective plant type. Each value is a mean of 6 replicates. Standard deviations presented in brackets.

legume litter would ultimately be expected to contribute more nitrogen to the soil nutrient pool. This would be particularly important in mine spoils where nitrogen can be the nutrient most limiting to plant growth. Lanning and Williams (1979) compared the decomposition rates of grass and clover in china clay sand wastes and, as in this study, found that clover decomposed more rapidly than grass. They also observed that nitrogen was released only from the decaying clover, but that this nitrogen did not accumulate in the sand waste. The loss of nitrogen from the sand wastes was attributed to rapid leaching of the mineralized nitrogen, loss of shoot material through wind and rain erosion, and volatilization of nitrogen as ammonia at a pH above 6.5 (Lanning and Williams, 1980b). The research of Lanning and Williams emphasizes the importance of gaining insight into the ultimate fate of nitrogen released from decaying plant material on mine spoils. Severe leaching losses, particularly during the early stages of reclamation, are possible in sandy or coarse textured spoils (Dancer, 1975), such as the tailing sand and subalpine spoil, and could eventually lead to the failure of the established plant cover. The decomposition results in this study have conclusively shown that alsike clover litter decay and presumably nitrogen release are significantly faster than that of grass litter. However, the beneficial aspects of more rapid nutrient release from this legume (i.e., faster return of nitrogen to the primary producers) would be lost if a reclamation strategy did not take into account the potential mechanisms for nutrient loss and include measures for managing this loss. Marrs and Bradshaw (1982) discussed in detail the significance of nitrogen accumulation and cycling in the development of a self-sustaining vegetative cover on nutrient-deficient mine wastes.

IV. CONCLUSIONS

The final goal of any reclamation program is the establishment of a self-sustaining vegetative cover that requires a minimum of maintenance. Successful establishment of the vegetation is dependent on climate, mine spoil topography, and chemical, physical, and biological properties of the mine wastes. Although much information relating chemical and physical characteristics of mine wastes to primary productivity is available, only recently has emphasis been given to the biological properties of these materials. Many reclamation strategies appear to have been developed on the assumption that vegetation establishment will occur without direct concern for soil microbiological processes that are necessary to improve soil fertility, maintain plant growth, and accelerate stable organic matter accumulation.

A wide variety of habitats are disturbed by surface mining for coal and oil; however, the effects of this disturbance on soil microbiological properties are very similar. In all cases, the surface soil horizons, which are often rich in organic matter and therefore harbor a highly diverse and active microbial community, are disrupted causing a drastic alteration of the bacterial and fungal populations, with a concomitant decrease in microbial activity and biomass. To raise the microbial activity levels in mine spoils, it is necessary to increase the level and input rates of soil organic matter. This is best achieved by applying an organic amendment, encouraging plant production by fertilization, or both. Plants exhibiting vigorous growth and high short-term productivity (e.g., grasses, legumes) are commonly used for initial mine waste revegetation since they are highly effective as erosion control mechanisms. However, unless these plant systems are grazed or harvested as in an agricultural system, a thick litter mat may develop. An additional point of interest from a long-term management standpoint is a very strong indication that although primary production may be stimulated by fertilization, microbial processes, including decomposition are not stimulated to the same degree. As a result, undecomposed plant residues may accumulate, creating a slow-turnover nutrient sink which can substantially reduce plant growth. Since plant and microbial productivity on mine spoils appear to be closely linked, a reduction in plant productivity could eventually lead to decreases in microbial activity to levels observed prior to revegetation. Decomposition rates and subsequent nutrient cycling processes will ultimately determine the success and maintenance of a particular plant community. For this reason the decomposition rates of the various litter types produced on a revegetated mine spoil should receive careful study when planning a revegetation program.

ACKNOWLEDGMENTS

Financial support for this study was provided by the Department of the Environment, Alberta, and the Natural Sciences and Engineering Research Council of Canada. The chemical analysis of the soil samples by Alberta Environment, Lethbridge, Alberta, is greatly appreciated. Expert statistical analysis was provided by C. L. Griffiths. This study would not have been possible without the support and assistance provided by R. M. Danielson, D. Parkinson, and J. C. Zak. A very special thanks to R. Danielson and D. Parkinson for critically reviewing the manuscript.

REFERENCES

Anderson, D. W. (1977). Early stages of soil formation on glacial till mine spoils in a semi-arid climate. *Geoderma 19*: 11--19.

Anderson, J. P. E. and Domsch, K. H. (1978). A physiological method for the quantitative measurement of microbial biomass in soils. *Soil Biol. Biochem. 10*: 215–221.

Ausmus, B. S., Edwards, N. T., and Witkamp, M. (1976). Microbial immobilization of carbon, nitrogen, phosphorus and potassium: Implications for forest ecosystem processes. In *The Role of Terrestrial and Aquatic Organisms in Decomposition Processes*, J. M. Anderson and A. Macfadyen, (Ed.). Blackwell Scientific Publications, London, pp. 397–416.

Barber, D. A., and Martin, J. K. (1976). The release of organic substances by cereal roots into soil. *New Phytol. 76*: 69–80.

Brady, N. C. (1974). *The Nature and Properties of Soils.* Macmillan, New York, 639 pp.

Bunnell, F. L., Tait, D. E. N., Flanagan, P. W., and van Cleve, K. (1977). Microbial respiration and substrate weight loss. I. A. general model of the influences of abiotic variables. *Soil Biol. Biochem. 9*: 33–40.

Carrel, J. E., Wieder, K., Leftwich, V., Weems, S., Kucera, C. L., Bouchard, L., and Game, M. (1979). Strip mine reclamation: production and decomposition of plant litter. In *Ecology and Coal Resource Development*, M. K. Wali (Ed.). Pergamon Press, New York, Vol. 2, pp. 670–676.

Christensen, P. J. (1977). The history, biology and taxonomy of the *Cytophaga* group. *Can. J. Microbiol. 23*: 1599–1653.

Cresswell, C. F. (1973). Changes in vegetational composition, nutritional status, soils and microbial populations with the establishment of vegetation on gold-mine dumps on the Witwatersrand. In *Ecology and Reclamation of Devastated Land,* R. J. Hutnik and G. Davis (Ed.). Gordon and Breach, Science Publ., New York, Vol. 2, pp. 335--358.

Cundell, A. M. (1977). The role of microorganisms in the revegetation of strip-mined land in the western United States. *J. Range Manag. 30*: 299–305.

Dancer, W. S. (1975). Leaching losses of ammonium and nitrate in the reclamation of sand spoils in Cornwall. *J. Environ. Qual. 4*: 499–504.

Domsch, K. H., Gams, W., Anderson, T.-H. (1980). *Compendium of Soil Fungi.* Academic Press, London, Vol. 1, pp. 540–611.

Eiland, F. (1981). The effects of application of sewage sludge on microorganisms in soil. Danish J. Plant and Soil Science, Report No. 1534; pp. 39–46.

Fresquez, P. R., and Lindemann, W. C. (1982). Soil and rhizosphere microorganisms in amended coal mine spoils. *Soil Sci. Soc. Am. J. 46*: 751–755.

Hedrick, H. G., and Wilson, H. A. (1956). The rate of carbon dioxide production in a strip mine spoil. *Proc. West Virginia Acad. Sci. 28*: 11--15.

Hersman, L. E., and Klein, D. A. (1979). Retorted oil shale effects on soil microbiological characteristics. *J. Environ. Qual. 8*: 520–524.

Hersman, L. E., and Temple, K. L. (1978). ATP as a parameter for characterizing strip mine spoils. *Soil Sci.* *126*: 350–352.
Hersman, L. E., and Temple, K. L. (1979). Comparison of ATP, phosphatase, pectinolyase, and respiration as indicators of microbial activity in reclaimed coal strip mine spoils. *Soil Sci.* *127*: 70–73.
Hersman, L. E., Molitoris, E., and Klein, D. A. (1981). Nitrogen fixation by legumes in retorted shale. *Soil Biol. Biochem.* *13*: 429–430.
Hons, F. M., and Hossner, L. R. (1980). Soil nitrogen relationships in spoil material generated by the surface mining of lignite coal. *Soil Sci.* *129*: 222–228.
Jones, P. C. T., and Mollison, J. E. (1948). A technique for the quantitative estimation of soil micro-organisms. *J. Gen. Microbiol.* *2*: 54–69.
Jurgensen, M. F. (1978). Microorganisms and the reclamation of mine wastes. In *Forest Soils and Land Use*, C. T. Youngberg (Ed.). Colorado State University, Colorado, pp. 251–286.
Langkamp, P. J., Swinden, L. B., and Dalling, M. J. (1979). Nitrogen fixation (acetylene reduction) by *Acacia holosericea* on areas restored after mining at Groote Eylandt, Northern Territory. *Aust. J. Bot.* *27*: 353–361.
Lanning, S., and Williams, S. T. (1979). Nitrogen in revegetated china clay sand waste—Part 1: Decomposition of plant material. *Environ. Pollut. (Ser. A)* *20*: 149–161.
Lanning, S., and Williams, S. T. (1980a). Nitrogen in revegetated china clay sand waste—Part 2: The effect of reduced leaching on decomposition and revegetation. *Environ. Pollut. (Ser. A)* *21*: 23–33.
Lanning, S., and Williams, S. T. (1980b). Nitrogen in revegetated china clay sand waste—Part 3: The use of clover in revegetation. *Environ. Pollut. (Ser. A)* *21*: 89–95.
Lawrey, J. D. (1977a). The relative decomposition potential of habitats variously affected by surface coal mining. *Can. J. Bot.* *55*: 1544–1552.
Lawrey, J. D. (1977b). Soil fungal populations and soil respiration in habitats variously influenced by coal strip-mining. *Environ. Pollut.* *14*: 195–205.
Lodhi, M. A. K. (1979). Inhibition of nitrifying bacteria, nitrification and mineralization in spoil soils as related to the successional stages. *Bull. Torrey Bot. Club* *106*: 284–289.
Macgregor, A. N., Naylor, L. M. (1982). Effect of municipal sludge on the respiratory activity of a cropland soil. *Plant Soil* *65*: 149–152.
Malik, K. A., and Azam, F. (1980). Effect of salinity on ^{14}C-labelled microbial biomass and its contribution to soil organic matter. *Pak. J. Bot.* *12*: 117–127.

Marrs, R. H., Granlund, I. H., and Bradshaw, A. D. (1980). Ecosystem development on reclaimed china clay wastes. IV. Recycling of above-ground plant nutrients. *J. Appl. Ecol. 17*: 803–813.
Marrs, R. H., and Bradshaw, A. D. (1982). Nitrogen accumulation, cycling and the reclamation of china clay wastes. *J. Environ. Manag. 15*: 139–157.
McKeague, J. A. (1976). Manual on Soil Sampling and Methods of Analysis. Prepared by Subcommittee of Canada Soil Survey Committee on Methods of Analysis, Agriculture Canada, Ottawa, 212 pp.
Minderman, G. (1968). Addition, decomposition and accumulation of organic matter in forests. *J. Ecol. 56*: 355--362.
Moore, T. R. (1981). Controls on the decomposition of organic matter in subarctic spruce-lichen woodland soils. *Soil Sci. 131*: 107–113.
Müller, K. (1973). The microflora of dumped soils in two opencast brown-coal mining regions of Poland. In *Ecology and Reclamation of Devastated Land,* R. J. Hutnik and G. Davis (Ed.). Gordon and Breach, Science Publishers, New York, Vol. 1. pp. 325–334.
Nelson, L. M., and Visser, S. (1978). Effect of spring thaw on microorganisms in an arctic meadow site. *Arctic Alpine Res. 10*: 679–688.
Nelson, L. M. (1978). Effect of temperature, growth rate, and nutrient limitation on the yield and composition of three bacterial isolates from an arctic soil grown in continuous culture. *Can. J. Microbiol. 24*: 1452–1459.
Nelson, L. M., and Parkinson, D. (1978). Effect of freezing and thawing on survival of three bacterial isolates from arctic soil. *Can. J. Microbiol. 24*: 1468–1474.
Olson, G. J., Turbak, S. C., and McFeters, G. A. (1979). Impact of western coal mining—II. Microbiological Studies. *Water Res. 13*: 1033–1041.
Olson, G. J., McFeters, G. A., and Temple, K. L. (1981). Occurrence and activity of iron and sulphur-oxidizing microorganisms in alkaline coal strip mine spoils. *Microb. Ecol. 7*: 39–50.
Park, D. (1976). Carbon and nitrogen levels as factors influencing fungal decomposers. In *The Role of Terrestrial and Aquatic Organisms in Decomposition Processes,* J. M. Anderson and A. Macfadyen (Ed.). Blackwell Scientific Publications, Oxford, pp. 41–59.
Parkinson, D. (1979). Microbes, mycorrhizae and minespoil. In *Ecology and Coal Resource Development,* M. K. Wali (Ed.). Pergamon Press, New York, pp. 634–642.
Parkinson, D., Visser, S., Danielson, R. M., and Zak, J. (1980). Restoration of fungal activity in tailing sand (oil sands). In *Soil Biology as Related to Land Use Practices. Proceedings of the VII International Colloquium of Soil Zoology,* D. L. Dindal (Ed.). Office of Pesticide and Toxic Substances, EPA, Washington, D. C., pp. 362–370.

Paul, E. A., and Johnson, R. L. (1977). Microscopic counting and adenosine 5'-triphosphate measurement in determining microbial growth in soils. *Appl. Environ. Microbiol. 34*: 263–269.

Paul, E. A., and Voroney, R. P. (1980). Nutrient and energy flows through soil microbial biomass. In *Contemporary Microbial Ecology*, D. C. Ellwood, J. N. Hedger, M. J. Latham, J. M. Lynch, and J. H. Slater (Ed.). Academic Press, New York, pp. 215–237.

Power, J. F., Ries, R. E., and Sandoval, F. M. (1978). Reclamation of coal-mined land in the Northern Great Plains. *J. Soil Water Conserv. 33*: 69–74.

Ratliff, R. D. (1980). Decomposition of native herbage and filter paper at five meadow sites in Sequoia National Park, California. *J. Range Manag. 33*: 262–266.

Ross, D. J., Molloy, L. F., Bridger, B. A., and Cairns, A. (1978). Studies on a climosequence of soils in tussock grasslands. 20. Decomposition of cellulose on the soil surface and in the topsoil. *N. Z. J. Sci. 21*: 459–465.

Russell, R. S. (1977). *Plant Root Systems: Their function and interaction with the soil.* McGraw-Hill, London, 298 pp.

Schafer, W. M., Nielsen, G. A., Dollkopf, D. J., and Temple, K. (1979). Soil Genesis, Hydrological Properties, Root Characteristics and Microbial Activity of 1- to 50-year old Stripmine Spoils. Interagency Energy/Environment R & D Program Report, USDA, EPA-600/7-79-100, 212 pp.

Schafer, W. M., Nielsen, G. A., and Nettleton, W. D. (1980). Minesoil genesis and morphology in a spoil chronosequence in Montana. *Soil Sci. Soc. Am. J. 44*: 802–807.

Skeffington, R. A., and Bradshaw, A. D. (1980). Nitrogen fixation by plants grown on reclaimed china clay waste. *J. Appl. Ecol. 17*: 469–477.

Smith, S. B. (1977). Repairing Disturbed Environments. Proceedings. Coal Industry Reclamation Symposium. The Coal Association of Canada, Calgary, Alta, pp. 92--97.

Stroo, H. F., and Jencks, E. M. (1982). Enzyme activity and respiration in minesoils. *Soil Sci. Soc. Am. J. 46*: 548–553.

Visser, S., Griffiths, C. L., and Parkinson, D. (1983). Effects of mining on the microbiology of a prairie site in Alberta, Canada. *Can. J. Soil Sci. 63*: 177–189.

Wainwright, M. (1982). Origin of fungal colonies on dilution and soil plates determined using nonanoic acid. *Trans. Br. Mycol. Soc. 79*: 178–179.

Walkley, A., and Black, I. A. (1934). An examination of the Degtjareff method for determining soil organic matter, and a proposed modification of the chromic acid titration method. *Soil Sci. 37*: 29–38.

Warcup, J. H. (1955). Isolation of fungi from hyphae present in soil. *Nature* London *175*: 953–954.

Williams, P. J., and Cooper, J. E. (1976). Nitrogen mineralization and nitrification in amended colliery spoils. *J. Appl. Ecol. 13*: 533–543.

Williams, S. T., McNeilly, T., and Wellington, E. M. H. (1977). The decomposition of vegetation growing on metal mine waste. *Soil Biol. Biochem. 9*: 271–275.

Wilson, H. A. (1965). The Microbiology of Strip-mine Spoil. West Virginia University Agricultural Experiment Station, Bulletin 506T, 44 pp.

Wilson, H. A., and Hedrick, H. G. (1957). Carbon dioxide evolution from some strip mine spoils. *Appl. Microbiol. 5*: 17–21.

Wilson, H. A., and Stewart, G. (1956). The number of bacteria, fungi, and actinomycetes in some strip-mine spoil. West Virginia University Agricultural Experiment Station, Bulletin 388T, 15 pp.

8

Microbial Technology for Aggregation and Dewatering of Phosphate Clay Slimes: Implications on Resource Recovery

CORALE L. BRIERLEY* / *New Mexico Bureau of Mines and Mineral Resources, New Mexico Institute of Mining and Technology, Socorro, New Mexico*

GUY R. LANZA / *University of Texas at Dallas, Richardson, Texas*

I. INTRODUCTION

Phosphate operations in Florida account for three-fourths of the phosphate production in the United States and are major contributors to world phosphate supplies (Terichow and May, 1973). In 1927 introduction of the flotation process for phosphate recovery revolutionized the industry but created a massive phosphate clay disposal problem. The slimes, or waste-clay suspensions, possess colloidal characteristics that make them difficult to dewater (Smelley and Feld, 1979). The fine particulate clays from the beneficiation process are discharged at a rate of 3.6 million metric tons per year per plant in an aqueous slurry of about 5% solids. These clay-wastes are stored in dams for settling; natural settling usually requires more than a decade to accomplish. These storage dams are an environmental hazard and an economic liability. Many laboratory techniques are available for dewatering of the slimes; however, none of these techniques is well-enough engineered to be economically practicable when applied in the field (Terichow and May, 1973; Lamont et al., 1975; Staff, 1975; Smelley and Feld, 1979). The primary settling problem with phosphatic slimes is attributed to attapulgite, a fibrous clay mineral with a cardboard-like structure.

This chapter reviews the mining and beneficiation processes applied to phosphate recovery, examines the nature of the clays associated with the Florida phosphate deposits, and discusses the use of biopolymers and microorganisms in flocculating and aggregating phosphatic clays.

**Current affiliation*: Advanced Mineral Technologies, Inc., Socorro, New Mexico

The clays associated with the phosphates of the central Florida district are unique among the world phosphate deposits, as they possess qualities that render them difficult to dewater and flocculate.

II. MINING AND BENEFICIATION PROCESSES FOR PHOSPHATE

Phosphate mining in central Florida is located in the land-pebble area near the towns of Lakeland, Bartow, Bradley Junction, and Fort Meade. The phosphate ore, called matrix, is composed of equal portions of quartz, sand, clays, and phosphate rock. The matrix is sometimes covered by 3.0–6.1 m of a "leached zone" and is nearly always covered by overburden (rock material overlying the deposit). The matrix is surface mined to depths of 15.2–18.3 m using large draglines. At the mine site, the phosphate ore, or matrix, is slurried with water and pumped several kilometers to the beneficiation plant where the phosphate is recovered. There are about 15 phosphate mining operations in the central Florida area, and all operations use essentially the same generalized mining and beneficiation processes (Lamont et al., 1975). The matrix received for beneficiation is reduced to 1.9 cm in size and the coarse phosphate rock (pebble phosphate) is separated from the sand, clays (called slimes), and fine phosphate using screens and "log washers" for washing the matrix. The material that is sized less than 1 mm is deslimed, i.e., the clays are removed by cyclones and hydroseparators. The sand and fine phosphate material are further processed by several flotation circuits for recovery of a phosphate concentrate.

Two phosphate products, pebble phosphate and phosphate concentrate, and two waste products, minus 1 mm plus 130-mesh (< 1 mm and > 0.1 mm) flotation tailings sand and minus 150-mesh (< 0.1 mm) clay-slimes, are generated from this processing. The tailings sand is effectively disposed of by backfilling into the open pit mining cuts, which can be revegetated. The phosphatic clays (slimes) that make up one-third of the matrix represent a major materials handling and ecological problem. The slimes are discharged from the beneficiation plants at a concentration of 2–6% suspended solids and pumped to settling basins excavated from mined-out areas (Lamont et al., 1975; Staff, 1975). The slow settling rates of the colloidal-like clay particles and the attendant lack of consolidation of the material necessitate long retention times in storage impoundments before the supernatant water can be reused for slurrying of the matrix. The above-ground storage impoundments are sometimes constructed to heights of 12.2–15.2 m above ground level and are nearly a 100 ha in size (Lamont et al., 1975). The slimes also represent a potential resource as nearly one-third of the phosphate in the matrix is lost to the slimes during the beneficiation process. Slimes stored through 1960 have an average analysis of 14% P_2O_5 representing 41,214,120 metric tons of P_2O_5 (Gary et al., 1963). The phosphate industry uses P_2O_5 to denote phosphate content of ores, products, and waste materials.

One major contemporary issue facing the phosphate industry in central Florida is the disposition of above ground-storage of clay wastes. Although major dam breakage is a rare occurrence, several severe breaks occurred in the mid-1960s and early 1970s. The damage to aquatic plants and animals was extensive. In response to these events, the industry now invests heavily in monitoring the dams and in better construction of the earthen structures (Blakey, 1973).

Because of the ecological and economical problems facing the phosphate industry, much research has been completed on physical, chemical, and mineralogical properties of the clays derived from the phosphate, schemes for economical dewatering of the slimes, and processes for recovery of phosphate values retained in the slimes during beneficiation. To date, no economical solution to the problem of slow settling rates and inability to consolidate the clays has been demonstrated. The development of microbial processes for flocculation and/or aggregation of clays generated by mining and beneficiation activities will be examined in this review.

III. CHARACTERIZATION OF PHOSPHATIC CLAYS

The slow settling characteristics of phosphatic clay-slimes are attributed to the attapulgite clay component, and water absorption by phosphatic clay-slimes is characteristic of montmorillonite. The nature of the slimes differs among mining operations, and considerable variation can exist in slimes from a single operation as mining progresses. The following discussion describes physical, chemical and mineralogical characteristics of selected samples of phosphatic clays.

A. Mineralogical Description

X-ray diffraction analyses of phosphatic slime samples indicate the presence of montmorillonite, kaolinite, attapulgite, illite, quartz, dolomite, wavellite, crandallite, and apatite. Most of the apatite is presumed to be fluorapatite; interstratified clay minerals and chlorites have not been detected (Gary et al., 1963; Lamont et al., 1975). Table 1 shows the range of net peak intensities for the nine clay minerals in 15 phosphatic slime samples. Attapulgite is found predominantly in the finest fraction (< 0.2 μm) of the slime sample.

Because attapulgite is believed to be the factor controlling settling rate, terminal solids (percent solids in precipitate), solids content of the filter cake (residue remaining after pressure filtration), flocculation, and viscosity (Lamont et al., 1975), the structure and characteristics of this clay are described. Attapulgite has a fibrous texture and a chain structure. Electron photomicrographs of attapulgite indicate

Table 1 Range of Net Peak Intensities for Clay Minerals in the <45-μm Size Fraction of 15 Phosphatic Slime Samples[a]

Clay	Range (Counts per second)
Montmorillonite	120–3200
Kaolinite	ND[b]–680
Attapulgite	ND–1270
Illite	ND–115
Apatite	180[c]–630
Quartz	160–830
Dolomite	ND–600
Wavellite	ND–190[c]
Crandallite	ND–180

[a]Values were obtained from the scan of the oriented glycerated sample except for wavellite, for which the oriented sample was used. The lines measured were as follows: montmorillonite (Montmorillonite-glycerol), about 18 Å: kaolinite, about 7 Å; attapulgite, about 10.6 Å; illite, about 10 Å; apatite (second strongest line), about 2.70 Å; quartz, about 3.34 Å; dolomite, about 2.89 Å; wavellite, about 8.7 Å; and crandallite, about 2.94 Å.
[b]ND, Not detectable.
[c]The diffraction line was either not completely resolved from a nearby line or was in a high background region, resulting in some inexactness of the value.
Source: Data from Lamont et al. (1975).

that this clay mineral is a fibrous, lath-like material. The thickness of the attapulgite varies from 0.011 to 0.046 micrometers (Gary et al., 1963); the length of the attapulgite laths may be many micrometers (Grim, 1968). The chemical formulation of attapulgite is proposed as $Mg_5Si_4O_{20}(OH)_2(OH_2)_4 \cdot 4H_2O$. Double silica chains run parallel to the *C* axis of the unit cell; these chains are linked through oxygen. The cavities between the chains will accommodate four molecules of water per unit cell. These four waters account for the dehydration observed when the clay is subjected to low temperatures. At the edges of the structure, hydroxyl groups are exposed; these are neutralized by protons, giving rise to bound water designated as (OH_2) in the formulation. Substitution for either the Mg(II) or the Si(IV) likely occurs in the attapulgite structure; however, the substitution of Al(III) for either the Si(IV) or Mg(II) weakens the structure of this clay. Ca(II), Na(I), and K(I) are also reported in attapulgite; these probably represent exchangeable cations (Caillere and Henin, 1972; Weaver and Pollard, 1973).

All clay minerals exhibit a cation exchange capacity (CEC). This results from (1) substitution of ions in the lattice structure of the clay, (2) broken bonds in the mineral, (3) conditions within the layers of the clay, and (4) conditions on the clay surface. CEC values reported in the literature for kaolinite and attapulgite range from 3 to 15 mEq/100 g and for montmorillonite from 80 to 150 mEq/100 g. The CEC values of Florida phosphatic clay-slimes are 17–45 mEq/100 g of slime solids; these values probably result from a mixture of attapulgite, kaolinite, and montmorillonite.

B. Chemical Composition

Extreme variations in chemical composition exist among phosphatic slime samples. Table 2 shows the average results of chemical analyses performed on 15 as-received (no size fractionation performed) slime solids. Table 3 lists the percent weight loss on ignition at four different temperatures. The weight loss is an indication of water bound by the clays. Some general characteristics of a clay-slime are tabulated in Table 4.

Table 2 Average Results of Chemical Analyses on 15 Phosphatic Slime Solids

Chemical analysis	Range %	Average %
CaO	8.3–26.9	16.2
P_2O_5	5.1–19.7	12.0
MgO	0.8–6.7	2.9
Al_2O_3	7.6–16.8	11.9
Fe_2O_3	1.6–8.3	3.4
SiO_2	23.1–49.1	36.5
Na_2O	0.1–1.2	0.3
K_2O	0.2–1.5	0.8
F	0.7–2.3	1.5
C	0.4–1.9	1.0
CO_2	0.7–5.3	2.3
S	0.1–0.4	0.2

Source: Data from Lamont et al. (1975).

Table 3 Results of Ignition Tests on Phosphatic Slime Solids

Ignition temperature (°C)	Weight loss (%)
105	3.5
250	6.0
500	10.3
1000	14.5

Source: Data from Lamont et al. (1975).

Table 5 illustrates the average chemical analyses of 15 different supernatant waters derived from slime samples collected at Florida phosphate operations.

C. Physical Characteristics

1. *Particle Size and Area*

Table 6 depicts the average percents of particles of various size distributions in 15 phosphatic slime samples; results are expressed in spherical diameters. The largest percentage of particles is in the smallest size range. The surface area of as-received phosphatic slimes ranges from 27.4 to 88.4 m^2/g with an average of 46.5 m^2/g (Lamont, et al., 1975).

2. *Sedimentation Characteristics*

Long-term sedimentation studies completed in 1000-ml glass soil-testing cylinders indicate that phosphatic slimes settle to about 84% terminal solids in 30 days. The data from this study indicate that the

Table 4 Some Characteristics of Phosphate Slime

Characteristic	Value
Solids content	9.1%
pH	6.55
Specific conductance	928 μmhos/cm
Color	Khaki
Specific viscosity	
Gravity	2.714 cp
Fann	14 cp
Brookfield	210 cp

Source: Data from Lamont et al. (1975)

Table 5 Average Chemical Analysis of Supernatent Waters from Phosphatic Slime Samples

Analysis	mg/liter[a]
Ca	57
Mg	22
Na	18
K	1.3
Fe (total recoverable)	0.119
Mn (total recoverable)	0.030
Al (total recoverable)	0.185
HCO_3^-	112
CO_3^{2-}	0
SO_4^{2-}	144
Cl^-	17
F^-	2.0
NO_2^- (as total N)	0.005
NO_3^-	1.06
Phosphate as HPO_4	ID
Phosphate as H_2PO_4	ID
Phosphorus, total as P	0.273
Phosphorus, total as ortho P	0.229
Silica as SiO_2	5.4
Dissolved solids	348
Total hardness as $CaCl_2$	234
Noncarbonate hardness as $CaCO_3$	142
Alkalinity as $CaCO_3$	92
pH	7.8
Specific conductance (μmhos/cm at 25 C)	523
Turbidity (Jackson Turbidity Units)	12

[a]ID, Insufficient data for averaging.
Source: Data from Lamont et al. (1975).

Table 6 Average Size Distribution of Phosphatic Slime

Sample size[a]	Distribution (%)
+104 μm	3.4
−104 μm + 43 μm	8.4
−43 μm + 20 μm	7.0
−20 μm + 10 μm	6.1
−10 μm + 5 μm	11.5
−5 μm + 1 μm	11.4
−1 μm + 0.5 μm	7.5
−0.5 μm + 0.2 μm	13.0
−0.2 μm	31.7

[a]+, Retained by sieve of designated size; −, passed through sieve of designated size.
Source: Data from Lamont et al. (1975).

settling rate is a function of terminal solids: the lower the terminal solids, the slower the settling rate. A comparison between the percent terminal solids and X-ray diffraction data indicates that, as the percent terminal solids decreases, the fraction of attapulgite in the sample increases (Lamont et al., 1975).

3. *Rheology*

Rheology is the study of the flow behavior or viscosity of a substance. An inverse relationship between the settling characteristics and viscosity of phosphatic slimes is thought to exist. Samples having a low viscosity at 3.7% solids settle the best, whereas slime samples with high viscosity at 3.7% solids settle poorly (Lamont et al., 1975).

4. *Filtration Qualities*

Filtration studies suggest that samples with poor settling characteristics also produce filter cakes (residue remaining after pressure filtration) of the lowest solids content. When comparing filtration data with attapulgite content, it is found that in samples with high attapulgite content the percent solids in the filter cake from pressure filtration decrease (Lamont et al., 1975).

5. *Flocculation Characteristics*

A screening test using anionic, cationic, and nonionic polyacrylamide flocculants on phosphatic clay-slime samples showed that an anionic polyacrylamide with a molecular weight of about 3 million is most effective in settling the clay-slime. Further tests were conducted in which increments of polyacrylamide were added; these tests indicate that clay-slime samples having the worst natural settling characteristics require the most flocculant for positive flocculating results. The solids content of the settled fraction after flocculation is a function of the solids content of the slimes without the flocculating agent (Lamont et al., 1975).

IV. COLLOIDAL AND SURFACE CHEMISTRY OF CLAYS

A. Colloidal Stability

Stumm and Morgan (1962) concluded that most colloids are dependent on chemical equilibria for stability. Acid-base, complex formation, solubility, and ion-exchange equilibria are involved. Organics and biological materials may become adsorbed on particulate matter such as clays and impart hydrophilic properties to these particulates. Many hydrophilic colloids are electrolytic (Stumm and Morgan, 1962); the charge generally arises from ionization of carboxylic, aliphatic, aromatic, hydroxyl, sulfate, phosphate, and amino groups attached to the particles. The addition of polyvalent cations to hydrophilic colloids results in fixation of the cations onto the negatively charged groups of the colloid and can lead to reversal of the charge of the colloid. The fixation may be an electrostatic or a chemical interaction (Stumm and Morgan, 1962).

Black et al. (1966) postulated that stability of a colloidal clay suspension can be attributed to two phenomena: (1) the force of repulsion between two similarly charged clay surfaces, and (2) the degree of hydration of the clay surfaces, which physically hinders the distance of closest approach between clay particle surfaces.

B. Electrical Nature of Colloidal Clays

Clays are believed to possess two oppositely charged double layers. One is associated with the clay faces; the other is associated with the edges of the plates. The positive edge results from exposure of lattice aluminum atoms at these locations. Therefore, the edge of one clay particle could be attracted to the face of another particle (Black et al., 1966; Marshall, 1969). Although they have a relatively short-range effect, van der Waal's forces are thought to be of considerable importance in attraction, since attraction between atom pairs is additive (Black et al., 1966).

Clays can obtain a net negative charge because of isomorphous substitutions of electropositive elements making up the clay by elements of a lower valence. The substitutive elements may be adsorbed by chemisorption or physical adsorption such as hydrogen bonding or van der Waal's forces. Charge may also be acquired by ionization of material adsorbed to a surface; this ionization is pH dependent. However, the entire system must remain electrically neutral, as counter-or gega-ions of the opposite charge (positive) distribute themselves around the clays. The distribution of these counter-ions around a charged surface constitutes the electrical double layer (Black et al., 1966; Shaw, 1966; Terichow and May, 1973). The closest layer to the surface is designated the Helmholtz layer. Some ions on the surface may attract ions of the opposite charge so strongly that an adsorbed dense layer results. This dense layer may partially neutralize the charge on the surface so that electrostatic attraction between the particles and ions in solution is diminished. A diffuse layer of ions is formed at a distance away from the particle surface; this diffuse double layer is referred to as the Stern double layer (Black, 1966; Shaw, 1966; Terichow and May, 1973).

In proposing the concept of a diffuse double layer, Stern also suggested that a mobility difference exists between the outer layer of adsorbed ions and the surface. This mobility difference creates a shear plane between the two layers. The potential related to this shear plane can be measured, and this potential can be considered a measure of the stability of the colloid. Electrophoretic mobility is a direct measure of the movement of charged particles relative to an applied gradient (Shaw, 1966; Terichow and May, 1973). The actual potential on particles is known as the zeta potential; it is calculated from the electrophoretic mobility (May, 1975).

Factors that compact the electrical double layer are an increase in the electrolyte concentration and an increase in the valency of the gega-ions. The valency of the co-ions (the ions affecting the charge on the colloidal particle) does not influence the double layer. The measured electrical potential will be highest near the charged surface and will decrease with distance at a greater than exponential rate (Shaw, 1966; Van Olphen, 1977). The repulsive forces that exist between particles surrounded by double layers of the same charge create colloidal stability. The addition of electrolytes, particularly to concentrations at which zero potential or reversal of charge is attained, may result in coagulation of clay-slimes (Terichow and May, 1973). Polyelectrolytes produced by microorganisms may induce coagulation. The aggregation of colloidal clay suspensions occurs only beyond a certain minimum electrolyte concentration; this is called the critical salt concentration (CSC). Factors affecting the CSC are the valency of the ions added, surface coatings on the clays, the degree of aluminum and iron hydroxide formation, and the mutual aggregation of clays by positive edge/face and negative surface interaction. Very low concentrations, in the order of 1 mEq/liter, calcium chloride or magnesium chloride are required to aggregate calcium- and magnesium-saturated clays (Arora and Coleman, 1979). The addition of aluminum sulfate or sand reduces the repulsion between similarly charged particles and results in attraction and coagulation of colloidal, phosphatic clay-slimes (Terichow and May, 1973). Research by May and Smelley (1979) on phosphatic clay-slimes indicates that increasing the chloride or hydroxide concentrations of the aqueous suspending medium increases the electrophoretic mobility of the slimes. The exact mechanism of this effect has not been determined; however, maximum mobility occurs near pH 11 corresponding to minimum viscosity of the clay-slimes. Increasing the hydrogen ion content to a pH of less than 4 also reduces the electrophoretic mobility.

C. Interactions between Clays and Organic Compounds

Several interactions between organic compounds and clays can occur. One reaction involves the adsorption of organic surface-active agents to clays. The surface-active agent then alters the properties of the interface by masking electrical charges and altering surface tension (Van Olphen, 1977). The other reaction is the interaction of polyelectrolytes (long-chain molecules with ionizing groups) with clays. In water functional groups (amino groups of cationic polyelectrolytes and carboxyl, sulfate, sulfonate, and other negatively charged groups of anionic polyelectrolytes) dissociate. The repulsion of similarly charged functional groups along the polyelectrolyte chain causes the stretched configuration and high viscosity of polyelectrolytes. The viscosity of polyelectrolytes is reduced by the addition of salts, since the salts form a diffuse layer around the charged functional group. As the charges are diminished, the polyelectrolyte will assume the statistically most probable configuration, a coiled configuration. At very high salt concentrations (molar quantities), the polyelectrolyte will precipitate (Van Olphen, 1977). Hydrophobic sols, like clays, become more sensitive to aggregation in the presence of polyelectrolytes even if the clay and the polyelectrolyte are of the same charge. This increased tendency toward aggregation is termed "sensitization." Bridging (the adsorption of the polyelectrolyte onto several hydrophobic soils at the same time) occurs in the region of sensitization. Salts must be present to establish particle bridging. The salts reduce the repulsion of the double electrical layer and enhance adsorption of the polyelectrolyte. An example of a strong aggregating electrolyte is calcium sulfate. Little is known about how nonionic polyelectrolytes function with hydrophobic sols or how such polymers influence ion distribution in the double electrical layer (Van Olphen, 1977).

Polyelectrolytes produced by bacteria bind large quantities of water, and it has been proposed (Forster and Lewin, 1972) that cations can exchange with the water. The polymer becomes more hydrophobic and the flocculating capacity of the polymer is improved (Brown and Lester, 1979). Two polyelectrolytes of a polysaccharide nature have been recognized—neutral polysaccharides and anionic polysaccharides. In an aqueous system, neutral polysaccharides have little affinity for cations; however, a bonding attraction may exist between the cation and the dipole of the donor molecule. Anionic polysaccharides strongly attract counter-ions, specifically monovalent cations. The bonding has been suggested to be largely ionic with some covalent character. Factors controlling binding of cations by anionic polysaccharides are: (1) linear charge density, (2) effective nuclear charge, (3) basicity and polarizability of the anionic group, (4) the nature of the solvent, (5) inner chelation, (6) intermolecular chelation, and (7) steric fit (Brown and Lester, 1979).

In addition to the reaction between clays and polyelectrolytes, other chemical reactions that can occur between organic matter and clays include: (1) reactions with Lewis acid sites (sites that accept electron pairs; in water these sites become Brönsted sites); (2) proton donor sites (Brönsted acid sites) where acid-treated clays and clays with other cations in exchange positions will react; (3) oxidizing sites where electrons are accepted; (4) reducing sites where electrons are donated; and (5) reactive hydroxyl groups at the edges of clays (Van Olphen, 1977).

Neutral and anionic polysaccharides and surface-active agents produced by microorganisms play an important role in flocculating clays. Negatively charged clays and anionic microbial polymers interact through a "bridging" mechanism resulting in flocculation. Neutral microbial polymers may interact with cations in solution through the formation of temporary dipoles in the polymers; this temporary charging enables interaction between the polymers and clays inducing flocculation. Microbially produced surface active agents adsorb to clays and alter electrical charge and surface tension of the colloid; these changes can enhance clay flocculation.

D. Destabilization of Colloids

The terminology used in aggregation of particles varies widely among disciplines. In this chapter, flocculation is used synonymously to describe the destabilization of particles by a chemical bridging mechanism that enmeshes the particles in a three-dimensional network (Black et al., 1966). Coagulation also involves particle destabilization, but is accomplished by reduction of the repulsive potential of similar electrical double layers surrounding two particles (Black et al., 1966).

1. Coagulation

Coagulation involves the concept of an electrical double layer and the participation of predominantly physical factors in the destabilization of colloids. These factors include counter-ion adsorption, reduction of zeta potential (electrokinetic potential resulting from the opposing effects of repulsive interactions between the electrical double layers and of attraction by van der Waal's forces), and ion-pair formation. Upon the addition of electrolytes, counter-ion adsorption and compaction of the diffuse part of the double layer cause coagulation.

The process can be visualized as a charge neutralization (Stumm and Morgan, 1962).

2. *Flocculation*

Considerable interest regarding the destabilization of colloids has resulted from the use of the activated sludge process for treatment of waste water. The activiated sludge process is dependent upon two characteristics: complete substrate utilization for synthesis and activation of a microbial mass and the aggregation of microorganisms and other colloidal components to produce a high-quality effluent. Despite the long use of the activated sludge system in waste water treatment, the flocculation phase has been ill-defined with no factual description of the process being available (Pavoni et al., 1972).

Zoogloea ramigera was first isolated by Butterfield in 1935 from activated sludge. For several years, it was widely believed that this organism was singularly responsible for both the flocculation and purification of water in activated sludge systems (Campbell, 1972; Pavoni et al., 1972); however, it is now known that other microorganisms play an equally important role.

Historical Development of the Flocculation Theory: A correlation between the biological production of extracellular polysaccharide material and bacterial flocculation was first observed by McKinney (1952). He postulated that polysaccharide production by organisms serves to displace the attractive forces between similar cells, thus reducing the effective critical potential on the cells so aggregation can result (Campbell, 1972; Pavoni et al., 1972). Following McKinney's work, several papers were published pointing out the potential importance of surface tension and interfacial tension in the process of floc formation in activated sludge. It was suggested that the common affinity of bacteria and surface-active agents for interfaces may be of considerable importance in aggregation of inorganic colloidal material and bacteria in waste treatment systems (Campbell, 1972). Little thought has been given to the importance of interfacial phenomenon until recently (Marshall, 1979).

Crabtree et al. (1956), working only with *Zoogloea ramigera*, showed that this organism accumulates large amounts of poly-β-hydroxybutyric acid (PHB) during the flocculation process, and it was hypothesized that PHB accumulation and flocculation were closely associated (Campbell, 1972). However, subsequent experimentation has shown that these events are independent (Deinema and Zevenhuizen, 1971; Deinema, 1972).

The theory that has received the most support and that is now used to define the bioflocculation phenomenon was developed by Tenney and Stumm (1965). This theory explains bioflocculation in terms of polyelectrolyte interaction. Polymers of bacterial origin are capable of flocculating inorganic, colloidal dispersions having a negative surface potential. Busch and Stumm (1968) suggested that these polymers form bridges between the bacterial cell and the colloidal solid causing coflocculation of the inorganic colloids and the bacterial cells. Mechanisms for the attachment of anionic functional groups on polymers to negatively charged surfaces on particles include: hydrogen bonding, anion interchange with adsorbed anions such as the hydroxyl group, or interactions with cations on or in the immediate vicinity of the particle (Busch and Stumm, 1968). The presence of polyvalent cations may enhance this aggregation by inducing complex or ion-pair formation between the functional groups of the polymer and the counter-ion in the vicinity of the particle. Coordination is possible near the negatively charged particle surface where, for electrostatic reasons, the cation concentration will be higher than in the solution (Busch and Stumm, 1968). Multivalent ions such as Al(III) and Fe(III) may be removed from solution by interaction with linear polymers produced by microorganisms (Tenney and Stumm, 1965). Busch and Stumm (1968) found evidence that bioflocs are not as sensitive to dispersion by shear forces as aggregates formed with synthetic polymers, and that natural flocs are not critically dependent on pH and polymer concentration.

Factors Affecting Polymer Bridging: The nature of clay minerals and polymer adsorption has been detailed by Theng (1979). Briefly described are the most important factors relative to microbial polymer destabilization of colloids through the bridging mechanism.

Polymer concentration: A stoichiometric relationship exists between clay concentration and optimal polymer dosage. The polymer adsorption data fit a Langmuirian model (the rate of adsorption is proportional to the pressure and the fraction of the surface available); as polymer dosage increases, an optimum zone of turbidity removal is noted. However, as the dose increases, a repeptized region occurs in which particle surfaces become saturated with polymer (Black et al., 1966; Harris and Mitchell, 1973). The optimum polymer concentration is a function of the total available surface area of the colloidal suspension (Harris and Mitchell, 1973). It is believed that as surface saturation is reached, polymers extending from the surface of one particle cannot find reactive surface sites on adjacent particles either because of the decreasing number of available sites or electrostatic or steric interferences between polymer segments (Harris and Mitchell, 1973).

Polymer configuration: The configuration of the polymer probably influences polymer bridging. Adsorption of the polymer to a colloidal surface is generally independent of molecular weight, but chain length does influence bridge formation; solution properties may actually influence chain elongation. Nonionic polysaccharides, which are effective flocculants over broad pH ranges and ionic conditions, may be linear polymers. The hydroxyl groups associated with these polymers adsorb by hydrogen bonding and/or ester formation to solid surfaces. Molecular weight does seem important if neutral polymers or polymers bearing the same charge as the colloid are used. Significant aggregation will occur only with polymer weights in excess of several hundred thousand (Harris and Mitchell, 1973).

Particle concentration: The destabilization of a clay suspension is dependent on the initial clay concentration. Polymer adsorption is very rapid, but the rate-controlling step in the flocculation process is the frequency of interparticle collision (Black et al., 1966). With low solid concentrations, interparticle collisions are so infrequent that polymer rearrangement on the surface of the solid leads to reduced polymer elongation before particle bridging occurs (Harris and Mitchell, 1973).

Agitation: To promote flocculation of colloids with polymers, interparticle repulsion must be overcome and distances among particles must be reduced to allow for interparticle bridging. Agitation of colloidal solutions promotes flocculation by bringing particles closer together (Hannah et al., 1967).

Solution pH and ionic properties: Cation moiety and pH affect the interaction between polymers and colloids. Factors influenced are particle properties such as surface potential, charge, and nature of the double layer surrounding each particle (Black et al., 1966). Also affected are charge and charge density of the polymer, as well as its elongation and solubility. Changes in pH that reduce repulsive potential between particles consequently increase particle contact (Harris and Mitchell, 1973).

Important factors in the flocculation of colloids by microbially produced polymers are polymer concentration, length of the polymer chain, concentration of the colloid, agitation to overcome intercolloid repulsion, cation concentration, and pH.

V. BIOPOLYMERS

Polymers produced by microorganisms have been studied for their ability to flocculate clays generated during mining and beneficiation activities. This section reviews the nature of these polymers and the factors affecting their formation.

A. The Nature of Biopolymers

Exocellular polymers all possess the same major constituents: polysaccharide, protein, RNA, and DNA (Pavoni et al., 1972). However, exocellular polymers can differ with the type of waste system; differences in composition of the polymers result when different energy sources are supplied to the organisms.

1. *Exopolysaccharides*

Exopolysaccharides exist in two forms: capsular polysaccharide attached to microbial cells as a distinct physical structure, and soluble slime secreted by the cell into the surrounding environment (Sutherland

and Ellwood, 1979; Geesey, 1982). Exopolysaccharides are identified as homopolysaccharides, synthesized from the substrate sucrose and constructed of a single structural unit, or heteropolysaccharides, composed of two or more monomers of neutral sugars and frequently uronic acid. The other common component of heteropolysaccharides is pyruvate (Sutherland and Ellwood, 1979; Smiley, 1976).

Many microorganisms produce exopolysaccharides. Nakamura et al. (1976) surveyed 214 strains of molds, bacteria, actinomycetes, and yeast, and identified 19 strains that produced flocculating compounds. Smiley (1976) discussed several polysaccharides having potential industrial application. Phosphomannan, produced by *Hansenula holstii* NRRL Y-2448 (a yeast), is composed of mannose and mannose-6-PO_4. *Arthrobacter viscosus* NRRL B-1973 produces a polymer consisting of nearly equal ratios of glucose, galactose and manuronic acid, when using glucose as a substrate. The polymer is stable over the pH range 5–10. *Xanthomonas campestris* NRRL B-1459, when using glucose, yields a very viscous polymer composed of D-glucose, D-mannose, and D-glucuronic acid in a ratio of 3:3:1; a wild-type strain of *X. campestris* yields a polymer of identical components with a ratio of 2:2:1 (Whitfield et al., 1981). The yeast *Aureobasidium (Pullularia) pullulans* produces a polysaccharide, D-glucan, in amounts up to 2–3 g/100 ml of culture medium when various carbohydrates are used as carbon sources (Ueda et al., 1963). Zajic and LeDuy (1973) reported 25.8% conversion of glucose to polysaccharide by this organism after 58 h of fermentation. This polymer (called pullulan) is characteristic of polyvinyl alcohol, and its applications include the flocculation of suspended clays from hydrometallurgical processes (Zajic, 1967; Slodki and Cadmus, 1978; Kang and Cottrell, 1979). Pullulan was patented for the flocculation of clay-slimes resulting from the beneficiation of uranium, potash, and aluminum (Zajic, 1967).

A darkly pigmented, extracellular polysaccharide is formed by the black yeast strain NRRL Y-6272 (Sanford et al., 1975). This acidic extracellular polysaccharide contains both *N*-acetyl-D-glucosamine and *N*-acetyl-D-glucosaminuronic acid; it has considerable industrial potential since it contains both acidic (carboxyl) and masked basic groups (*N*-acetyl) and disperses readily in water to produce a viscous solution. The black pigmentation has some characteristics of a protein, but cannot be removed from the polysaccharide by dialysis, suggesting that the pigment is either too large to pass through the dialysis membrane or is small but strongly associated with the polysaccharide. The black yeast NRRL Y-6272 has not been classified (Sanford et al., 1975). However, there may be some relationship between the black pigment-forming yeast-like fungus *Aureobasidium (Pullularia) pullulans* (Ueda et al., 1963; Zajic and LeDuy, 1973) and the black yeast NRRL Y-6272 (Sanford et al., 1975).

Xanthan, dextran, and glucan are produced commercially by microbial processes. Xanthan, produced by *Xanthomonas* species, contains glucose, mannose, and glucuronic acid. Xanthan is stable with most salts; however, polyvalent metals cause gelation or precipitation of xanthan at high pH values (Slodki and Cadmus, 1978; Kang and Cottrell, 1979). Dextran is produced by *Leuconostoc* and is composed of D-glucose. Dextran has no charged groups and is compatible with salts, acids, and bases (Kang and Cottrell, 1979). The microbe *Sclerotium* produces glucan, which consists of β-D(1-3)-glucopyranose units. The most notable property of scleroglucan is its synergetic viscosity increase in the presence of bentonite (Kang and Cottrell, 1979).

2. *Polymers Other Than Exopolysaccharides*

Nucleic acids and proteins can also play an important role in aggregation (Harris and Mitchell, 1973). When grown under nutrient-deficient conditions, *Micrococcus halodenitrificans* and *Vibrio costicola* produce a sticky extracellular DNA; similar results are observed with *Pasteurella pestis (Yersinia pestis)*. RNA and protein accumulations are noted accompanying *Staphylococcus aureus* growth (Harris and Mitchell, 1973).

B. Factors Affecting Biopolymer Formation

The use of biopolymers for flocculation of clays produced in phosphate beneficiation would involve either in situ cultivation of organisms that produce polymers or polymer production in separate fermentation vats with subsequent harvesting of the polymer.

Campbell (1972) outlined four conditions necessary for biopolymer formation: (1) the presence of sufficient oxygen and other nutrients to initiate and sustain growth of the organism, (2) a change in the environment inducing the endogenous phase, (3) the excretion of a polymer during the endogenous phase, and (4) a genetic make-up permitting the synthesis and excretion of polymers during the endogenous phase.

Many investigators (Tenney and Stumm, 1965; Busch and Stumm, 1968; Campbell, 1972; Sutherland and Ellwood, 1979) have observed that natural polymers are produced by microbes under conditions of declining or endogenous growth. Some microorganisms synthesize extracellular polymers under conditions in which nutrients other than carbon are limited (for example, restricted nitrogen) (Busch and Stumm, 1968). There is considerable evidence that exopolysaccharide production is more evident in cultures when the media has a high carbon:nitrogen ratio (Deinema and Zevenhuizen, 1971). The synthesis of polymers by organisms under nutrient limitation is a paradox, since it is generally thought that bacteria under such conditions would use all available nutrients for growth (Sutherland and Ellwood, 1979). However, polymer synthesis must be less energy demanding than might be expected because polymers are readily produced in the absence of growth (Busch and Stumm, 1968; Sutherland and Ellwood, 1979). It has also been observed that microbial polysaccharide production is greater when the organisms are grown on carbohydrates rather than hydrocarbons and alcohols derived from petrochemicals (Slodki and Cadmus, 1978). Aeration is required for all polysaccharide production of commercial significance, but it must be noted that as the viscosity of the medium increases, aeration efficiency decreases. All microbial polymers of commercial importance are produced by mesophilic organisms but, to combat the inefficiency of mass transfer of air due to increased viscosity, hot air and increased agitation are applied during commercial production (Kang and Cottrell, 1979).

Sutherland and Ellwood (1979) detailed the disadvantages in using biopolymers for industrial applications; these disadvantages include the high cost of installation and start-up of fermentation equipment together with solvent requirements and the associated need for large amounts of energy. However, these problems are only critical when a purified polymer for food use is to be produced. Such purified polymer probably would not be essential for nonfood uses. However, one important concern is the suitability and availability of an inexpensive energy source. Most organisms noted for polymer production are heterotrophic, requiring an organic energy source. Glucose in some cases enhances polymer production (Zajic and LeDuy, 1973; Smiley, 1976; Balkwill and Casida, 1979). For an industrial application of the magnitude of phosphatic slime dewatering, the use of microorganisms would dictate the need for inexpensive, readily available waste products as energy sources or the use of photosynthetic organisms (Tischer and Davis, 1971). Use of inorganic energy sources has little appeal, as there is relatively little that can be oxidized in the phosphatic slime wastes; however, such possibilities should not be totally overlooked.

VI. MICROBIAL INTERACTIONS AT INTERFACES

In recent years, there has been considerable interest in the activities occurring at interfaces between microorganisms and liquids, gases, and solids. Understanding these interactions could be a key in industrial application of microorganisms in flocculation and wastewater clarification.

A. Biosurfactants

Some biopolymers are considered biosurfactants because they have a tendency to concentrate at interfaces and reduce surface tension. Surfactants are molecules that consist of both hydrophilic (polar) and hydrophobic (nonpolar) moieties. These agents affect the physical characteristics of interfaces between immiscible phases. Surfactants are responsible for reducing surface tension (air-water), reducing interfacial tension in liquid-liquid and liquid-solid systems, and altering foaming properties (Gerson and Zajic, 1979; Margaritis et al., 1979).

Biosurfactants can influence the initial attachment of bacteria to solid surfaces by altering the wettability of the surface (Weaver and Dugan, 1972). It has also been suggested that organic films may form on clay particles; however, aggregates induced by such films are large and unstable (Sarma and Dakshinamurti, 1971). In contrast, bacterial polymers bind clays into smaller but more stable aggregates. The first monolayer of adsorbed organic conditioning film that forms on a solid surface will mask the critical surface tension of the original surface. It is thought to be the conditioning film that controls the attachment process. Conditioning films studied in marine biofouling are usually composed of glycoprotein (Dexter et al., 1975).

B. Bacterial Surface Adhesion and Flocculation

Parameters affecting the attachment of microorganisms to solid surfaces include: (1) van der Waal's forces, (2) shape of the electrical double layer, (3) chemotaxis, (4) ability of the organism to produce polymeric fibrils, (5) surface texture, (6) surface charge, (7) wettability of the substrate (Dexter et al., 1975), (8) pH, (9) temperature (Fletcher and Floodgate, 1976), (10) time allowed for microbial attachment, (11) growth phase of the culture (Fletcher, 1977), and (12) nutrient concentrations (Fletcher and Floodgate, 1976; Fletcher, 1977). The adsorption of marine bacteria to surfaces is thought to involve an instantaneous reversible phase and a time-dependent irreversible phase (Marshall et al., 1971b).

1. Reversible Adsorption

Reversible adsorption is an instant attraction of the bacteria to the surface; bacteria so attracted are held weakly near the surface. This phase can be interpreted as a balance between the electrical double-layer repulsion energies at different electrolyte concentrations and van der Waal's attractive energies (Marshall et al., 1971a,b; Marshall and Cruickshank, 1973; Marshall, 1979). The initial attachment of the bacteria to a surface can be modeled on the basis of physicochemical adsorption, suggesting that nonbiological processes may play a major role in the initial events of adhesion (Fletcher, 1977).

The primary colonization of surfaces by microorganisms is probably limited by the availability and concentration of nutrients on the surface (Corpe, 1972). However, it is known that nutrients accumulate at a variety of solid-solid, gas-liquid, and liquid-liquid interfaces (Marshall, 1979). Motile bacteria exhibit a chemotactic response along the nutrient gradient established at interfaces; however, other attractive responses must exist besides motility, since nonmotile bacteria will also accumulate at interfaces (Marshall, 1979).

Most interfaces are negatively charged and, as a consequence, attract a variety of cations and macromolecules (such as biosurfactants) to form a surface microlayer, or conditioning film. This film can alter the surface free energy and therefore can influence the behavior of bacteria at the surface (Marshall, 1979). Cell-surface hydrophobicity is a major factor influencing the behavior of bacteria at interfaces

(Marshall, 1979). Both perpendicular and rosette formations of bacteria are observed attached to surfaces (Marshall and Cruickshank, 1973; Marshall, 1979).

Initial attachment of an organism to a surface may be due to the formation of a "primary polysaccharide." A "secondary polysaccharide" is formed after adhesion, and this probably strengthens the cell's attachment (Fletcher and Floodgate, 1976). In studies involving the adsorption of bacteria from seawater (Marshall et al., 1971a), small rod-shaped bacteria are found to be the primary colonizers of a glass surface. Their production of extracellular polymeric fibrils may be important in their selective sorption (Marshall et al., 1971a). These primary colonizers may form a monolayer that alters the critical surface tension of the original surface (Dexter et al., 1975).

2. *Irreversible Adsorption*

Irreversible adsorption involves firm adhesion of the bacteria to the surface. The cells may be anchored to the surface by extracellular polysaccharides or glycoproteins involving polymer bridging (Marshall et al., 1971a,b; Marshall, 1979, 1980). *Arthrobacter globiformis* other soil isolates attach to soil particles by producing extracellular polysaccharide (Balkwill and Casida, 1979). *Azotobacter chroococcum*, *Derxia* sp. and a *Rhizobium* sp. also bind soil particles using an extracellular polysaccharide (Yadev et al., 1970).

Costerton et al. (1978) describe the tangled fibers of polysaccharide associated with many bacteria as a "glycocalyx." This material is negatively charged and can form polar bonds with higher cells polysaccharides by means of divalent positive ions in the medium (Costerton et al., 1978). Calcium and magnesium are important in the structure of intercellular matrix (Fletcher and Floodgate, 1976). The glycocalyx acts similarly to an ion-exchange resin because it can also bind nutrient ions and molecules in the environment. It also offers protection for the organism from predatory bacteria and viruses. Therefore, this structure gives the organism a competitive advantage in the natural environment (Costerton et al., 1978).

C. Adsorption of Inorganic Materials to Microbial Structures

Microorganisms, particularly fungi, have the ability to bind soil particles into stable masses. This ability has been attributed to both polymer production and charge interactions. Lynch (1981) reported that microbial products rather than the cells are the principal agents responsible for promoting or inhibiting soil aggregate stability.

Fungal mycelia are effective sorbents of inorganic materials. Mycelia have been used to remove uranium from solution (Jilek et al., 1975), and fungi can aggregate clay-slimes produced by phosphate beneficiation (Davis, 1965). In the latter study, the mechanism of solids removal is believed to be the adsorption of the particles onto the fungal mycelia followed by mechanical binding, since clearing of the solution is related to the available mycelial surface. Zvyagintsev et al. (1969) reported that clay particles attach to cell walls, fimbriae, and flagella of microorganisms by processes other than binding to extracellular polymers. However, it is known that some fungi, namely *Sclerotium* (Sutherland and Ellwood, 1979), *Aureobasidium (Pullularia) pullulans* (Ueda et al., 1963; Zajic, 1967; Zajic and LeDuy, 1973; Sutherland and Ellwood, 1979), and the black yeast (NRRL Y-6272) (Sanford et al., 1975) produce exocellular polysaccharides.

Biosurfactants that alter the surface properties of colloids, the attraction of microorganisms to interfaces, and the adsorption of microorganisms to solid particles have been identified as important factors in effecting particle removal from water. However, the relative importance of each of these phenomena has yet to be assessed.

VII. EFFECT OF CLAYS ON MICROORGANISMS

When considering the use of microorganisms to clarify industrial wastewaters of suspended clay minerals, the effect of these clays on the growth, respiration, and activity of the organisms must be addressed.

Bacterial respiration is stimulated by montmorillonite clay (Stotzky and Rem, 1966); this is believed to be the result of the clay maintaining the pH of the environment at suitable levels for bacterial development. Kaolinite and montmorillonite serve equally well as sources of minerals for bacterial nutrition. In addition, clay surfaces probably provide a greater permanence of water film, concentrate energy sources for the organisms, and inactivate microbial inhibitors (Stotzky and Rem, 1966).

Respiration of bacterial cells increases as the cation exchange capacity (CEC) and surface area of clay particles increase (Stotzky, 1966). The CEC probably aids in sustaining a moderate pH by replacing hydrogen ions produced during metabolism with basic cations from the exchange complex. Particle size does not appear to be a critical factor in bacterial respiration. Based on these observations Stotzky (1966) concluded that montmorillonite is the dominant clay influencing microbial activity.

The addition of low concentrations of K-montmorillonite can flocculate *Klebsiella aerogenes* (Britton et al., 1976). The polysaccharide of this organism contains a carboyxl moiety as the sole ionogenic group, and this probably attaches to and neutralizes positive charges on the edges of the clay particle. Marshall (1969) observed that *Rhizobium* sp., which possess exclusively acidic (carboxyl) surface ionogenic groups, adsorb approximately twice as much sodium-montmorillonite or sodium-illite as cells with more complex carboxyl-amino surface groups. Using *Bacillus subtilis* and montmorillonite, Marshall (1969) found that the orientation of clay particles at bacterial cell surfaces is edge-to-face;

the positively charged edges of the clay particles are attracted to the negatively charged groups on the bacterial surface. Zvyagintsev et al. (1969) noted that cells can adsorb onto particles larger than the cells; adsorption is by the lateral surfaces or poles of the cells. Adsorption of cells onto particles smaller than the cells causes the entire cell surface to be almost entirely covered by the particle.

In industrial applications in which microorganisms are cultured with the clays to be flocculated, the effect of clays on the growth and polymer-producing activity of the organisms is important. Research suggests that microbial respiratory activities are enhanced by the presence of clays and that microorganisms will effectively adsorb to clay surfaces.

VIII. MICROBIAL PROCESSES FOR WASTE WATER CLARIFICATION AND RESTORATION

A. Flocculation of Clays by Microbial Polymers

An attractive industrial application for microorganisms possessing extracellular polymers is the restoration of waste water by removal of heavy metals associated with particulates and flocculation or aggregation of suspended clays (Morris, 1971; Dugan and Pickrum, 1972; Dugan, 1975; Gale and Wixson, 1979; Hodge et al., 1979; Kelly et al., 1979; Marshall, 1979; Norris and Kelly, 1979).

In recent years, several processes have been patented for the flocculation of clays. Microbial polysaccharides from *Pullularia*, *Xanthomonas*, *Arthrobacter*, *Cryptococcus*, *Hansenula*, and *Plectania* species can flocculate finely divided inorganic solids suspended in an aqueous medium (Goren, 1968). Fermentation processes for preparation of microbial polysaccharides and steps for recovery of the flocculating agent are described in Goren's 1968 patent. In 1972 Bomstein patented the use of alkaline-treated microbial nucleoprotein for flocculating organic and inorganic wastes. Used in the process were nucleoproteins from microbial genera *Polyangium*, *Myxococcus*, *Sorangium*, *Flavobacterium*, *Leuconostoc*, *Micrococcus*, and *Alcaligenes*. The nucleoprotein derived from these organisms is treated with any one of a variety of alkaline compounds [$Ca(OH)_2$, KOH, NaOH, NH_3, Na_3PO_4, or quaternary ammonium compounds] which increase the pH causing the microbial material to lyse and form a sol. Flocculation of suspended waste results when the concentration of alkaline-treated microbial material is present in concentrations of 1--500 ppm (Bomstein, 1972).

Species of the algae *Anabena*, *Chlamydomonas*, and *Chlorella* are effective in flocculating bentonite in the presence of an electrolyte (Avnimelech et al., 1982). The flocculation is attributed to the large amounts of polysaccharide and other polymers produced by the algae. Algae also accumulate polonium and plutonium from sea water (Hodge et al., 1979); these metals are probably removed from solution in a particulate state, and uptake is related to the organic matrix on the surface of the organisms. Algae are currently used to adsorb heavy metals from mine waste water (Gale and Wixson, 1979). The process is thought to involve physical entrapment of particulate minerals followed by chelation of the metals to anionic sites on the cell walls of the algae. The chelation may entail displacement of di- and monovalent cations associated with the cell wall. Selected algal cultures exhibit cation exchange capacities of 640 μEq/g (dry wt.) (Gale and Wixson, 1979).

Stabilization and dewatering of phosphatic clays were attempted using the polymer-producing organisms *Leuconostoc mesenteroides*, *Xanthomonas* sp., *Azotobacter vinelandii*, and *Beijerinckia indica* (Brierley et al., 1981; Brierley and Lanza, 1982). The flocculation capacity of clays by whole cells, determination of respiratory activity of whole cells in the presence of phosphatic clays, observation of clay flocculation rates by polymers derived from these organisms, and clarity of polymer-flocculated clay suspensions were examined. The polymer-producing organisms tested exhibit little capacity for flocculating phosphate clays.

A total of 350 Texas soil isolates were also screened for their ability to flocculate phosphatic clays. Of these isolates, 13 did demonstrate some flocculation capacity (Brierley and Lanza, 1982).

B. Fungal Aggregation of Phosphatic Clays

Agitation of fungal spores of *Cladosporium cladosporioides* (isolated from a phosphate clay impoundment) with phosphatic clays results in the formation of clay-fungal pellets. Studies with *C. cladosporioides* to assess the organism's suitability for aggregating clays included the determination of suitable energy substrates and supplements for growth, determination of the time required for bioflocculant production by the organism, and an elucidation of the conditions required for the pelletization of clays by the fungus.

Little flocculation of phosphatic clays by *Cladosporium cladosporioides* is observed if the fungus is incubated with clays under static conditions; however, if sterile clays are supplemented with sugar and yeast extract and the suspension agitated, spores or newly germinated spores bind the clays in tightly compacted pellets yielding a clear or slightly discolored supernatant. Although *C. cladosporioides* can grow on starch, gelatin, rosin, or wax, fructose yields the greatest biomass (Eveleigh 1970). When the organism is provided with starch as an energy source, growth of the organism is excellent; however, clay-fungal pelletization is poor. If the starch is supplemented with fructose, pelletization improves. The mechanism that *Cladosporium cladosporioides* uses for pelletizing the clays is not known, although the phenomenon has been observed with phosphatic clays and in soil (Davis, 1965; Lynch, 1981). Clay pelletization occurs with 2–3 days following the addition of spores to the phosphatic clay, and this correlates well with production of bioflocculant by the organism.

Factors other than bioflocculant production by *Cladosporium cladosporioides* may be important in clay-fungal pelletization. Adsorption of clay by mycelial structures may be a method of physical entrapment of the clays with the agitation inducing pelletization. A mutual flocculation between selected clay minerals and some kinds of asexual spores from soil borne fungi can occur (Horikawa et al., 1979). Moreover, *Cladosporium cladosporioides* can induce bacteria to respond in ways similar to positive and negative taxis; this response is substrate dependent (Diem, 1975). Since both organisms and clays are negatively charged, it would be expected that there would not be an attraction between the particles. However, the biopolymer produced by *C. cladosporioides* may "mask" the negative charge of the fungus itself and enhance clay aggregation in this way.

The role of yeast extract in bringing about pelletization of germinating fungal spores and clay is unknown; perhaps a specific nutritional component (or components) is responsible for the phenomenon. The yeast extract component(s) may induce a positive taxis of the germinating spore toward the clay, as noted by Diem (1975). Alanine and glycine or $Ca(NO_3)_2$ could substitute for yeast extract with little effect on clay pelletization by *C. cladosporioides*. *Xanthomonas campestris* exhibits similar nutritional requirements for xanthan production. A 4% sucrose or glucose medium yields the highest xanthan titer, and certain amino acids and nitrate salts are superior to ammonium salts for xanthan production (Souw and Demain, 1979).

Studies of the clay-fungal pellet produced by *Cladosporium cladosporioides* show the pellet to have a network of mycelia throughout suggesting that physical entrapment of the clay may be an important factor in the pellet formation. Since the density of the pellets approximates water (1.1 g/ml), a screening procedure may be required to remove the pellets from solution.

The bioflocculant produced by *C. cladosporioides* may be important in the clay pelletization process. A protein was identified in the 70--75 $\times 10^3$ molecular weight range, and the presence of a structural component indicates that the bioflocculant possesses two compounds with similar chemical structure (possibly members of a homologous chemical series) but different molecular weights. The bioflocculant also contains a polysaccharide component.

Cladosporium cladosporioides spores are unable to pelletize clays when the clays are not sterile. Davis (1965) concluded that competition from other organisms within the clay prohibits sufficient fungal development; however, it has not been resolved whether this competition results from antifungal agents produced by other organisms or whether it is substrate competition. Bacterial inhibition of fungi can be influenced by clay mineralogy, pH, nutrient levels, type of fungi present, and the spatial relations of the organisms. Fungal spores are more sensitive to inhibition by bacteria than mycelial filaments (Rosenzweig and Stotzky, 1979). The inability of *C. cladosporioides* to pelletize nonsterile clay could be a major setback in commercial implementation of this microbial technology unless the problem can be overcome.

IX. FUTURE RESEARCH THEMES

Scientists, engineers, and managers developing microbial technology for various reclamation processes face three common challenges: (1) dealing with the increasing problem of water supply as a limiting factor, (2) identifying the best approaches to minimize energy requirements for reclaiming redistributed resources, and (3) combining process water-waste treatment with methods to extract redistributed resources such as mineral values and water. All three are extremely important in developing efficient, cost-competitive processes for resource reclamation in the phosphate industry, and they should form the basis of future research programs. A brief discussion of each follows.

A. Water as a Limiting Factor

Increased demand on water for all beneficial uses is not unique to the mining and beneficiation of phosphate. Table 7 illustrates the trend of major beneficial water use in the United States between 1900–1975. A dramatic increase in water demand is clearly evident in each selected category. Self-supplied industrial (and miscellaneous) water uses alone increased almost 12-fold.

Phosphate mining-beneficiation operations are water intensive and draw almost exclusively from groundwater sources. Mine planners must consider all facets of water use to minimize waste, maximize recycle volumes, and avoid polluting local water supplies. In spite of a relatively high efficiency of water recycling (between 85–95% of all water used by the phosphate industry), considerable volumes are lost to the slime (Hoppe, 1976). As a result, one research area of extreme importance continues to be water reclamation from phosphatic waste clays to alleviate limited water supply.

The few processes for water reclamation by microbial flocculation-dewatering discussed in this chapter are in preliminary stages and not without shortcomings. However, it is important to recognize that microbial systems have received limited research effort in the area of inorganic particulate wastes. Relatively little information exists on the indigenous microflora of low-nutrient mine-beneficiation waste streams and clay-waste storage lagoons. Basic information on the microbial community in both systems could provide the applied microbiologist with valuable insights for developing useful microbial reclamation systems. Data should be generated for specific use in calibrating microbial reclamation systems. Research should stress the factors regulating microorganism metabolism and growth in inorganic particulate systems. Of paramount importance in this regard are a better understanding of factors influencing organism-substrate interaction including chemical tolerance, and pH, Eh, nutrients, surface tension, critical gas, temperature, light, and pressure optima. New genetic engineering approaches should be evaluated as to their applicability in expanding use-

Table 7 Trend of Major Beneficial Use of Water in the United States, 1900--1975, in Millions of Cubic Meters Per Day

Year	Irrigation	Public water supplies	Domestic	Self-supplied, industrial and miscellaneous	Steam-electric power	Total
1900	76.8	11.4	7.6	38.0	19.0	152.8
1910	148.2	17.9	8.4	53.2	24.7	252.4
1920	212.4	22.8	9.1	68.4	38.0	350.7
1930	228.8	30.4	11.0	79.8	69.9	419.9
1940	269.8	38.4	11.8	110.2	84.4	514.6
1944	306.3	45.6	12.2	212.8	136.4	713.3
1945	315.8	45.6	12.2	182.4	109.4	665.4
1946	328.3	45.6	13.3	148.2	102.2	637.6
1950	380.0	53.6	17.5	175.0	145.9	772.0
1955	455.2	64.6	20.5	228.0	227.2	995.5
1960	513.0	83.6	22.8	273.2	294.9	1187.5
1965	562.8	95.0	24.7	333.3	350.4	1366.2
1970	604.2	105.6	26.2	391.4	409.6	1537.0
1975	644.9	113.2	27.4	438.5	497.8	1712.8

Source: U.S. Department of Commerce (1976).

ful organism-substrate interactions. The application of this type of information could produce efficient microbial dewatering techniques geared to reclaiming 5–15% of the process water lost in mineral waste slimes.

B. Minimizing Energy Requirements

In general, physicochemical processes are more costly to operate and energy intensive for treatment and/or reclamation than are conventional biological systems. A decade ago, some authorities predicted that complete physicochemical treatment of conventional wastewater using chemical precipitation and carbon adsorption would completely replace biological treatment processes (Bunch, 1982). It is now recognized that the high costs of maintaining the external (nonmicrobial) energy supply required by physicochemical approaches to treatment and/or reclamation are prohibitive.

Research is needed to identify the minimal nutrient requirements to stimulate microbial processes for reclamation in inorganic particulate systems. The feasibility of integrating (blending) different organic and inorganic waste streams to provide an inexpensive carbon source for microbial growth should be examined. Transformation of the "universal energy currency," ATP, during combined treatment of organic-inorganic wastes should be clarified. The energetics of carbon utilization, biopolymer synthesis, and the microbiological dewatering process itself should be stressed.

Research on microbial reclamation processes using microbial assemblages of several species would also be extremely useful. These assemblages could be structured to contain several different microbes in a synergistic relationship providing a combined metabolic attack on inorganic waste materials. Finally, more should be learned about the antagonistic relationships among natural microbial assemblages so as to permit useful manipulations in treatment situations.

C. Resource Recovery

Combining microbial processes to treat fine particulate inorganic wastes with methods to extract redistributed resources is another area worthy of more research effort. In the case of phosphatic waste-clays, methods to release simultaneously water and recover lost phosphate mineral values should be emphasized. Fluoride and uranium harvesting from phosphatic waste clays during the microbial dewatering process could also be studied in more detail.

The development of combined processes would be heavily dependent on the successful integration of organic and inorganic waste streams. Unfortunately, there is a tendency toward research that isolates physicochemical treatment from biological treatment. A lack of effort to develop integrated waste water treatment systems will most certainly result in relatively ineffective and energy-inefficient waste water treatment and management in future decades. The feasibility of incorporating biological units in existing physicochemical treatment processes, to decrease process water demand, enhance treatment efficiency, and produce or salvage useful products following pollutant removal, should be carefully studied. Perhaps the single major reason for the failure of biological treatment processes to process many nondomestic waste materials lies in their incompatibility with the abiotic or physicochemical properties of the waste material itself. The biotic portion of the system often fails because the microbial community is not "environmentally programmed" to a new set of physicochemical conditions and cannot function under those conditions. But this is not an insurmountable problem and part of the solution could result from integration of a biological segment at the proper level of treatment. In the treatment of petroleum waste waters for example, oily waste material at concentrations above 50–100 ppm is effectively treated by physicochemical processes largely involving gravity separation (Davis, 1967). Remaining oily residuals in the order of 50–100 ppm should be low enough to treat with a biological segment if carefully integrated into the treatment system.

Full integration of the physicochemical and biological treatment segments should produce a treatment system operating as a continuum of the two treatment types. Integration of the two treatment types would require a careful recalibration of each to insure compatibility; optimizing the biological component necessitates some adjustment of the physicochemical component since it provides the basis or habitat for biological metabolism. Since the physicochemical segment provides the habitat for the organism, major parameters critical to organism survival and efficient function should form the framework of integration.

The potential of existing microbial metal treatment processes as part of a larger integrated waste water treatment scheme should be examined. The potential for microbially produced polymers to dewater simultaneously fine particulate wastes (e.g., separate clays from mine process water) and scavenge critical minerals lost in the process waters must be further studied before the bioprocess can be implemented on a commercial scale.

Substitution of natural polymers for polymers based on petroleum feedstocks should be actively examined using waste products of other treatment systems as the carbon source for biopolymer production (e.g., starches from beet, citrus, and sweet potato waste waters). Research in this direction should help to optimize the treatment system to reclaim the recycle water, produce an effective treatment of exotic wastes using other production wastes, and allow the harvest of valuable byproducts such as minerals otherwise lost in treatment residuals. The end result could be effective, energy efficient treatment of waste water with substantial recovery of reclaimed water and minerals. Much of the technology base for both the physicochemical and biological treatment already exists. Scientists, engineers and managers need only examine the feasibility of restructuring these methods into experimental integrated systems and test the integrated components for effectiveness.

X. SUMMARY

Mining and beneficiation of phosphate mineral values produce large volumes of waste-clay suspensions. The chemical, physical, and rheological properties of phosphatic waste-clays result in colloidal suspensions with poor sedimentation, filtration, and flocculation characteristics. The difficulties associated with dewatering colloidal waste-clay suspensions result in storage problems with substantial economic and environmental liabilities.

Destabilization of colloidal waste-clays through coagulation or flocculation has been the major focus of research on methods to dewater phosphatic slimes. The bioflocculation theory offered by Tenney and Stumm (1965) has received the most support in explaining the mechanism of biologically mediated destabilization. This theory explained bioflocculation in terms of polyelectrolyte interaction and was later expanded by Busch and Stumm (1968) to include the concept of polymer bridging. Bridging between microorganism polymers (biopolymers) and clay colloidal solids results in coflocculation of organisms and clay. Biopolymers usually have major constituents as polysaccharides, proteins, and nucleic acids with specific composition strongly regulated by the microorganism's energy source.

Understanding the interactions between microorganisms and liquids, gases, and solids is basic to the development of microbial technology for waste-clay aggregation. Major microbial interactions include activity as biosurfactants, adhesion by reversible and irreversible adsorption, and the adsorption of inorganics. Clay effects on the microorganism are also important, especially as they influence growth, respiration and activity largely through pH, cation exchange, and surface charge phenomena.

Recent research has produced mixed success in developing microbial processes to flocculate and/or remove metals and suspended clays from various types of waste water. Processes examined to date have included the use of bacteria, fungi, and algae as treatment organisms, and some processes have been patented (Goren, 1968; Bomstein, 1972; Zajic, 1967); none is commercially applied. One approach specifically applied to phosphatic waste-clays has met with some success (Brierley and Lanza, 1982; Brierley et al., 1981). *Cladosporium cladosporioides*, a fungus isolated from a phosphatic waste-clay impoundment, has been examined as an aggregation organism. Agitation of fungal spores of *C. cladosporioides* with sterile, nutrient-amended phosphatic slimes produced waste aggregates as clay-fungal pellets. The mechanism used by *C. cladosporioides* to pelletize clay and the specific reasons(s) for the inability of the microbe to produce pellets in nonsterile waste-clay are not known. Clarification of both points is essential to the successful development of *C. cladosporioides* as a technological tool in mine waste reclamation.

Scientists, engineers, and managers developing microbial technology for various reclamation processes face three common challenges: (1) dealing with the increasing problem of water supply as a limiting factor, (2) identifying the best approaches to minimize energy requirements for reclaiming redistributed resources, and (3) combining process water-waste treatment with methods to extract redistributed resources such as mineral values and water. All three are extremely important in developing efficient, cost-competitive processes for resource reclamation in the phosphate industry; these points should form the basis of future research programs.

REFERENCES

Anon. (1976). Water uses in the United States. U.S. Dept. of Commerce (BSB-136), Washington, D.C.

Arora, H. S., and Coleman, N. T. (1979). The influence of electrolyte concentration on flocculation of clay suspensions. *Soil Science 127*: 134–139.

Avnimelech, Y., Troeger, B. W., and Reed, L. W. (1982). Mutual flocculation of algae and clay: Evidence and implications. *Science 216*: 63–64.

Balkwill, D. L., and Casida, L. E. Jr. (1979). Attachment to autoclaved soil of bacterial cells from pure cultures of soil isolates. *Appl. Env. Microbiol. 37*: 1031–1037.

Black, A. P., Birkner, F. B., and Morgan, J. J. (1966). The effect of polymer adsorption on the electrokinetic stability of dilute clay suspensions. *J. Colloid Interface Sci. 21*: 626–648.

Blakey, A. F. (1973). *The Florida phosphate industry: A History of the Development and Use of a Vital Mineral.* Wertheim Committee, Harvard University Press, Cambridge.

Bomstein, R. A. (1972). Waste treatment with microbial nucleoprotein flocculating agent. U.S. Patent No. 3,684,706.

Brierley, C. L., and Lanza, G. R. (1982). The microbial flocculation of phosphate and potash slimes. U.S. Dept. of the Interior, Bureau of Mines Minerals Research Contract Report J0199150.

Brierley, C. L., Lanza, G. R., and Scheiner, B. (1981). Microbial processes for removal of suspended clays from selected industrial wastewaters. In *Biotechnology and Bioengineering Symposium No. 11*, C. D. Scott (Ed.). John Wiley & Sons, Inc., New York.

Britton, G., Henis, Y., and Lahav, N. (1976). Effect of montmorillonite and the capsular polysaccharide on the flocculation of *Klebsiella aerogenes*. *Plant Soil 44*: 37–47.

Brown, M. J., and Lester, J. N. (1979). Metal removed in activated sludge: The role of bacterial extracellular polymers. *Water Res. 13*: 817–837.

Bunch, R. L. (1982). Status of physicochemical treatment in the U.S.A. In *Physicochemical Methods for Water and Wastewater Treatment*, L. Pawlowski (Ed.). Elsevier, Amsterdam.

Busch, P. L., and Stumm, W. (1968). Chemical interactions in the aggregation of bacteria—bioflocculation in waste treatment. *Env. Sci. Tech. 2*: 49–53.

Caillere, S., and Henin, S. (1972). Palygorskite. In *X-ray Identification and Crystal Structures of Clay Minerals*, G. Brown (Ed.). Mineralogical Society (Clay Minerals Group), London.

Campbell, L. A. (Aug. 1972). The initiation of bioflocculation. *Water and Pollution Control*, pp. 14–17.

Corpe, W. A. (1972). Periphytic marine bacteria and the formation of microbial films on solid surfaces. In *Effect of the Ocean Environment on Microbial Activities*, R. R. Colwell and R. Y. Morita (Ed.). Universities Park Press, Baltimore.

Costerton, J. W., Geesey, G. G., and Cheno, K. -J. (1978). How bacteria stick. *Sci. Am. 238*: 86–95.

Crabtree, K., McCoy, E., and Boyle, W. C. (1965). Isolation, identification, and metabolic role of the sudanophilic granules of *Zoogloea ramigera*. *Appl. Microbiol. 13*: 218–226.

Davis, C. A. (1965). Microbiological beneficiation of phosphate slime. M. S. Thesis, University of Alabama.

Davis, J. B. (1967). *Petroleum Microbiology*. Elsevier Publishing Company, New York.

Deinema, M. H. (1972). Poly-beta-hydroxybutyrate. *Appl. Microbiol. 24*: 857–858.

Deinema, M. H., and Zevenhuizen, L. P. T. M. (1971). Formation of cellulose fibrils by gram-negative bacteria and their role in bacterial flocculation. *Arch. Mikrobiol. 78*: 42–57.

Dexter, S. C., Sullivan, J. D., Jr., Williams, J. III, and Watson, S. W. (1975). Influence of substrate wettability on the attachment of marine bacteria to various surfaces. *Appl. Environ. Microbiol. 30*: 298–308.

Diem, H. G. (1975). Attraction and repulsion of bacteria by spores of *Cladosporium cladosporioides* during germination. *Can. J. Bot. 53*: 1092–1096.

Dugan, P. R. (Sept. 1975). Bioflocculation and the accumulation of chemicals by floc-forming organisms. Environmental Protection Agency Report EPA-600/2-75-032.

Dugan, P. R., and Pickrum, H. M. (1972). Removal of mineral ions from water by microbially produced polymers. *Proceedings of the 27th Purdue Industrial Waste Conference*, pp. 1019–1038.

Eveleigh, D. E. (1970). Fungal disfigurement of paper and soft rot of cedar shingles. *Appl. Microbiol. 19*: 872--874.

Fletcher, M. (1977). The effects of culture concentration and age, time, and temperature on bacterial attachment to polystyrene. *Can. J. Microbiol. 23*: 1–6.

Fletcher, M., and Floodgate, G. D. (1976). The adhesion of bacteria to solid surfaces. In *Microbial Ultrastructure: The Use of the*

Electron Microscope, R. Fuller and D. W. Lovelock (Eds.). Academic Press, London.

Forster, C. F., and Lewin, D. C. (1972). Polymer interactions at activated sludge surfaces. *Effluent Water Treatment Journal 12*: 520–525.

Gale, N. L., and Wixson, B. G. (1979). Removal of heavy metals from industrial effluents by algae. *Dev. Ind. Microbiol. 20*: 259–273.

Gary, J. H., Feld, I. L., and Davis, E. G. (1963). Chemical and physical beneficiation of Florida phosphate slimes. U.S. Bureau of Mines Report of Investigations 6163.

Geesey, G. G. (1982). Microbial exopolymers: Ecological and economic considerations. *ASM News 48*: 9–14.

Gerson, D. F., and Zajic, J. E. (1979). Microbial biosurfactants. *Process Biochemistry*, pp. 20–22.

Goren, M. B. (1968). Process for flocculating finely divided solids suspended in an aqueous medium with a microbial polysaccharide. U.S. Patent No. 3,406,114.

Grim, R. E. (1968). *Clay Mineralogy*, Second Edition. McGraw-Hill Book Co., New York.

Hannah, S. A., Cohen, J. M., and Robeck, G. G. (1967). Measurement of floc strength by particle counting. *J. Am. Water Works Assoc. 59*: 843–858.

Harris, R. H., and Mitchell, R. (1973). The role of polymers in microbial aggregation. *Ann. Review Microbiol. 27*: 27–50.

Hodge, V. F., Koide, M., and Goldberg, E. D. (1979). Particulate uranium, plutonium and polonium in the biogeochemistries of the coastal zone. *Nature 277*: 206–209.

Hoppe, R. W. (1976). Phosphates are vital to agriculture and Florida mines for one-third the world. *Eng. Min. J. 177*: 79–89.

Horikawa, Y., Terai, T., and Ogura, H. (1979). Mutual flocculation between selected clay minerals and some kinds of asexual spores from soil-borne fungi. *Soil Sci. Plant Nutr. (Tokyo) 25*: 357–364.

Jilek, R., Prochazka, H., Stamberg, K., Katzer, J., and Nemec, P. (1975). Properties and development of cultivated biological sorbent. *Rudy (Prague) 23*: 282–286.

Kang, K. S., and Cottrell, I. W. (1979). Polysaccharides. *Microbial Technology 1*: 417–481.

Kelly, D. P., Norris, P. R., and Brierley, C. L. (1979). Microbiological methods for the extraction and recovery of metals. In *Microbial Technology: Current State, Future Prospects*, A. T. Bull, D. C. Ellwood, and C. Ratledge (Ed.). Soc. for Gen. Microbiol. Symp. 29, Cambridge University Press, Cambridge.

Lamont, W. E., McLendon, J. T., Clements, L. W., Jr., and Feld, I. L. (1975). Characterization studies of Florida phosphate slimes. U.S. Bureau of Mines Report of Investigations 8089.

Lynch, J. M. (1981). Promotion and inhibition of soil aggregate stabilization by micro-organisms. *J. Gen. Microbiol. 126*: 371–375.

Margaritis, A., Zajic, J.E., and Gerson, D. F. (1979). Production and surface-active properties of microbial surfactants. *Biotechnol. Bioeng. 21*: 1151–1162.

Marshall, K. C. (1969). Orientation of clay particles sorbed on bacteria possessing different ionogenic surfaces. *Biochem. Biophys. Acta 193*: 472–474.

Marshall, K. C. (1979). Theoretical and practical significance of bacteria at interfaces. *Dev. Ind. Microbiol. 20*: 1–7.

Marshall, K. C. (1980). Microorganisms and interfaces. *BioScience 30*: 246–249.

Marshall, K. C., and Cruickshank, R. H. (1973). Cell surface hydrophobicity and the orientation of certain bacteria at interfaces. *Arch. Mikrobiol. 91*: 29–40.

Marshall, K. C., Stout, R., and Mitchell, R. (1971a). Selective sorption of bacteria from seawater. *Can. J. Microbiol. 17*: 1413–1416.

Marshall, K. C., Stout, R., and Mitchell, R. (1971b). Mechanism of the initial events in the sorption of marine bacteria to surfaces. *J. Gen. Microbiol. 68*: 337–348.

May, A. (1975). Electrophoretic mobilities and cation exchange capacities of Florida phosphate slimes. U.S. Bureau of Mines Report of Investigations 8046.

May, A., and Smelley, A. G. (1979). Effect of electrolytes on the electrophoretic mobilities of Florida phosphatic clay wastes. U.S. Bureau of Mines Report of Investigations 8398.

McKinney, R. E. (1952). A fundamental approach to the activated sludge process. II. A proposed theory of floc formation. *Sewage Ind. Wastes 24*: 280–287.

Morris, A. W. (1971). Trace metal variations in sea water of the Menai Straits caused by a bloom of *Phaeocystis*. *Nature 233*: 427–428.

Nakamura, J., Miyashiro, S., and Hirose, Y. (1976). Screening, isolation, and some properties of microbial cell flocculants. *Agr. Biol. Chem. 40*: 377–383.

Norris, P. R., and Kelly, D. P. (1979). Accumulation of metals by bacteria and yeasts. *Dev. Ind. Microbiol. 20*: 299–308.

Pavoni, J. L., Tenney, M. W., and Echelberger, W. F., Jr. (1972). Bacterial exocellular polymers and biological flocculation. *J. Water Pollution Control Fed. 44*: 414–451.

Rosenzweig, W. D., and Stotzky, G. (1979). Influence of environmental factors on antagonism of fungi by bacteria in soil: Clay minerals and pH. *Appl. Environ. Microbiol. 38*: 1120–1126.

Sanford, P. A., Burton, K. A., Watson, P. R., Cadmus, M. C., and James, A. (1975). Extracellular polysaccharide from the black yeast NRRL Y-6272: Improved methods for preparing a high viscosity, pigment-free product. *Appl. Microbiol. 29*: 769–775.

Sarma, V. A. K., and Dakshinamurti, C. (1971). Relative efficiency of bacterial gum and organic matter extract as aggregating agents in sand-clay mixtures. *Curr. Sci.* *2*: 35–36.

Shaw, D. J. (1966). *Introduction to Colloid and Surface Chemistry.* Buttersworth, London.

Slodki, M. E., and Cadmus, M. C. (1978). Production of microbial polysaccharides. *Adv. Appl. Microbiol.* *23*: 19–54.

Smelley, A. G., and Feld, I. L. (1979). Flocculation dewatering of Florida phosphatic clay wastes. U.S. Bureau of Mines Report of Investigation 8349.

Smiley, K. L. (1976). Microbial polysaccharides—a review. *Env. Tech.* *20*: 112–116.

Souw, P., and Demain, A. L. (1979). Nutritional studies on xanthan production by *Xanthomonas campestris* NRRL B1459. *Appl. Environ. Microbiol.* *37*: 1186–1192.

Staff (1975). The Florida phosphate slimes problem—a review and a bibliography. U.S. Bureau of Mines Report of Investigations 8668.

Stotzky, G. (1966). Influence of clay minerals on microorganisms. III. Effect of particle size, cation exchange capacity, and surface area on bacteria. *Can J. Microbiol.* *12*: 1235–1246.

Stotzky, G., and Rem, L. T. (1966). Influence of clay minerals on microorganisms. I. Montmorillonite and kaolinite on bacteria. *Can. J. Microbiol.* *12*: 547–563.

Stumm, W., and Morgan, J. I. (1962). Chemical aspects of coagulation. *J. Am. Water Works Assoc.* *54*: 971–994.

Sutherland, I. W., and Ellwood, D. C. (1979). Microbial exopolysaccharides—industrial polymers of current and future potential. In *Microbial Technology: Current State, Future Prospects*, A. T. Bull, D. C. Ellwood and C. Ratledge (Ed.). Soc. of Gen. Microbiol. Symp. 29, Cambridge University Press, Cambridge.

Tenney, M. W., and Stumm, W. (1965). Chemical flocculation of microorganisms in biological waste treatment. *J. Water Pollution Control Fed.* *37*: 1370–1388.

Terichow, O., and May, A. (1973). Electrophoresis and coagulation studies of some Florida phosphate slimes. U.S. Bureau of Mines Report of Investigations 7816.

Theng, B. K. G. (1979). *Formation and Properties of Clay-Polymer Complexes.* Elsevier Scientific Publishing Co., Amsterdam.

Tischer, R. G., and Davis, E. B. (1971). The effect of various nitrogen sources upon the production of extracellular polysaccharide by the blue-gree alga *Anabena flos-aquae* A-37. *J. Exp. Bot.* *22*: 546–551.

Ueda, S., Fujita, K., Komatsu, K., and Nakashima, Z. (1963). Polysaccharide produced by the genus *Pullularia* I. Production of polysaccharide by growing cells. *Appl. Microbiol.* *11*: 211–215.

Van Olphen, H. (1977). *An Introduction to Clay Colloid Chemistry*, Second Ed. John Wiley & Sons, New York.

Weaver, T. L., and Dugan, P. R. (1972). Enhancement of bacterial methane oxidation of clay minerals. *Nature 237*: 518.
Weaver, C. C., and Pollard, L. D. (1973). *The Chemistry of Clay Minerals. Developments in Sedimentology 15.* Elsevier Scientific Publishing Co., Amsterdam.
Whitfield, C., Sutherland, I. W., and Cripps, R. E. (1981). Surface polysaccharides in mutants of *Xanthomonas campestris*. *J. Gen. Microbiol. 124*: 385–392.
Yadav, N. K., Abrol, I. P., and Vyas, S. R. (1970). Note on the effect of gum-producing bacteria on soil aggregation. *Ind. J. Agric. Sci. 40*: 623–625.
Zajic, J. E. (1967). Process for preparing a polysaccharide flocculating agent. U.S. Patent No. 3,320,136.
Zajic, J. E., and LeDuy, A. (1973). Flocculant and chemical properties of a polysaccharide from *Pullularia pullulans*. *Appl. Microbiol. 25*: 628–635.
Zvyagintsev, D. G., Pertsovskaya, A. F., Duda, V. I., and Nikitin, D. I. (1969). Electron-microscope study of the adsorption of microorganisms on soil and minerals. *Mikrobiologiya 38*: 1091–1095.

9

Low-Level Radioactive Wastes in Subsurface Soils

AROKIASMY J. FRANCIS / *Brookhaven National Laboratory, Upton, New York*

I. INTRODUCTION

Low-level radioactive wastes are generated from nuclear power plants, research institutions, hospitals, universities, radiopharmaceutical manufacturers, fuel fabrication industry, government energy research, and military programs. In addition to radionuclides, the bulk of the waste consists of natural and synthetic organic materials. These wastes are usually disposed of in shallow-land trenches and pits. Of particular concern in the disposal of solid wastes is the contamination of surface and ground waters by waste leachates containing radionuclides and other toxic substances. Some problems encountered over the past 35 years of shallow-land disposal of low-level radioactive wastes on both government and commercial sites are biota intrusion, biodegradation of waste forms, trench cover subsidence, migration of radionuclides due to the presence of chelating and complexing agents, contamination of ground surfaces, release of radioactive gases, flooding, fire, erosion, siting in areas that are geologically very complex, water accumulation in disposal trenches and disposal in areas of high topographic relief, which make surface water movement a particular site concern.

Microorganisms play a significant role in the transformation of the radioactive waste and waste forms disposed of at shallow-land burial sites. Various microbial processes may bring about (1) changes in pH and Eh (oxidation-reduction reactions), which affect the valence or ionic state of the radionuclides and enhance their mobility in the environment by retarding the soil-binding characteristics; (2) chelation, solubilization, and leaching of certain elements by microbial metabolites or decomposition products; (3) bioaccumulation, biomethylation, and production of volatile and/or toxic alkylated metal compounds; (4) production of radioactive, toxic, and explosive gases which result in container pressurization and explosion in burial sites; and (5) enhancement of radionuclide uptake by crop plants.

In this paper the various microbial processes that significantly affect the long-term storage, stability, movement, and isolation of the radionuclides from the disposal environment are discussed.

A. Classification of Radioactive Waste

Radioactive wastes are classified as high-level wastes, transuranic wastes, and low-level wastes (DOE, 1979). High-level wastes (HLW) are either intact fuel assemblies that are discarded after serving their useful life in a nuclear reactor (spent fuel) or the portion of the wastes generated in the reprocessing of spent fuel that contain virtually all of the fission products and most of the actinides remaining after reprocessing. These wastes are being considered for disposal in geologic repositories or by other technical options designed to provide long-term isolation of the wastes from the biosphere.

Transuranic (TRU) wastes result predominantly from spent fuel reprocessing, the fabrication of plutonium to produce nuclear weapons, and plutonium fuel fabrication for recycle to nuclear reactions. TRU waste is currently defined as material containing more than 10 nCi of transuranic activity per gram of material. These wastes will be disposed of in a manner similar to that used for high-level waste disposal.

Low-level wastes (LLW) contain less than 10 nCi of transuranic contaminants per gram of material, or they may be free of transuranic contaminants, require little or no shielding, and have low but potentially hazardous concentrations or quantities of radionuclides. LLW are generated in almost all activities involving radioactive materials and are presently being disposed of by shallow-land burial.

Uranium mine and mill tailings are the residues from uranium mining and milling operations that contain low concentrations of naturally occurring radioactive materials. The tailings are generated in very large volume and are currently stored at the site of mining and milling operations.

B. Shallow-Land Burial of Low-Level Waste

Low-level solid wastes are buried in shallow rectangular trenches and pits with varying dimensions, ranging from 60 m to 150 m long, 8 m to 15 m wide, and 6 m deep. The trench floor is sometimes sloped toward one or more sumps, and riser pipes are provided for routine observation of water levels and for removal of water. In the past, the long-term performance of these sites was not considered, and several problems developed as a result of improper disposal, site selection, and design of the trenches. Consequently, improved engineering design, site selection, construction, and operation of the future burial sites are being considered to alleviate the problems encountered with past burial practices (NRC, 1982). About 16 million ft^3 and 51 million ft^3 of low-level radioactive wastes from commercial and defense activities, respectively, are buried at shallow-land disposal sites. The half-lives of selected radionuclides commonly buried at shallow-land burial sites are given in Table 1. Many of the radionuclides are extremely radioactive and have long half lives such as fission product strontium-90 (^{90}Sr) (25 years) and transuranic isotope plutonium-239 (24,000 years).

An analysis of the history of low-level waste disposal at both commercial and government disposal sites (Clancy et al., 1981); a summary of the characteristics and waste volumes projected to the year 2000 and the treatment options of the waste (Wild et al., 1981); and the methodology for analysis of the impacts of handling and disposing of low-level waste based upon consideration of alternative waste forms, disposal facility design and operating practices, disposal facility environmental characteristics, and institutional control considerations (Oztunali et al., 1981) have been discussed in great detail.

Low-level radioactive wastes produced by nuclear industry, hospitals, universities, and research institutions consist of contaminated equipment, protective clothing, gloves, small tools, plastics, rags, paper, scintillation liquids, organic solvents, aqueous liquids containing radionuclides, decontamination solutions, radioactive carcasses of experimental animals, spent-ion exchange resins, and sludges. Many of the facilities producing these wastes have recently chosen to dispose of the bulk of the waste by incineration because of the high cost of shallow-land disposal and the relaxation of state and federal standards concerning disposal of these materials.

Commercially produced low-level wastes are buried at Barnwell, South Carolina; Beatty, Nevada; Hanford, Washington; Maxey Flats, Kentucky; Sheffield, Illinois; and West Valley, New York. The volume and radioactivity of the wastes buried at commercial sites are given in Table 2. Three of the sites, Maxey Flats, Sheffield, and West Valley, have been closed and are not currently accepting wastes.

Low-level wastes resulting from government energy research and military programs are generated by Department of Energy research in new applications of nuclear energy, reactor safety, and waste treatment techniques; production and use of nuclear fuel for submarines and ships; and veterans' hospitals and government-operated research clinics. These wastes are disposed of at the government-owned and -operated disposal sites located at Oak Ridge National Laboratory (ORNL), Tennessee; Los Alamos National Laboratory (LANL), New Mexico; Hanford Reservation (HR), Washington; Idaho National Engineering Laboratory (INEL), Idaho; and Savannah River Plant (SRP), South Carolina. Other disposal

Table 1 Half-lives of Selected Radionuclides Disposed of at Shallow-Land Burial Sites

Radionuclide	Type of radiation emitted	Half-life (in years)
Americium-241	alpha	433
Carbon-14	beta	5730
Cesium-134	beta, gamma	2.06
Cesium-137	beta, gamma	30
Cobalt-60	beta, gamma	5.26
Curium-244	alpha	18.11
Hydrogen-3 (tritium)	beta	12.3
Nickel-63	beta	100
Plutonium-238	alpha	87.74
Plutonium-239	alpha	2.41×10^4
Plutonium-240	alpha	6.57×10^3
Radium-226	alpha	1600
Strontium-90	beta	28.8
Thorium-232	alpha	1.40×10^{10}
Uranium-234	alpha	2.45×10^5
Uranium-235	alpha	7.04×10^8
Uranium-238	alpha	4.47×10^9

facilities include Nevada Test Site (NTS), Pantex Plant (Texas), Sandia Laboratory (New Mexico), the Feed Materials Production Center (Ohio), the National Lead Company (Niagara Falls, New York), the Paducah Gaseous Diffusion Plant (Kentucky), the Portsmouth Gaseous Diffusion Plant (Ohio), Oak Ridge Gaseous Diffusion Plant (Tennessee), Oak Ridge Y-12 Plant (Tennessee), Weldon Springs Site (Missouri), and Brookhaven National Laboratory (New York).

C. Problems Encountered at the Shallow-Land Disposal Sites

Several problems have been encountered at both commercial and government sites because of inadequate criteria used in site selection, characterization, and design. These problems are catagorized as bathtub effect, loss of trench cap integrity, erosion, high water table, hydrogeologic complexity, flooding, leachate chemistry, and rapid radionuclide migration. Several radionuclides such as americium-241, barium-133, carbon-14, cesium-134 and 137, cobalt-60, iodine-129, nickel-63, plutonium-238, 239, and 240, strontium-90, thorium-232, tritium, and uranium-234, 235, and 238 have been identified in leachate samples. Concentrations of radionuclides found in leachate samples are given in Table 3. Radionuclides can be released from the disposal trenches through (1) surface-water runoff, (2) atmospheric fallout from the

Table 2 Radioactive Wastes Buried at Commercial Sites

Site	Waste volume (ft^3)	Radioactivity (Curies)	Plutonium (kg)
Maxey Flats, KY	4.8×10^6	2.4×10^6	64
West Valley, NY	2.4×10^6	7.1×10^5	4
Sheffield, IL	3.2×10^6	6.0×10^4	13
Barnwell, SC	11.4×10^6	1.7×10^6	NA
Richland, WA	2.2×10^6	9.5×10^5	37
Beatty, NV	3.2×10^6	4.6×10^5	14

evaporator used to concentrate the accumulated water samples in the trenches, (3) lateral movement from trenches through the soil zone, and (4) movement from the trenches through fractures in surrounding rocks.

The problems encountered at the Maxey Flats site are brought about by a number of interrelated factors, including site characteristics, waste form, site design and operation, and institutional considerations. The low permeability of the soils, the humid environment, and the site operational practices have resulted in water accumulation leading to a "bathtub" effect in many of the trenches. In addition, several fractured formations exist in the subsurface media. The waste form is one of the major contributors to the problem. Most of the waste disposed of at the site is made up of decomposable material as received so that large void spaces exist within the waste or between the waste and the packaging. The presence of other chemicals increased biodegradation of waste and leaching of radionuclides. This led to settlement of the disposal trench contents, followed by subsidence of the trench covers that increased water percolation into the disposal trenches.

Major problems at the West Valley facility are the accumulation of leachate within the disposal trenches and the erosion of a portion of the disposal site. Water levels within some trenches started to rise, and reached the ground level from one trench and broke the ground cover and started to flow at a rate of 4 liters per day. Consequently, pumping of the leachate from the problem trenches was employed as a remedial action. This site was closed in 1975, after 12 years of operation.

The three major problems encountered at the Sheffield, Illinois, disposal site were erosion, subsidence, and elevated radionuclide concentrations in some ground water samples collected from some on-site wells and disposal trench sumps. Trench subsidence occurred in the form of several large sinkholes, one of which was approximately 10 ft deep and 8 ft X 12 ft in area. Tritium was the major radionuclide detected in water samples from on-site monitoring wells.

Table 3 Radiochemical Analysis of Trench Leachate from Commercial Low-Level Radioactive Disposal Sites

	Range (pCi/liter)	
Analysis	Maxey Flats, KY	West Valley, NY
Gross Alpha	140--640,000	<200–29,000,000
Gross Beta	1,500--57,000,000	91,000--31,000,000
Gross Gamma	<10–16,000 cpm[a]	—
Tritium	250,000–7,400,000,000	220,000--5,600,000
Sodium-22	23–130	—
Manganese-54	170--190,000	—
Cobalt-60	19–840,000	<110--10,000
Strontium-90	1,900–9,900,000	—
Cesium-134	<100–22,000	<110–330,000
Cesium-137	<20–170,000	<160–900,000
Plutonium-238	<2–126,000	—
Plutonium-239/240	<1--21,000	—
Americium-241	<20–28,000	<170–1,110

[a]cpm, Counts per minute.
Source: Clancy et al. (1981).

At the ORNL disposal site the principal radiocontaminant released from the solid waste disposal area (SWDA)-4 is ^{90}Sr; concentrations of this isotope in the seeping liquid are approximately 5×10^{-6} μCi/ml. The calculated discharge of ^{90}Sr from the site in the mid-1970s was between 1 and 2 Ci per year. Thirteen different surface seeps of contaminated water have been observed at SWDA-5. These seeps are the result of the "bathtub" effect, which is due to filling of a disposal trench with water which occurs where the rate of infiltration into the trench is much greater than the rate of percolation out of the trench over long time periods. After the trench fills with water, the height differential between the two ends of the trench provides a driving force for surface seepage.

High concentrations of ^{90}Sr and significant concentration levels of ^{238}Pu and ^{244}Cu were observed in one of the SWDA-5 seeps. In SWDA-6 only, contaminated trench leachate containing ^{90}Sr (880 pCi/ml) has been observed within individual disposal trenches. Mitigative measures to prevent seepage from the trenches through precipitation reactions with the application of sodium carbonate appear to be successful on a short-term basis (Spalding, 1981).

There have been a number of incidents related to low-level waste management operations at the Hanford Reservation site in Washington. Two of them involved fires in waste disposal trenches, and a number of incidents resulted from contamination of ground surfaces due to leaks or spills of both high-level and low-level radioactive waste liquids. A number of contaminated areas still remain on the reservation, although off-site impacts from the contaminated soils are believed to be minimal.

Intrusion of plants and animals into disposed waste have been recorded. For example, burrowing animals have, on occasion, burrowed into liquid waste disposal cribs in an effort to obtain salts deposited by the percolating liquids. Radioactive salts thus consumed were then dispersed by the burrowing animals and the predators. Swallows have been observed to obtain radioactive mud from settling basins for use in constructing nests. Plants growing over disposal trenches and cribs have accumulated fission products and transuranic elements in shoot tissues. In arid and semiarid low-level radioactive waste disposal sites, only minor problems were encountered.

II. ORGANIC COMPOUNDS IN WASTE LEACHATE

The presence of organic materials in the waste creates great concern because they are responsible for most of the problems encountered at burial sites. Many of the organic compounds are capable of forming stable complexes with radionuclides or of increasing the solubilization and leaching of the buried radionuclides. The distributions of organic compounds in trench leachate and monitoring well water samples were studied by Francis et al. (1980b). Leachate samples collected from several low-level radioactive waste disposal sites contained high levels of dissolved organic carbon (Table 4). These values may be compared with the value for unpolluted groundwater, which is reported by most investigators to range from 5 and 30 ppm and usually less than 10 ppm.

A comprehensive list of all organic compounds that have been identified by methylene chloride extraction and gas chromatographic-mass spectrometric (GC/MS) analysis of trench leachates from the West Valley and Maxey Flats sites is given in Table 5. It should be noted that not all compounds can be extracted by methylene chloride, and probably many other organic compounds are present in the leachates. Several straight- and branched-chain aliphatic acids, aromatic acids, alcohols, aldehydes, ketones, amines, aromatic hydrocarbons, esters, ethers, and phenols were detected.

Typically, the acidic fraction of the samples contained several compounds, the neutral fraction contained fewer compounds, and the basic fraction, except in the case of one or two trench leachate samples, yielded no detectable compounds. This may be due to the limitation of using a gas chromatographic column not specifically selected for the detection of basic compounds. Quantitative results on the numerous compounds detected in Maxey Flats and West Valley samples are shown in Table 6. In many cases, samples of a trench leachate taken a year apart contained identical and similar types of major compounds, although some quantitative variations between the two samplings were generally observed.

Table 4 Distribution of Water Samples from Low-level Radioactive Waste Disposal Sites According to Dissolved Organic Carbon Content

Sample source	Total number of samples	Number of samples DOC Content (mg/liter) 2–10	10–100	100–1000	1000–7000
Maxey Flats, KY					
Trench	46	2	10	23	11
Wells	10	10	0	0	0
West Valley, NY					
Trench	13	0	0	4	9
Wells	10	8	2	0	0
Sheffield, IL					
Trench	1	0	1	0	0
Wells	26	0	4	13	9
Barnwell, SC					
Trench	9	4	2	3	0
Wells	3	3	0	0	0

Source: Francis et al. (1980b).

The types of organic compounds identified in the trench leachates reflect both the nature of the buried waste and the products of biodegradation. For example, *p*-dioxane, toluene, and xylene may be discarded solvents or scintillation fluids. Tributyl phosphate may originate from the process of solvent extraction of metal ions from solutions of reactor products, and low-molecular-weight straight- and branched-chain aliphatic acids are due mainly to microbial degradation of complex natural and synthetic materials in the buried waste.

Some mechanisms of radionuclide migration have been examined by de Marsily et al. (1977), including transport in ground water by convection, diffusion, and sorption phenomena. These mechanisms are complicated by the ability of radionuclides to form stable complexes with natural or microbially produced organic compounds that are buried with the radioactive waste. Studies with heavy metals and naturally occurring humic acid have shown that soluble metal complex formation may be important in mobilization (Zurino and Martin, 1977; Takamatsu and Yoshida, 1978). Likewise, microbial metabolites and waste degradation products or intermediates may be an important source of agents that affect the long-term solubility characteristics of transuranic elements. Bolter et al. (1975) reported that organic acids from decaying leaf litter in soil increased the solubility of heavy metals deposited from smelters. Ethylenediaminetetraacetic acid (EDTA), an extremely strong complexing

Table 5 Compounds Identified in Trench Leachates from Maxey Flats and West Valley Disposal Sites

Acids	*Aldehydes and ketones*
Benzoic acid	p-Hydroxybenzaldehyde
Butanoic acid	Paralydehyde
Decanoic acid	Vanillin
2-Ethylhexanoic acid	Acetovanillon
Hexanoic acid	Camphor
Hydroxybenzoic acid	Diacetone alcohol
3-Methoxy-4-hydroxybenzoic acid	Dibutyl ketone
2-Methylbutyric acid	Fenchone
3-Methylbutyric acid	Methyl ethyl ketone
2-Methylhexanoic acid	Methyl isobutyl ketone
2-Methylpentanoic acid	*Amines*
3-Methylpentanoic acid	
2-Methylpropionic acid	Aniline
Nonanoic acid	Cyclohexylamine
Octanoic acid	Dicyclohexylamine
Pentanoic acid	Methyldicyclohexylamine
Phenylacetic acid	*Aromatic hydrocarbons*
Phenylbutyric acid	
Phenylhexanoic acid	Benzene
Phenylpropionic acid	Biphenyl
Toluic acid (isomers)	Dimethylnaphthalene
Alcohols	Naphthalene
	Toluene
Borneol	Xylene (isomers)
2-Butanol	*Esters*
Cyclohexanol	
2-Ethylhexanol	Butyl phthalate
3-Ethylhexanol	Diethyl phthalate
2-Hexanol	Several unidentified phthalates
3-Hexanol	Tributyl phosphate
2-Methyl-2-butanol	Triethyl phosphate
Methylcyclohexanol	Triphenyl phosphate
Octanol	*Ethers*
2-Phenylcyclohexanol	
(α-Propylene glycol)$_n$	Anisole
α-Terpineol	Bis(2-Chloroethyl) ether
3,3,5-Trimethylcyclohexanol	Bis(2-chloroethoxy) methane
Phenols	Bis(2-Ethoxyethyl) ether
	1,1-Diethoxyethane
Cresol (isomers)	1,1-Diethoxy-2-chloroethane
Octylphenol	1,4-Dioxane
Phenol	Tetrahydrofuran
4-*t*-Butylphenol	Tripropylene glycol methyl ether
Tetramethylbutylphenol	

Source: Francis et al. (1980b).

Table 6 Organic Compounds Identified in Leachate Samples from Commercial Low-Level Radioactive Waste Disposal Sites

Compound	Concentration (mg/liter)			
	Maxey Flats, KY (5 Trenches)		West Valley, NY (5 Trenches)	
	Range	(mean)[a]	Range	(mean)[a]
Dissolved organic carbon (mg/liter)	540–1900	(940)	630–2900	(1966)
Acids				
Benzoic acid	0.22–1.9	(0.83)	0.28–13	(5.66)
C_8 branched acids[b]	0–1.4	(0.29)	–	
Decanoic acid	–		0–2.6	(1.02)
2-Ethylhexanoic acid	5.6–27	(12.4)	12–216	(106)
Hexanoic acid	1.2–4.7	(2.26)	0.15–65	(26.1)
Hydroxybenzoic acid	–		+	
2-Methylbutyric acid	0–19	(7.52)	1.0–189	(77.2)
3-Methylbutyric acid	0–5.8	(1.94)	0–17	(4.39)
2-Methylhexanoic acid	0.30–3.2	(1.32)	0.33–14	(7.05)
2-Methylpentanoic acid	0–1.4	(0.72)	0–20	(10.8)
3-Methylpentanoic acid	0–4.2	(2.26)	0–2.1	(1.01)
2-Methylpropionic acid	0–3.6	(1.32)	0–21	(9.66)
Nonanoic acid	–		0–6.4	(3.00)
Octanoic acid	0–1.3	(0.54)	4.6–20	(12.5)
Pentanoic acid	0–4.7	(2.52)	0–73	(31.8)
Phenylacetic acid	0.08–3.4	(1.16)	0.41–7.5	(4.06)
Phenylhexanoic acid	+		+	
Phenylpropionic acid	0.70–9.8	(2.73)	0.56–15	(8.09)
Toluic acid	–		0–0.78	(0.40)

Others				
Anisole	–		0–2.7	(0.54)
Biphenyl	–		+	
Bis(2-ethoxyethyl) ether	+		–	
Butyl phthalate	–		+	
Camphor	+		–	
Cresol (isomers)	0.28–4.2	(2.35)	0.56–4.2	(2.83)
Cyclohexanol	0–1.1	(0.34)	0–0.59	(0.12)
Dibutyl ketone	+		–	
Diethyl phthalate	0–0.08	(0.016)	–	
p-Dioxane	+		+	
2-Ethyl-1-hexanol	+		0–8.1	(4.8)
Fenchone	0–0.03	(0.01)	0–0.14	(0.3)
Methyl isobutyl ketone	+		–	
Naphthalene	0–0.12	(0.17)	0–1.7	(0.43)
Octyl phenol (isomer)	–		+	
Phenol	+		0–9.5	(2.2)
Phthalate (unknown)	+		+	
α-Terpineol	0–0.49	(0.16)	0–0.64	(0.16)
Toluene	0.39–6.9	(3.42)	0.31–18	(5.22)
Tributyl phosphate	0–0.36	(0.10)	0–0.86	(0.40)
Triethyl phosphate	0–0.03	(0.008)	–	
Tripropylene glycol methyl ether	–		+	
Vanillin	0–8.6	(1.72)	–	
Xylene (isomers)	0–0.48	(0.12)	–	

[a]–, Not detected; +, compound present but not quantified.
[b]Quantified by using 2-ethylhexanoic acid standard.
Source: Francis et al. (1980b).

agent commonly used in decontamination operations at nuclear facilities, was the dominant mobilizing agent of ^{60}Co and uranium from intermediate-level liquid waste disposal pits and trenches at the Oak Ridge National Laboratory burial grounds (Means et al., 1978). They also showed that organic ligands can have pronounced effects by reducing the absorption capacity of soil for radionuclides. Other constituents detected in trench leachates include palmitic acid, phthalic acid, and other mono- and dicarboxylic acids, which are much weaker complexing agents than EDTA. These ligands may also contribute to the mobilization of other radionuclides.

Analysis of trench leachates at the Maxey Flats radioactive waste disposal site show that plutonium exists as dissolved species (0.05 μm filtration) in the tetravalent state complexed with strong organic ligands such as EDTA (Cleveland and Rees, 1981). These complexes are not sorbed well by sediment and are only partly precipitated by ferric hydroxide. Gel permeation chromatographic data indicate that plutonium was distributed through several widely separated organic fractions, although the organic species were not identified (Cleveland and Rees 1981). Similar results were reported by Kirby et al. (1982). The latter study reported that EDTA was the major organic chelating agent in sump water from Maxey Flats disposal site. Other polar water-soluble organics were also detected but were not identified. A variety of organic acids and other water-soluble polar compounds coeluted with ^{90}Sr and ^{137}Cs. Kirby et al. (1982) concluded that chelation of radionuclides by polar organic species, such as organic acids, provide a mechanism for the migration of radionuclides at the waste disposal sites. In the presence of water containing strongly complexing inorganic species, plutonium can be mobilized to a significant degree as a soluble organic complex that can withstand submicrometer filtration, ferric hydroxide carrier precipitation, and surface adsorption on sediments (Cleveland and Rees, 1981). Significant amounts of the synthetic chelating agent EDTA have been found in the leachate samples (Table 7) and have been responsible for radionuclide mobilization from the disposal sites. Nitrilotriacetic acid (NTA) and diethylenetriamine pentaacetic acid (DTPA) were not detected in any of the leachate samples analyzed.

III. MICROBIAL POPULATION AND ACTIVITY

Low-level radioactive wastes buried in shallow-land disposal sites contain solids and liquids generated from fuel-cycle and nonfuel-cycle sources. In general, much of the waste consists of organic materials such as contaminated paper, clothing, plastics, packing materials, cleanup and decontamination solutions, cardboard boxes, and radioactive carcasses of experimental animals. Microorganisms play an important role in the decomposition of natural and man-made organic materials

Table 7 Concentration of EDTA in Trench Leachate Samples from Low-Level Shallow-Land Radioactive Waste Disposal Sites

Waste disposal site	EDTA (mg/liter)	Reference
Maxey Flats, KY		
Trench 19S	0.11	Cleveland and Rees (1981)
Trench 26	0.5	Cleveland and Rees (1981)
Trench 27	0.1	Cleveland and Rees (1981)
Trench 27	0.99	Kirby et al. (1982)
Trench 32	0.2	Cleveland and Rees (1981)
Trench 33–L–4	0.1	Cleveland and Rees (1981)
Oak Ridge National Laboratory Burial Site		
Trench 7 - RS7	0.1	Means et al. (1978)
Pit 4 - RS3	+	Means et al. (1978)
Trench 5 - RS9	+	Means et al. (1978)

and are responsible for the transformation of various elements in nature. Some of the problems encountered at the commercial and government disposal sites have been largely due to microbial activity such as decomposition of the waste material resulting in subsidence of the trench cover, gas generation, and the interaction of microbial metabolites with radionuclides which results in the complexation, solubilization, and leaching of radionuclides.

A. Microbial Population Distribution

Little information dealing with the microbiological decomposition of radioactive waste is available; however, studies with leachate samples collected from several disposal sites strongly indicate that significant microbial activity is occurring at these sites (Francis et al., 1980a). Trench leachate samples collected anoxically from four commercial disposal sites were analyzed for total aerobic and anaerobic populations and the distribution of denitrifying, sulfate reducing, and methanogenic bacteria (Table 8). Among the several aerobic and anaerobic bacteria isolated, only *Bacillus* sp., *Pseudomonas* sp., *Citrobacter* sp., and *Clostridium* sp. were identified. Several unidentified aerobic Gram-negative isolates with different biochemical properties were present. The radionucludes and the organic compounds present in the trench leachates may be acting as mutagenic agents on these bacteria.

Barnhart et al. (1980) enumerated microorganisms in core samples from a Los Alamos National Laboratory (LANL) low-level transuranic waste burial trench. They found 1.5×10^6 aerobic bacteria, 2.8×10^3

Table 8 Bacterial Population Distribution in Leachate Samples from Commercial Low-Level Radioactive Waste Disposal Sites

Sample	Collection date	Aerobic (CFU/ml)	Anaerobic (CFU/ml)	Denitrifiers (MPN/ml)	Sulfate reducers (MPN/ml)	Methanogens (MPN/ml)
Maxey Flats, KY						
Trench 2	7/77	1.2×10^3	1.0×10^2	NA	NA	NA
26	7/77	4.7×10^3	4.1×10^2	NA	NA	NA
32[a]	7/77	4.8×10^4	1.2×10^4	NA	NA	NA
19	5/78	2.2×10^2	3.2×10^2	3.3×10^1	4.0×10^0	4.9×10^0
Well UBIA	5/78	3.4×10^3	NA[b]	4.6×10^2	ND[c]	1.0×10^0
West Valley, NY						
Trench 3	10/78	5.0×10^4	4.0×10^3	1.3×10^4	7.0×10^1	2.3×10^1
4	10/78	2.3×10^3	3.3×10^3	2.3×10^3	4.9×10^2	1.7×10^0
5	10/78	1.6×10^3	3.5×10^2	3.3×10^2	1.1×10^1	ND
8	10/78	1.4×10^3	7.6×10^2	7.9×10^2	1.7×10^2	1.0×10^0
9	10/78	5.0×10^2	7.3×10^3	1.3×10^2	3.5×10^2	4.5×10^0

Barnwell, SC						
Trench 8D2	3/79	2.0×10^8	1.0×10^4	2.3×10^5	1.1×10^0	0.8×10^0
6D1	3/79	3.3×10^3	1.3×10^2	1.1×10^3	ND	ND
25/21D1	3/79	3.5×10^4	2.2×10^3	1.3×10^4	1.3×10^2	0.2×10^0
3D1	3/79	1.5×10^5	1.2×10^3	5.4×10^4	ND	ND
Sheffield, IL						
Trench 14A	4/79	1.7×10^5	4.4×10^4	2.4×10^5	ND	0.2×10^0
18	4/79	7.1×10^2	6.9×10^1	9.5×10^2	4.9×10^1	ND
525	4/79	6.3×10^2	4.2×10^2	1.7×10^3	2.3×10^0	ND

[a]Sample analyzed 7 days after collection.
[b]NA, Not analyzed.
[c]ND, None detected.
Source: Francis et al. (1980a).

anaerobic bacteria, and 3.5 X 10^3 fungi at a depth of 5–15 cm; 1.3 X 10^5 aerobic bacteria, 1.4 X 10^2 anaerobic bacteria, and 5.8 X 10^2 fungi at a depth of 40 to 50 cm; and 4.5 X 10^4 aerobic bacteria, no anaerobic bacteria, and 6.5 X 10^1 fungi at a depth of 100–110 cm. The radioactivity measured as dpm/g of soil was as follows: at 40–50 cm, 4 X 10^4 γ, 6 X 10^4 β; at 100 to 110 cm 2 X 10^3 γ, 2 X 10^3 β, and no activity was detected at 5–15-cm depth. The γ-emitter was identified as ^{137}Cs; the β-emitters were not identified; and there was no detectable α-activity. The levels of radioactivity found in the soil samples did not eliminate the microorganisms.

Microorganisms detected in radioactive LANL transuranic waste burial site soil and flammable waste contaminated with ^{239}Pu (Barnhart et al., 1980) are viable and metabolically active. Potential microbial effects include radionuclide mobilization by microbially produced chelates which can mediate dissolution of essentially insoluble plutonium dioxide, and by possible gas pressurization of radioactive waste storage containers or enclosures as the result of microbially produced gases such as carbon dioxide, methane, NO_x, and N_2. An analysis of the experimental results of waste degradation studies shows that microorganisms produce far more gas than that produced by physical and chemical processes (Molecke, 1979).

Bacterial populations in a reactor-spent resin (cation, anion, and mixed resin) and in influent and effluent cooling water samples from the reactor at Brookhaven National Laboratory (BNL) were analyzed (Table 9) by Francis and Quinby (1981). Bacteria present in the resin and water samples when incubated with the resin produced CO_2, albeit at a slow rate, indicating that these organisms may play a significant role in the degradation of the resin, the radiolytic decomposition products of resin, or other organic compounds present in the resin.

B. Effect of Radiation and Radionuclides on Microorganisms

1. *Effect of Radiation on Microorganisms*

Microorganisms in general are tolerant or resistant to ionizing radiation. Among bacteria, *Micrococcus radiodurans* is resistant to γ-radiation (<10^6 rad) and has been studied in detail. Much of the information on radiation resistance of microorganisms comes from the food industry, because ionizing radiation has been proposed as a means of preserving foods and is being used in many countries. By exposing foods to low doses of ionizing radiation (<1 mrad), one may substantially reduce the number of microorganisms responsible for food spoilage and thereby extend the shelf life of the food. The radiation dose, the moisture content, and various other parameters including free radical formation effects on microbial survival have been studied.

γ-radiation has been used in the disinfection and treatment of waste water effluents. Ridenour and Armbruster (1956) reported that γ-radiation from a ^{60}Co source at a dosage of 8.4 X 10^4 rad was sufficient to

Table 9 Population of Bacteria in BNL Reactor-Spent Resin and Water Samples

Sample	pH	Number of bacteria
Resin	6.00	6.59×10^8 g^{-1} dry wt
Influent H_2O	5.65	3.64×10^5 ml^{-1}
Effluent H_2O	5.90	1.72×10^5 ml^{-1}

reduce the total bacterial population in effluent from primary waste water treatment plants by 99%. Lowe et al. (1956) found that 99% microbial inactivation dosage was 1.5×10^5 rad, whereas that for sterilization was 5.0×10^6 rad. When a secondary waste water treatment plant effluent was treated with 10^5 rad, a 99.9% reduction in both total bacterial plate count and coliform numbers was observed (Touhill et al., 1969).

The effect of moisture content on bacterial inactivation by ionizing radiation in waste water sludge has been studied (Ward et al., 1981). Most bacteria were found to be somewhat protected from ionizing radiations at reduced moisture levels. However, no excessively large D_{10} values (the amount of irradiation required to inactivate 90% of a population of a particular organism or group of organisms) were found for the bacterial species tested. Similar results were observed with soil and soil isolates.

Many radioresistant strains of bacteria have been isolated from soil. For example, *Arthrobacter radiotolerans* was isolated from soil (Yoshinaka et al., 1973), and Yano et al. (1975) made a survey of radioresistant asporogenous bacteria in natural environments. Arunachalam et al. (1974) reported that γ-radiation up to 2×10^6 rad was required to eradicate bacteria from soil. Several cultures of bacteria were isolated by Kiselov et al. (1961) from a spring containing radium, and these cultures were 3–10 times more radioresistant than cultures from a nonradioactive source. Matsuyama et al. (1973) found that bacterial strains isolated from high-background radioactive areas in Japan possessed high radioresistance to γ-rays.

Although the effect of radiation on microorganisms present at radioactive waste disposal sites may be similar, these organisms are continuously exposed not only to low levels of radiation but also to the radionuclides. In such instances one might expect to see both radiation as well as radionuclide toxicity effects. This could lead to selection of radiation- and radionuclide-resistant organisms which might then become the dominant species in that environment. In such samples, enumeration of total microbial population indicates that the microorganisms found in that environment have changed radioresistant patterns, and as such the data should be interpreted with caution in relation to the effects of radiation and radionuclides on microorganisms in general.

Studies with soil microflora irradiated with long-term γ-irradiation in an open-air facility at 15, 150, 800, and 1500 krad/18 months show that radiation resistance increased at all doses (Table 10) when compared with the radiation resistance of the soil microflora from soil shielded from the irradiation with lead bricks (Eriksen and Emborg, 1978). Barnhart et al. (1980) found that about 68% of the bacterial isolates from radioactive shallow land burial site soil exhibited greater resistance to γ- and β-radioactivity than did the common soil bacterium *Bacillus subtilis* to ionizing radiation. Approximately 25% of the isolates from nonradioactive soil were more resistant than *B. subtilis*, but all these bacterial isolates were less radioresistant than *Micrococcus radiodurans*, which is resistant to γ-radiation (10^5 rad). However, radio-resistant bacteria are constantly being enriched in an environment containing nanocurie levels of α- and β-activity (Barnhart et al., 1980). In the latter study, the radio-resistance of 10 isolates of bacteria from a soil core 40 to 50 cm depth containing 20–30 nCi of radioactivity from the shallow-land burial site at Los Alamos National Laboratory was also studied. Several of the isolates were resistant to 5×10^4 rad γ-radiation. At 7.5×10^4 rad, none of the isolates survived. Radiation resistance of bacteria with reference to the studies dealing with radioactive waste is shown in Table 11. These data are in good agreement with values reported for sewage, food, and sterilization processes in the literature.

2. *Effect of Radionuclides on Bacteria*

The radioactivity of the waste in the trenches is several orders of magnitude higher than the levels of activity detected in the trench leachate. In addition to radionuclides, a variety of low-molecular-weight organic acids and alcohols are present in the leachates, the result primarily of the microbial decomposition of complex organic materials. Although aerobic and anaerobic bacteria have been detected in leachate samples and have been shown to be metabolically active, it is not known whether these bacteria can grow in the presence of higher levels of radioactivity. Therefore, the threshold level of radioactivity beyond which the trench water bacteria cannot survive or contribute to the degradation of the buried wastes was determined (Francis et al., 1980a). A mixture of radionuclides consisting of ^{60}Co, ^{85}Sr, and $^{134,137}Cs$ was prepared in 0.5 M HCl and added to growth media, using the isotope levels shown in Table 12. The effect of a mixture of the radionuclides on the growth of a mixed bacterial culture isolated from trench 32 at Maxey Flats is shown in Fig. 1. There was no significant difference in effects on bacterial growth between the control, containing no radionuclides, and the media, containing 2.6×10^2 and 2.7×10^3 pCi/ml of radioactivity. The radionuclide levels added to these bacterial growth media were of the same order of magnitude as those radionuclides found in Maxey Flats trench leachates. At a concentration of 2.7×10^4 pCi/ml, inhibition of growth

Table 10 Radioresistance of Soil Microorganisms Previously Exposed to Long-Term Gamma Irradiation

Soil sample previously exposed to gamma radiation (krad)	Number of tubes with visible growth (%)			
	Irradiation dose (Mrad)			
	2	3	4	5
0 (Control)	100	47	0	0
15	100	93	3	0
150	100	100	33	0
800	100	100	40	7
1500	100	100	100	40

Source: Erikson and Emborg (1978).

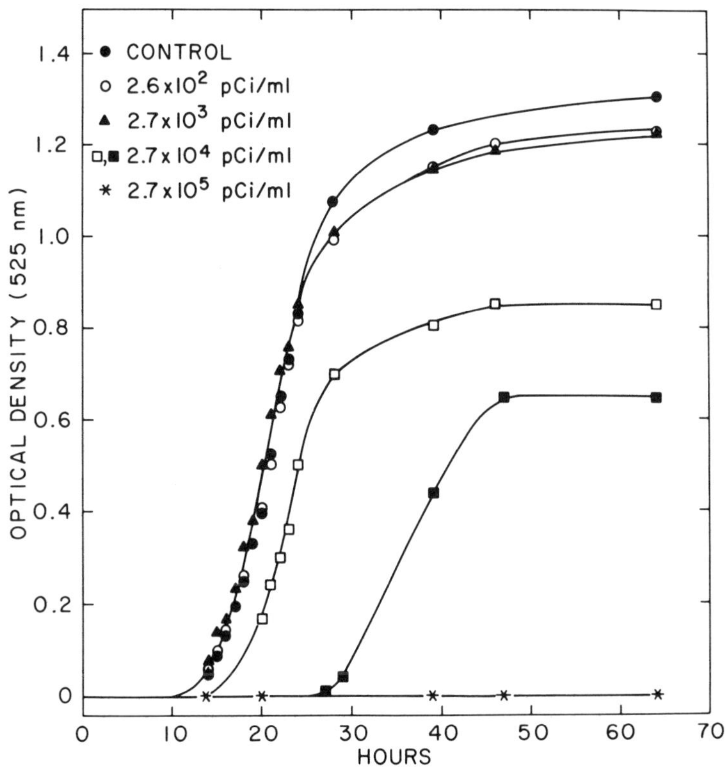

Figure 1 Effect of ^{60}Co, ^{85}Sr, and $^{134,137}Cs$ on growth of mixed culture bacteria from trench 32 at the Maxey Flats, Kentucky, disposal site. (From Francis et al., 1980a.)

Table 11 Gamma Radiation Resistance of Bacteria Isolated from Various Sources

Organism/source	Radiation dose (rad)	Reference
Micrococcus radiodurans		
ATCC 13939 (wet)	10^5	Barnhart et al. (1980)
(dry)	2.4×10^6	Barnhart et al. (1980)
Bacterial isolate QB2 from		
BNL reactor spent resin (wet)	10^4	Francis and Quinby (1981)
Strains K.3.24 and K.3.25		
from control soil (dry)	1.5×10^6; 2.5×10^6	Erikson and Emborg (1978)
Strains 30.5.11 and 30.5.16		
from irradiated soil (dry)	5×10^6; 8×10^6	Erikson and Emborg (1978)
Bacillus subtilis ATCC 6051	$<2.5 \times 10^4$	Barnhart et al. (1980)
Several isolates from		
radioactive waste disposal sites	7.5×10^4	Barnhart et al. (1980)

Table 12 Levels of Radionuclides Added to Bacterial Growth Media

Isotope	Activity (pCi/ml)			
^{60}Co	1.1×10^2	1.0×10^3	1.4×10^4	1.1×10^5
^{85}Sr	8.9×10^0	1.3×10^2	1.4×10^3	1.4×10^4
^{134}Cs	3.8×10^1	3.9×10^2	3.4×10^3	3.4×10^4
^{137}Cs	1.0×10^2	1.2×10^3	1.1×10^4	1.1×10^5
Total	2.6×10^2	2.7×10^3	2.7×10^4	2.7×10^5

Source: Francis et al. (1980a).

of bacteria was observed, resulting in two distinct growth curves presumably due to different strains, possibly by selection of radioresistant strains or mutants of bacteria (Fig. 1). Bacterial growth was completely inhibited at 2.7×10^5 pCi/ml of radioactivity.

The effect of plutonium on soil microorganisms was studied by Wildung and Garland (1982). The toxicity of plutonium to microorganisms depends on plutonium solubility in soil. Soil microorganisms are generally resistant to plutonium; toxicity was due principally to radiation rather than to chemical effects. Several fungal and actinomycete cultures isolated from soil were capable of growth at 100 μg/ml of plutonium added as the soluble complex. In a related study, additions of thorium nitrate or uranyl nitrate (0, 10, 100, 1,000, and 10,000 ppm) to a defined medium inoculated with *Pseudomonas aeruginosa* CSU (known to bioaccumulate plutonium and uranium) increased the lag period of the organism with increase in concentration of uranium or thorium (Francis et al., unpublished results). Cultures grew slowly at first and then at a rate nearly equal to that of control with no thorium or uranium. Retardation of growth initiation was a fairly common feature of sublethal concentrations of metals on bacteria and it was also possible that the solubility and bioavailability of thorium or uranium in the medium was a major limiting factor.

C. Bioaccumulation

Microorganisms are known to bioaccumulate, extracellularly or intracellularly, several heavy metals and radionuclides. "Microbial uptake," "biosorption," and "bioconcentration" are terms often used with reference to the bioaccumulation process, but specific mechanisms are involved in each case. Bioaccumulation by microorganisms has been reported for such metals as lead (Aicken and Dean, 1973; Tornabene and Edwards, 1973), silver (Charley and Bull, 1979), platinum, palladium, gold, mercury (Chakrabarty, 1976), gallium (Menon et al., 1978), and radionuclides such as ^{60}Co, ^{137}Cs (Grossbard and Stranks, 1959), uranium (Shumate et al., 1978; Strandberg et al., 1981(a,b); Tsezos and

Volesky, 1981), thorium (Tsezos and Volesky, 1981), radium (Strandberg et al., 1981b), americium (Giesey and Paine, 1977), and plutonium (Johnson et al., 1974; Beckert and Au, 1976). The potential use of microorganisms in metals and radionuclide removal from waste process streams has been described (Shumate II et al., 1978; Brown and Lester, 1979; Strandberg et al., 1981b). The ability of mixed bacterial cultures isolated from low-level radioactive waste leachate to bioaccumulate a mixture of radionuclides present at varying concentration levels in a mineral salts medium containing dextrose and yeast extract was investigated. As shown in Table 13, the bacteria accumulated $^{85}Sr > ^{60}Co > ^{137}Cs > ^{134}Cs$ from a nutrient-rich medium. With increased radionuclide concentration in the medium, radionuclide uptake increased, but the percent uptake decreased. Among the radionuclides tested, ^{85}Sr was preferentially taken up by the bacteria followed by ^{60}Co and $^{134,137}Cs$. Lesser concentrations of either ^{134}Cs or ^{137}Cs were accumulated. Similarly, ^{137}Cs and ^{226}Ra were concentrated to a considerably lesser extent than uranium by several microbial cultures (Strandberg et al., 1981). The absorption rate of strontium by higher plants has been found to be 10–100 times that of cesium and 10–1000 times that of cobalt (Chapin and Cooper, 1979). Strontium and cesium are structural analogs of calcium and potassium, respectively.

Since low-level radioactive wastes disposed of at the shallow-land burial sites contain a variety of radionuclides and organic compounds, the microflora present in the waste or leachate can indeed preferentially bioaccumulate certain radionuclides. It is possible that these microorganisms, when transported in large numbers by water, may deposit the radionuclides elsewhere in the environment upon cell lysis. Such transport phenomena, although not investigated, may be important on a localized basis in natural situations.

D. Microbial Degradation of Organic Constituents in Waste

1. *Growth of Bacteria in Leachates*

Chemical and biological degradation of solidifed wastes can reduce the integrity of the waste form, which can lead to increased waste leachability primarily due to an increase in the exposed surface of the materials. Aerobic, anaerobic, and facultative organisms are present in trench leachates, and it is not known whether these organisms are capable of growth in trench leachates containing several radionuclides and a variety of organic compounds. Mixed-cultures of bacteria, isolated under anaerobic conditions from leachate samples taken from Maxey Flats trenches 26 and 32 and West Valley trenches 3 and 5, were tested for their ability to grow in the respective filter-sterilized trench leachates under anaerobic conditions (Francis et al., 1980a). Addition of mineral salts medium (Francis et al., 1980a) without any carbon source to the trench leachates did not significantly increase the growth rates of these bacteria (Fig. 2). The growth curve fluctuations may be due to the

Table 13 Bioaccumulation of Radionuclides by Mixed-Culture Bacteria Isolated from Maxey Flats Trench Leachate

	Radionuclide mixture[a]								
	1			2			3		
	Radioactivity (x 10^2 pCi)			Radioactivity (x 10^3 pCi)			Radioactivity (x 10^4 pCi)		
Isotope	Added to medium	Found in cells	Removed by cells (%)	Added to medium	Found in cells	Removed by cells (%)	Added to medium	Found in cells	Removed by cells (%)
^{60}CO	55	2.07	4	50	2.02	4	55	1.09	2
^{85}Sr	4.5	0.42	9	6.5	0.393	6	7.0	0.268	4
^{134}Cs	19	ND[b]	0	19	0.010	0.06	17	0.007	0.04
^{137}Cs	50	0.059	0.12	60	0.053	0.09	55	0.025	0.05

[a]Total activity levels of mixture 1 = 1.28×10^4; mixture 2 = 1.36×10^5; and mixture 3 = 1.34×10^6 pCi.

[b]ND, No activity detected.

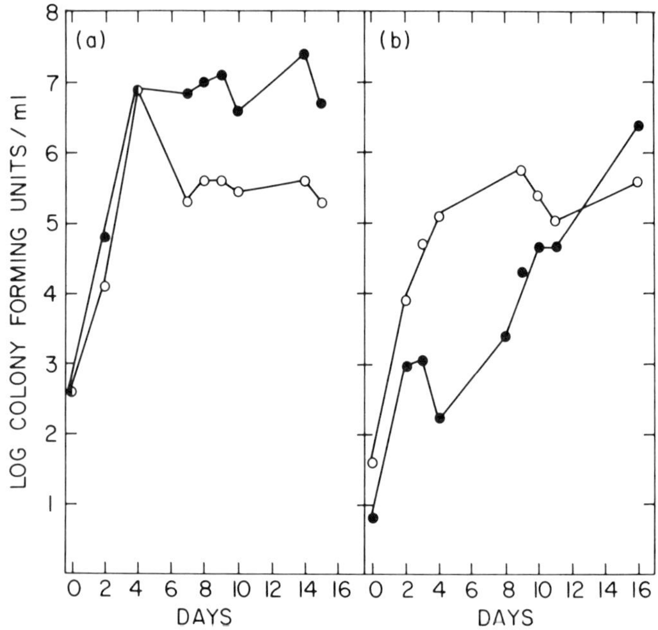

Figure 2 Anaerobic growth of mixed-culture bacteria in trench leachates. (a) Maxey Flats, trench 26 (○), trench 32 (●); (b) West Valley, trench 3 (○), trench 5 (●). (From Francis et al., 1980a.)

presence of mixed-culture bacteria and differences in the types and quantities of the organic constituents in the leachates. Nevertheless, these bacteria were able to grow under anoxic conditions using the nutrients present in the leachates, which indicates that the organic compounds and radionuclides present in the leachates did not inhibit bacterial growth.

2. *Carbon Dioxide Production from Leachates*

Carbon dioxide production from the degradation of organic compounds present in the leachate samples under aerobic conditions was studied by Francis (1982) and is shown in Fig. 3. Leachate samples initially contained dissolved CO_2, primarily resulting from anaerobic microbial activity, and leachate samples supplemented with nitrogen produced more CO_2 than unamended samples. This effect was more pronounced with West Valley trench 5 leachate than with Maxey Flats trench 19S leachate. Differences in CO_2 production rates by the leachates are probably due to the nature and type of organic compounds, the microbial community composition and the varying concentrations of different radionuclides present in the leachates.

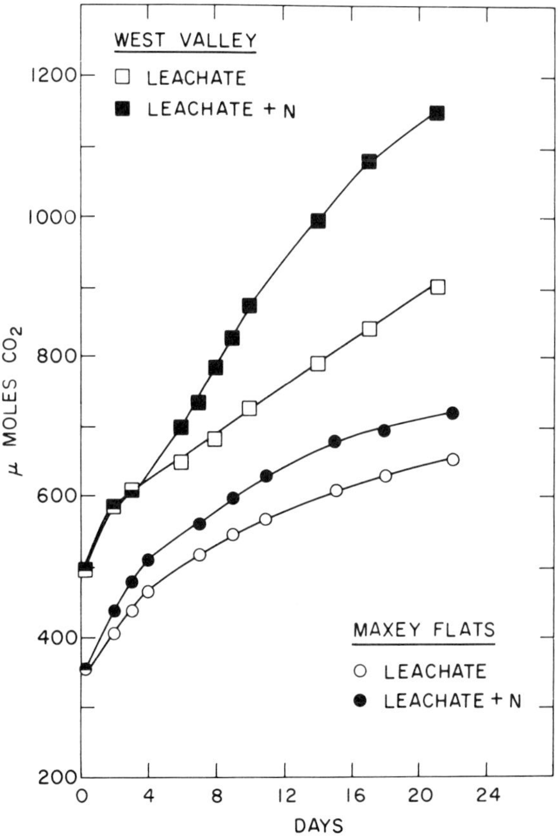

Figure 3 Microbial production of CO_2 from low-level radioactive waste leachate sample. (From Francis, 1982.)

3. *Degradation of Organic Compounds Present in Leachates*

Microbial degradation products of organic wastes may influence the transport of buried radionuclides by leaching, solubilization, and formation of organic radionuclide complexes. Degradation of selected organic compounds, such as *p*-cresol, 2-ethylhexanoic acid, phenylacetic acid, phenylpropionic acid, α-terpineol, and tributyl phosphate by mixed-culture bacteria isolated from a leachate sample was studied under aerobic conditions (Francis, 1982). These compounds were generally found in many leachate samples and may have been part of the radioactive waste, or some may have originated from microbial decomposition of the waste. Tributyl phosphate, a compound used in the extraction of metal ions from solutions of reactor products and α-terpineol, as sole carbon source, was not degraded by the mixed-culture bacteria.

The ability of the indigenous microflora of the radioactive waste to degrade the organic compounds under aerobic and anaerobic conditions has been reported (Francis, 1982). Trench leachate samples collected under anoxic conditions from the low-level radioactive waste disposal sites were incubated aerobically in Erlenmeyer flasks fitted with cotton plugs on a rotary shaker at 28°C. Anaerobic samples were incubated in serum bottles in an atmosphere of 85% N_2, 10% CO_2, and 5% H_2. Acidified control samples were incubated under aerobic and anaerobic conditions. At the end of the incubation period, samples were extracted with methylene chloride and analyzed by gas chromatography and mass spectrometry (Francis, 1982). Maxey Flats trench 19S leachate sample incubated aerobically for 21 days showed changes in the concentrations of several compounds (Table 14). Some compounds were degraded, whereas several compounds showed increased concentrations, presumably due to breakdown of complex materials. However, samples incubated under anaerobic conditions for 60 days showed increased concentrations of all constituents detected except for valeric and hexanoic acids (Table 15).

West Valley trench 3 leachate samples incubated under aerobic conditions for 30 days showed degradation of almost all the compounds detected. Many were decomposed to undetectable levels (Table 16). This sample, when collected, contained 300 ppm NH_3, with little NO_3^- and a pH of 7.3. Ammonium nitrogen present in the sample may have enhanced the decomposition rate under aerobic conditions.

West Valley trench 5 leachate sample incubated under aerobic conditions showed reductions in many compounds (Table 17), and addition of nitrogen enhanced the degradation of all compounds detected. Significant increases in the degradation of phenylacetic acid, phenylpropionic acid, phenylhexanoic acid, cyclohexanol, 2-ethyl-hexanol, tripropylene glycol methyl ester, and tributyl phosphate were observed in samples amended with nitrogen. Although tributyl phosphate was not degraded as the sole carbon source by mixed-culture bacteria, possibly it is being used either as sole carbon source or cometabolized in the presence of other organic compounds by the trench leachate bacteria. Degradation of tributyl phosphate by naturally occurring mixed-microbial populations present in activated sludge and river water has been reported (Saeger et al., 1979). Samples incubated under anaerobic conditions showed little degradation of organics, and several compounds showed increases due to breakdown of complex organic materials and accumulation of their degradation products (Table 18). Production of branched-chain volatile fatty acids such as isobutyric, isovaleric, and 2-methylbutyric acids and phenylacetic acid by anaerobic bacteria from culture media containing trypticase and glucose (Allison, 1978), production of phenol, *p*-cresol, and phenylpropionic acid from anaerobic degradation of tyrosine (Spoelstra, 1978), and production of cyclohexane carboxylic acid and 1-cyclohexene-1-carboxylic acid and a number of volatile fatty acids such as heptanoic, butyric, propanoic, and acetic from anaerobic breakdown of

Table 14 Aerobic Degradation of Organic Compounds Present In Maxey Flats Trench 19S Leachate Sample

Compound	Initial concentration (mg/liter)	Percent change
2-Methylpropionic acid	1.1	+175
2-Methylbutanoic acid	10.4	+ 46
3-Methylbutanoic acid	3.7	+ 37
Valeric acid	4.0	− 58
2-Methylpentanoic acid	5.3	+ 5
3-Methylpentanoic acid	0.9	− 9
Hexanoic acid	1.3	−100[c]
2-Methylhexanoic acid	2.0	− 4
Cresol	3.0	− 80
2-Ethylhexanoic acid	5.1	− 4
C_8 branched acid[a]	NQ[b]	− 100
C_8 branched acid[a]	NQ	+ 61
C_8 branched acid[a]	NQ	+ 8
Benzoic acid	0.13	+ 80
Octanoic acid	0.5	− 100
Phenylacetic acid	0.93	− 100
Toluic acid	0.77	− 34
Phenylpropionic acid	1.3	0
Phenylhexanoic acid[a]	NQ	− 17

[a]Percent change in concentration was determined on the basis of the ratio of the compound to the internal standard.
[b]NQ, Not quantified.
[c]Quantities present after incubation were below the detection limit of the assay procedure.
Source: Francis (1982).

benzoic acid (Keith et al., 1978) have been reported. Addition of nitrogen to the leachate under anaerobic conditions allowed degradation of several compounds. Since nitrogen in the form of NH_4NO_3 and $(NH_4)_2SO_4$ was added to the anaerobic samples, NO_3^- and SO_4^{2-} were available for use as electron acceptors during decomposition of the organic materials. Tributyl phosphate was not degraded under anaerobic conditions even with the addition of nitrogen.

These studies show that several of the organic compounds in the leachates were degraded under aerobic conditions, and that nitrogen amendments increased the decomposition rates. Under anaerobic conditions, the degradation of the organics was very slow, and changes in concentrations of several acidic compounds were observed. Several low-

Table 15 Anaerobic Degradation of Organic Compounds Present in Maxey Flats Trench 19S Leachate Sample

Compound	Initial Concentration (mg/liter)	Percent change
2-Methylpropionic acid	5.9	+ 4
2-Methylbutanoic acid	18.3	+ 3
3-Methylbutanoic acid	7.2	+ 5
Valeric acid	6.7	− 51
2-Methylpentanoic acid	0.9	+ 13
3-Methylpentanoic acid	6.3	+ 27
Hexanoic acid	1.7	− 69
2-Methylhexanoic acid	1.9	+ 23
Cresol	3.0	+ 13
2-Ethylhexanoic acid	4.5	+ 27
C_8 branched acid[a]	NQ[b]	+ 55
C_8 branched acid[a]	NQ	+ 62
C_8 branched acid[a]	NQ	+ 48
Benzoic acid	0.5	+ 12
Phenylacetic acid	2.0	+ 8
Toluic acid	0.6	+ 8
Phenylpropionic acid	1.6	+ 14
Phenylhexanoic acid[a]	NQ	+ 17

[a]Percent change in concentration was determined on the basis of the ratio of the compound to the internal standard.
[b]NQ, Not quantified.
Source: Francis (1982).

molecular-weight organic acids are formed by breakdown of complex organic materials and are further metabolized by microorganisms; hence, these compounds are in a dynamic state, being both synthesized and destroyed. Addition of a nitrogen source had only a slight effect on these degradation rates. Degradation of tributyl phosphate, a compound used in the extraction of metal ions from solutions of reactor products, was not observed under anaerobic conditions. The formation of straight- and branched-chain aliphatic acids and their long residence times in an anaerobic environment could significantly affect the migration of radionuclides from the disposal sites.

Table 16 Aerobic Degradation of Organic Compounds Present in West Valley Trench 3 Leachate Sample

Compound	Initial concentration (mg/liter)	Percent change
Propionic acid[a]	NQ[b]	−100[c]
2-Methylpropionic acid	25.3	− 97
2-Methylbutanoic acid	100	− 76
3-Methylbutanoic acid	17.9	−100
Valeric acid	39.7	−100
2-Methylpentanoic acid	16.8	− 72
3-Methylpentanoic acid	1.5	−100
Phenol	0.7	−100
Hexanoic acid	19.7	−100
2-Methylhexanoic acid	7.1	− 84
Cresol	5.2	−100
2-Ethylhexanoic acid	59.5	− 50
C_8 branched acid[a]	NQ	−100
C_8 branched acid[a]	NQ	−100
Benzoic acid	6.5	−100
Octanoic acid	4.3	−100
Phenylacetic acid	4.9	−100
Nonanoic acid	1.6	− 75
Phenylpropionic acid	8.0	− 94
Decanoic acid	0.3	−100
Phenylhexanoic acid[a]	NQ	− 58

[a]Percent change in concentration was determined on the basis of the ratio of the compound to the internal standard.
[b]NQ, Not quantified.
[c]Quantities present after incubation were below the detection limit of the assay procedure.
Source: Francis (1982).

4. *Degradation of Chelating Agents*

Chelating agents are widely used in the nuclear industry for decontamination of reactors and equipment and in cleanup operations. Several metal-complexing agents, along with solvents such as decalin or diisopropylbenzene, are used in metal separation chemistry. Examples of the types of complexing agents used are polyphosphates, mono-, di-, and tributyl phosphates, amino carboxylic acids such as ethylenediaminetetraacetic acid (EDTA), diethylenetriaminepentaacetic acid (DTPA), nitrilotriacetic acid (NTA), and *N*-hydroxyethylenediaminetriacetic acid (HEDTA), dihexyl-*N*,*N*-diethylcarbamyl methylene phosphonate

Table 17 Aerobic Degradation of Organic Compounds Present in West Valley Trench 5 Leachate Sample

Compound	Initial concentration (mg/liter)	Percent change: Addition: None	Percent change: Addition: Nitrogen
Acetic acid[a]	NQ[b]	+ 98	−100[c]
2-Methylpropionic acid	36.8	+ 1	− 18
2-Methylbutanoic acid	112	+ 2	− 4
3-Methylbutanoic acid	38.6	− 5	− 13
Valeric acid	82.4	− 8	− 17
2-Methylpentanoic acid	20.8	− 4	− 10
3-Methylpentanoic acid	2.6	− 4	− 15
Phenol	6.0	− 19	− 22
Hexanoic acid	83.0	− 21	− 28
2-Methylhexanoic acid	15.0	− 1	− 10
Cresol	5.2	− 93	− 28
2-Ethylhexanoic acid	67.8	− 16	− 24
Benzoic acid	16.8	− 10	− 19
Octanoic acid	26.2	− 96	−100
Phenylacetic acid	13.6	− 13	− 20
Toluic acid	2.4	− 5	− 10
Nonanoic acid	4.9	− 100	−100
Phenylpropionic acid	10.3	− 13	− 21
Decanoic acid	1.4	− 100	−100
Phenylhexanoic acid[a]	NQ	− 5	− 24
p-Dioxane[a]	NQ	0	− 10
Cyclohexanol	1.2	− 17	− 49
2-Ethyl-1-hexanol	2.2	− 31	− 52
Tripropylene glycol methyl ester[a]	NQ	− 15	− 38
Tributyl phosphate	0.7	− 13	− 38

[a]Percent change in concentration was determined on the basis of the ratio of the compound to the internal standard.
[b]NQ, Not quantified.
[c]Quantities present after incubation were below the detection limit of the assay.
Source: Francis (1982).

Table 18 Anaerobic Degradation of Organic Compounds Present in West Valley Trench 5 Leachate Samples

Compound	Initial concentration (mg/liter)	Percent change: Addition: None	Percent change: Addition: Nitrogen
Acetic acid[a]	NQ[b]	0	− 8
2-Methylpropionic acid	36.0	− 3	− 21
2-Methylbutanoic acid	113	+ 3	− 12
3-Methylbutanoic acid	38.0	0	− 15
Valeric acid	79.0	+ 4	− 14
2-Methylpentanoic acid	20.2	+ 9	− 5
3-Methylpentanoic acid	2.6	+ 9	− 6
Phenol	5.8	+ 13	− 16
Hexanoic acid	78.2	+ 11	− 4
2-Methylhexanoic acid	14.0	+ 18	+ 3
Cresol	6.4	− 7	− 44
2-Ethylhexanoic acid	62.6	+ 20	+ 5
C_8 acid[a]	NQ	− 42	− 48
Benzoic acid	17.0	0	− 17
Octanoic acid	22.6	+ 33	+ 6
Phenylacetic acid	13.6	0	− 16
Toluic acid	2.6	+ 2	− 14
Nonanoic acid	4.2	+ 52	+ 22
Phenylpropionic acid	10.6	+ 1	− 15
Decanoic acid	0.2	+ 66	+ 1
Phenylhexanoic acid[a]	NQ	+ 1	− 17
Benzene[a]	NQ	− 8	− 17
p-Dioxane[a]	NQ	− 4	− 13
Toluene	13.3	− 17	− 20
Cyclohexanol	1.2	+ 33	+ 10
2-Ethyl-1-hexanol	13.2	+ 4	− 5
Tripropylene glycol methyl ester[a]	NQ	0	− 6
Tributyl phosphate	0.7	0	0

[a]Percent change in concentration was determined on the basis of the ratio of the compound to the internal standard.
[b]NQ, Not quantified.
Source: Francis (1982).

(DHDECMP), and carboxylic acids such as citric, hydroxyacetic, oxalic, and tartaric acids. Of these, EDTA and DTPA are widely used in decontamination as well as in the therapy of removal of lead and plutonium intoxication in humans. They form extremely strong complexes with rare earths and actinides. Although these multidentate chelates are effective in the decontamination of radionuclides by complexation, by virtue of their chemical reactions they increase the solubility of the metals and are easily transported and are readily bioavailable in soil.

In addition to the use of synthetic chelating agents, complexing agents are also found in decomposing organic matter introduced as waste, and water mobile organic constituents from soils, primarily humic and fulvic acids. These soil organic ligands and decomposing organic waste contain complex mixtures composed of oxygenated organic degradation products. Soil organics are generally composed of amino acids, polysaccharides, polyfunctional aromatic compounds, and porphyrins. The most abundant and well-characterized fraction of soil fulvic acid is composed of a mixture of organic acids such as formic, acetic, benzoic, 3-hydroxy-5 methylbenzoic, *p*-hydroxybenzoic, protocatechuic, vanillic, gallic, propionic, citric, fumaric, malic, oxaloacetic, butyric, glycolic, lactic, and tartaric acids (McLaren and Petersen, 1967; Schnitzer and Kahn, 1972). These compounds are generally excellent bi- and polydentate chelating ligands for a variety of metals.

Complexes may be formed before waste burial as a result of laboratory procedures, process operation, or decontamination activities or at the burial site when species interact with buried or soil-derived organic matter.

When the metal and metal-organic complex species contact soil, interaction mechanisms include ion exchange, precipitation, and mineral replacement, and major variables influencing these interactions include temperature, pH, clay content, organic carbon content, and cation exchange capacity of the soil. Minor variables in the soil environment can also affect these interactions among which are competing cation content and microorganisms. For example, Co^{2+} and Mg^{2+} depress iron complexation, and stereospecific organic complexing can occur in microorganisms (McLaren and Petersen, 1967).

Chelating agent persistence in the disposal environment is a major concern because of the potential of increasing the transport and bioavailability of radionuclides and toxic metals. In general, information on microbial degradation of chelating agents is scarce. Many of these chelates either are poorly biodegraded under aerobic conditions or are found to undergo little biodegradation under anaerobic conditions. This is indicated by the isolation of EDTA-^{60}Co complexes in seepage after several years from the Oak Ridge National Laboratory burial site. The migration of ^{60}Co was attributed primarily to the chelating agent EDTA. Other leachate organic constituents such as organic acids and phthalates might have contributed to the mobilization of radionuclides, but to a lesser degree (Means et al., 1978). Recently, Means et al.

(1980) reviewed the biodegradation of NTA, EDTA, and DTPA in light of their widespread use in various industries, particularly the nuclear industry, and because of their ability to solubilize, complex and mobilize toxic heavy metals and radionuclides in the environment. They also studied the relative rates of biodegradation of NTA, EDTA, and DTPA in flasks inoculated with soil. Rates of both biological and nonbiological degradation decrease in the order: DTPA > EDTA > NTA over the short term, and NTA ∿ DTPA > EDTA over the long term. However, photolysis appeared to account for a significant proportion of DTPA degradation. Therefore, considering only biodegradation: NTA > EDTA ∿ DTPA. Degradation of chelates, particularly in anaerobic environments, is important because in many of the humid waste disposal sites, the conditions are anaerobic or highly reduced.

The degradation of NTA in soils incubated under air or argon was studied by Tiedje and Mason (1974). They found that under anaerobic conditions NTA degradation was inhibited. NTA is decomposed more rapidly under aerobic conditions and to a variable extent under anaerobic conditions by facultative organisms in the presence of alternate electron acceptors such as NO_3^- (Enfors and Molin, 1973; Claesson, 1971). Recently, Kirk et al. (1982) reported that the removal of NTA observed during anaerobic digestion was not biological and in part was the result of adsorption onto the solid phase of sewage sludge.

Biodegradation of EDTA in soils, sediments, and aerated lagoons has been reported (Belley et al., 1975; Tiedje, 1975; Means et al., 1980). The influence of various environmental factors on EDTA biodegradation in soils and sediments has been investigated by Tiedje (1977). EDTA was degraded slowly by cometabolism under aerobic conditions in a variety of soils of different geographical origin, texture, use, and pH, while no significant degradation was detected under anaerobic conditions. Degradation of EDTA was observed to occur at initial levels as high as 1000 ppm (the highest concentration tested), and first-order kinetics were followed from 0.4 to 90 ppm. The extent of degradation among soils was variable; common values for 2–4 ppm of added [^{14}C]EDTA mineralized were 13–45% after 15 weeks and 65–70% after 45 weeks (Tiedje, 1977).

Little information is available on the biodegradation of DTPA, however. Means et al. (1980) reported that DTPA was biodegraded in laboratory experiments, but photodegradation of DTPA was very significant. No information is available concerning possible DTPA degradation under anaerobic conditions. However, in sewage, no microbial degradation of DTPA for carbon or energy was observed (Bunch and Ettinger, 1967), suggesting the possible cometabolic degradation.

Tributyl phosphate, a compound used in the extraction of metal ions from solutions of reactor products, was not degraded under aerobic conditions as sole carbon source by a mixed-culture bacteria isolated from low-level radioactive waste leachate sample (Francis, 1982). However, tributyl phosphate in leachate rich in other organic compounds was being either metabolized as a sole carbon source or cometabolized by the bacteria (Francis, 1982). Aerobic degradation of tributyl phosphate by naturally occurring mixed-microbial populations present in activated sludge and river water have been reported (Saegar et al., 1979).

Microbial degradation studies of citrate, EDTA, and NTA chelates of europium indicate that the average half-lives for citrate, NTA, and EDTA under aerobic conditions were 3.2, 8.0, and 28 years, respectively (Barnhart et al., 1980).

Decomposition of chelates leached from sanitary landfill waste disposal sites was determined by the potential chelation capacity of solutions (regardless of the types of chelating agents present) on the solubility of $CaCO_3$ (Avnimelech and Raveh, 1982). Leachate samples were incubated aerobically and anaerobically with clay loam, calcareous clay loam, and $CaCO_3$ for up to 120 days. Chelate decomposition was slow and insignificant under anaerobic conditions whereas under aerobic conditions it was rapid, leading to a residual chelation capacity of only 0.1–1% of the initial value after a 74-day incubation. Physical adsorption of the chelating agents on clay or $CaCO_3$ particles was insignificant. These results suggest chelating agent persistence under anaerobic conditions and the potential for the transport of toxic metals in the soil environment. Further, it emphasizes the importance of aerobic pretreatment of the decontamination solutions prior to disposal to reduce the mobilization of toxic metals from the disposal environment (Avnimelech and Raveh, 1982).

5. *Effect of Chelating Agents on Microorganisms*

Although chemical structure imparts resistance or determines the recalcitrant nature of organic compounds, degradation of chelating agents may in part be determined by their ability to bind essential metals such as calcium or iron and render them unavailable for microbial growth. Alternatively, toxicity may arise as the result of increased bioavailability or release of toxic metals from the degradation of metal chelates. It has been shown that EDTA and DTPA do affect the growth of *Pseudomonas aeruginosa* in culture media. For example, Eagon and Carson (1965) have shown that *P. aeruginosa* strain 64 was extensively lysed by EDTA. They found that 0.2 mg EDTA/3 ml of cell suspension had no lytic effect even after the addition of lysozyme. With higher amounts, up to 0.8 mg EDTA increased lysis was observed. EDTA was replaceable by a structurally related polyacetic and chelating agent, diaminocyclohexanetetraacetic acid. DTPA was a poor replacement of EDTA. EDTA was not replaced by other chelating agents such as 2,2′-dipyridyl, 8-hydroxyguinoline, iminodiacetic acid, *o*-phenanthroline, citric acid, and cysteine. EDTA, on the other hand, was not replaced by detergents, such as alkyl benzene sulfonic acid, cetyl trimethyl ammonium chloride, deoxycholate, and sodium dodecyl sulfate.

There is evidence that calcium was chelated in greatest quantity and that the binding of divalent metals, which may function to form cross-linkages, is essential for the integrity of cell walls of *P. aeruginosa*. Humphrey and Vincent (1962) reported that calcium may be involved structurally in cell walls of *Rhizobium trifolii*, stabilizing lipoprotein and mucopeptide components. Similarly, Repaske (1958) suggested that EDTA functions in lysis by chelation of metals associated with cell walls of *Escherichia coli* and *Azotobacter vinelandii*.

Ethylenediamine-di-(*O*-hydroxyphenylacetic acid) (EDDA) has been used as an inhibitor of the growth of a variety of bacteria on the basis of its iron-chelating behavior (Miles and Khimji, 1975). EDDA at 1.25 mg/ml in a casamino medium (pH 7.5) retarded the onset of growth of *P. aeruginosa* by 15 hr, but allowed final growth comparable to that in the control medium. Addition of both iron-free pyochelin (an iron-chelating compound produced by *P. aeruginosa* in a medium containing low iron) and iron-bound pyochelin to the medium containing EDDA reversed this compound's growth inhibition. Neither citrate (0.05 mM) nor NTA (0.05 mM) reversed EDDA inhibition. Theoretically, the iron in the medium exists in the EDDA complex, and initiation of growth depends upon accumulation of sufficient pyochelin to sequester iron in a usable form for the pseudomonads (Cox and Graham, 1979). In a related study with *P. aeruginosa* (CSU), which bioaccumulates the radionuclides plutonium and uranium, DTPA at 4×10^{-3} M concentration completely inhibited growth of this organism in a defined medium (Fig. 4). DTPA at concentrations 4×10^{-5} and 4×10^{-4} M also inhibited growth with a long lag period, and when growth resumed it reached half the maximum growth in medium containing no DTPA. The growth inhibition of *P. aeruginosa* by DTPA may be due to chelation of divalent cations, particularly calcium or chelation of iron or both, depending upon the concentrations and stability constants of these ions in the growth medium. Whether similar inhibition of other organisms occurs at higher concentrations in nature is not known, but the presence of excess cations in natural samples should have little effect on bacterial growth. Nevertheless, it is particularly important from the waste management viewpoint if one chooses to biologically pretreat the waste containing chelating agents.

Although several types of chelating agents are buried at the low-level radioactive waste disposal site, to date only EDTA has been clearly identified (Table 7). Other chelating agents such as DTPA or NTA have not been identified, but several other unidentified organic compounds with molecular weights higher than EDTA associated with ^{60}Co, ^{90}Sr, or ^{137}Cs have been reported. More information on the nature of other organic chelating agents present in the leachates is needed so that mitigative measures can be planned in light of chemical persistence and mobility under various environmental conditions normally expected at the burial sites.

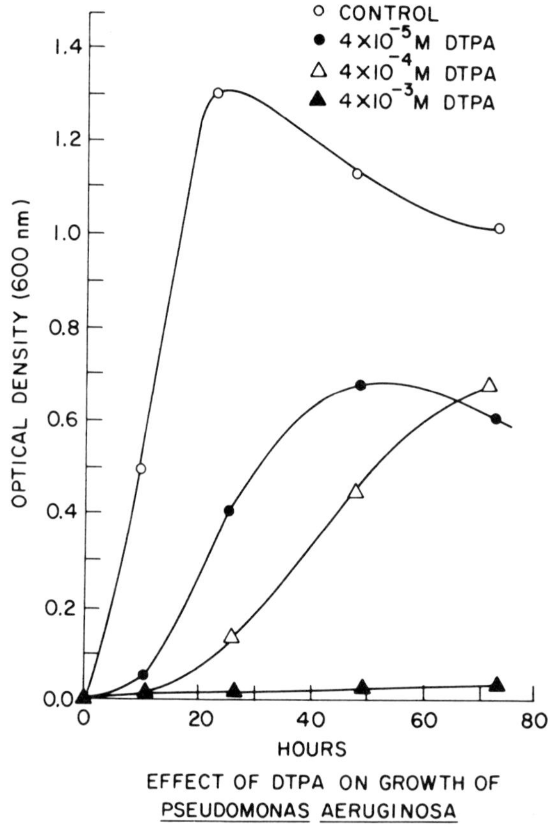

Figure 4 Effect of DTPA on growth of *Pseudomonas aeruginosa*.

E. Microbial Generation of Radioactive Gaseous Compounds

Radioactive gaseous compounds such as CH_3T, HTO, ^3H, and other tritiated hydrocarbons, as well as ^{85}Kr, ^{222}Rn, $^{14}CH_4$, and other ^{14}C hydrocarbons have been identified as evolving from burial trenches at the West Valley, New York, disposal site (Lu and Matuszek, 1978; Husain et al., 1979). Often, tritiated methane is one of the most abundant compounds, and it has been estimated that 200–6000 mCi/yr of CH_3T are released to the environment from various trenches at the West Valley disposal site (Matuszek, 1980). Microorganisms may play a significant role in the generation of radioactive gases directly through their metabolic activity, or they may indirectly enhance the release of trapped gases such as radon which result from the radioactive decay of radium. Particular attention has been given to methanogenic bacteria because of anoxic conditions that prevail in the trenches and their possible contribution to the release of tritiated methane. The population distributions of methanogenic bacteria found in trench leachates collected from several low-level radioactive waste disposal sites (Table 8) are shown to range from 20 to 2300 methanogens/ml of leachate. These populations, however, may be much higher in waste that is undergoing decomposition and can be influenced by various environmental conditions including substrate availability.

1. Methanogenesis from Leachate Samples

The rates of methane production by inoculated and uninoculated leachate samples incubated in different atmospheres was studied by Francis et al. (1980c). Uninoculated samples incubated under a $N_2 + CO_2 + H_2$ gas phase produced much less CH_4 than inoculated samples (Fig. 5). Leachate samples incubated under $H_2 + CO_2$ produced more methane at an accelerated rate than samples incubated under $N_2 + CO_2 + H_2$. Initially, the uninoculated sample incubated under $H_2 + CO_2$ followed the same rate as the uninoculated sample incubated under $N_2 + CO_2 + H_2$. After 15 days of incubation, the rate of methane production by the uninoculated sample under $H_2 + CO_2$ initially followed the same rate as the inoculated sample under $N_2 + CO_2 + H_2$. After 30 days, in comparison the inoculated samples incubated in an atmosphere of $H_2 + CO_2$ produced methane at a faster rate and reached a rate almost equal to that of the uninoculated sample under $H_2 + CO_2$. In these studies, differences in the rates and total production of methane from the leachate samples by the methanogens were observed. These differences resulted from the presence of other microflora consisting of clostridia and other unidentified bacteria, and the generation and availability of substrates such as acetate, CO_2, and H_2 for the growth of methanogens. A threefold increase in methane production was observed from a defined medium inoculated with the mixed methanogenic culture incubated in an atmosphere of 80% H_2 + 20% CO_2 over that in an atmosphere of 85% N_2 + 10% CO_2 + 5% N_2.

2. Generation of Tritiated Methane

Tritiated and [^{14}C]methane were detected in leachate samples inoculated with methane bacteria and incubated under $N_2 + CO_2 + H_2$ or $CO_2 + H_2$ (see Table 19). The control sample had some tritium and ^{14}C activity in the methane fraction. Inoculated samples, incubated under $N_2 + CO_2 + H_2$ evolved approximately 18,000 nmol of CH_3/30 ml of leachate, whereas a fourfold increase in total methane was detected in samples incubated under a $CO_2 + H_2$ atmosphere. Similarly, the sample incubated under $CO_2 + H_2$ exhibited a 20- and 57-fold increase, respectively, in total $^{14}CH_4$ and CH_3T activities over that of the sample incubated under $N_2 + CO_2 + H_2$ atmosphere.

Table 19 Microbial Production of $^{14}CH_4$ and CH_3T from Maxey Flats Trench 19S Leachate

Leachate sample (30 mi)	CH_4 produced (nmol)	Total activity (pCi)	
		$^{14}CH_4$	CH_3T
Control (Formaldehyde treated)	980	0.05	0.03
Inoculated ($N_2 + CO_2 + H_2$)	18,000	0.59	1.0
Inoculated ($CO_2 + H_2$)	68,000	12	57

Source: Francis et al. (1980c).

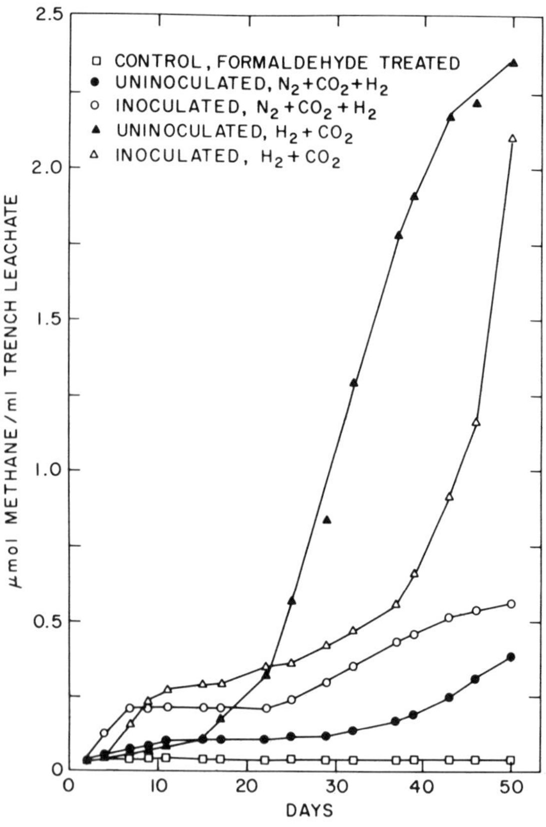

Figure 5 Microbial generation of methane from low-level radioactive waste leachate sample incubated in different atmospheres. (From Francis et al., 1980c.)

Tritiated methane production from synthetic media containing 2 mCi of tritium as HTO or tritiated acetate (CH_2TCOO^-) was studied (Francis et al., 1980c). The tritium levels (2 mCi) used in the study had no apparent effect on methanogenesis, and the production of CH_3T increased proportionally with the increase in concentration of HTO added to the medium (Francis, 1984). Increased levels of tritiated methane were produced from media containing tritiated acetate (Fig. 6). This is mainly due to preferential utilization of acetate by methanogens via methyl group reduction to methane with carboxyl group oxidation to CO_2 and H_2O (Mah et al., 1977). The mechanism of incorporation of tritium from HTO to the methane fraction by methanogens is not clearly understood. Nevertheless, tritium from HTO can exchange with exchangeable hydrogen atoms of organic compounds with functional groups such as $-OH$, $-COOH$, $-NH_2$, and $=NH$ and can be subsequently metabolized by microorganisms in the course of methane production. As an additional tritium source, bacteria are able to use tritium from tritiated water by enzymatic reactions in methane biosynthesis.

Gaseous compounds such as CO_2 and H_2, generated by biological decomposition and/or radiolytic degradation of wastes, may be reduced to CH_4 by methane bacteria in anaerobic environments, or if the radioactivity is sufficient to inhibit bacterial activity, then the gases generated from radiolytic decomposition may migrate from the source of radioactivity and finally escape as CO_2 and H_2 or be reduced to CH_4 by methanogens elsewhere in the environment. If these gases contain ^{14}C and/or ^3H activity, then one would expect the radioactivity in the methane fraction.

These results also suggest that under appropriate conditions significant amounts of ^{14}C- and tritium-containing methane can be released into the environment by microbial activity of the radioactive waste containing ^{14}C and ^3H organic compounds and tritiated water. It has been reported that 200–6000 mCi/yr of tritiated methane is released from various trenches from West Valley disposal site (Matuzek, 1980), primarily due to microbial activity. Tritium and ^{14}C released as tritiated and ^{14}C methane from the waste may also be oxidized to $^{14}CO_2$ and HTO once they reach the aerobic layer in the soil by microbial communities and thus become readily bioavailable to other life forms in the biosphere.

3. *Microbial Oxidation of Molecular Tritium*

The microbial oxidation of gaseous tritium to tritiated water has been recently studied by many investigators because tritiated water is more readily bioavailable than molecular tritium, and may result in significant contamination of food and water. Consequently, maximum permissible concentrations of HTO are about 100 times lower than for gaseous tritium. These studies have been undertaken in light of the expectation of increasingly large atmospheric tritium discharges from nuclear fuel reprocessing plants, development of nuclear fusion technology, radioactive waste leachate evaporation practices at the disposal sites, and the nuclear industry.

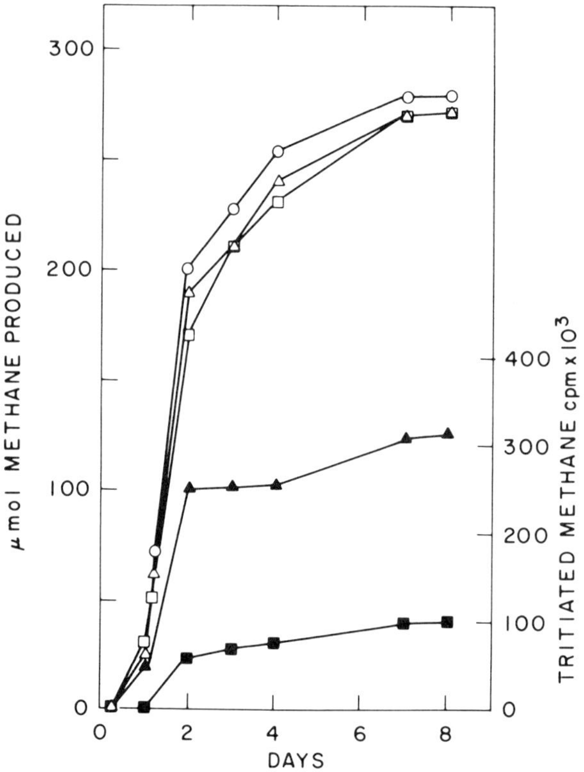

Figure 6 Microbial production of tritiated methane from HTO and tritiated acetate (CH_2TCOO^-) from synthetic media incubated under 85% N_2 + 10% CO_2 + 5% H_2. Open symbols represent methanogenesis; closed symbols represent tritiated methane. (○) Control; (□,■) 2 mCi HTO; (△,▲) 2 mCi tritiated acetate/30 ml media. (From Francis et al., 1980c.)

Soils exposed to molecular tritium under laboratory conditions showed that molecular tritium was converted to tritiated water within the soil column (Murphy et al., 1976), and soil conditions that promoted soil microbial activity increased tritiated water production. Soil sterilization greatly decreased the conversion of molecular tritium, whereas reinoculation with soil microorganisms restored the conversion of molecular tritium to tritiated water. When closed plant-soil systems were exposed to molecular tritium, tritiated water was found in the soil and plant water, suggesting microbial as well as plant metabolism of tritium (Murphy et al., 1976). McFarlane et al. (1978) found that exposure of intact and excised leaves, sterilized clay loam soil, and various extractable nonliving soil components to molecular tritium resulted in less than 4% conversion to tritiated water in 48 hr. However, exposure of natural (unsterilized) clay loam or of sterilized soil inoculated with a soil water extract yielded over 95% conversion of tritium to tritiated water. Bacteria isolated from the soil were able to catalyze this reaction in solution. Molecular tritium oxidation rates in various soils by soil microorganisms ranged from 12% to 66% per hour, and were generally independent of the soil type or soil chemical properties (McFarlane et al., 1979). The authors suggest that these rates are sufficiently rapid to account for significant oxidation of HT if it were present in the environment.

Fallon (1982) studied the incorporation of gaseous tritium (T_2, HT, or both) by soils. The optimal temperature and pH values ranged from 20°C to 50°C and 4 to 7, respectively, and the tritium metabolism declined at the wet and dry extremes in soil moisture content. Hydrogenase enzymes have been suggested to play a role in the metabolism of tritium (Fallon, 1982).

Rapid exchange of gaseous tritium into soil water can occur and should be taken into consideration in soil uptake studies. T_2-HT uptake in soils can occur by oxidation $2HT + O_2 \rightarrow 2HTO$, or through an exchange reaction involving no net transfer of electrons, $H_2O + HT \rightarrow HTO + H_2$ (Anand and Krasna, 1965; Adams et al., 1981; Klibanov and Huber, 1981). In vivo studies with *Methanobacterium thermoautotrophium* (Daniels et al., 1980; Spencer et al., 1981) show that the exchange reaction can be up to 40 times more rapid than the net oxidation reaction. The exchange reaction has an equilibrium constant, $K = (HTO)(H_2)/(H_2O)(HT) = 6.25$ (Anand and Krasna, 1965; Klibanov and Huber, 1981).

It is probable that significant quantities of tritium would be found in the soil following release or leakage of HT. The amount would obviously depend not only on microbial activity but also on air mixing near the soil surface, duration of exposure, and, of course, the molecular tritium concentration.

4. *Release of Other Radioactive Gases by Microbial Activity*

Radon-222 is one of the naturally occurring radioisotopes and is a member of the ^{238}U decay series. When ^{226}Ra decays by α-emission, it transmutes to its daughter ^{222}Rn an inert gas having a half-life of 3.82 days. Radon-222, a radioactive gas, may be formed in aquatic systems through an enrichment of ^{226}Ra, which is dissolved naturally in waters. The occurrence of ^{222}Rn associated with biologically produced gases such as N_2, CH_4, CO_2 from sediments in a production reactor cooling pond at the Savannah River Plant (SRP), Aiken, was studied (Fliermans et al., 1978). Radon-222 is transported into the atmosphere by both physical and biological mechanisms and the rates of ^{222}Rn flux by biological gassing appear to be less important than that released by physical transport processes from terrestrial systems. In contrast, the biological production and transport processes may be significant in the overall flux or release of radioactive gases from the low-level waste disposal sites, particularly from those sites located in humid regions.

IV. MICROBIAL MOBILIZATION AND IMMOBILIZATION OF METALS

The mobility of the elements in the geosphere is controlled by several environmental parameters. For example, the mobilities of cadmium, lead, nickel, zinc, copper, and chromium are enhanced with increasing acidity, while elements such as molybdenum, selenium, tin, barium, and arsenic tend to be enhanced with increasing alkalinity. Oxidation of sulfur such as pyritic or elemental sulfur in the waste results in the production of sulfuric acid. In addition, the iron oxidizing bacterium *Thiobacillus ferrooxidans* generates ferric iron which is an effective oxidant. The oxidizing and reducing conditions present in the landfill area to a large extent influence the mobilization and immobilization of metals. Eh tends to have little direct influence on metal mobility but if microbial activity leads to reducing conditions yielding sulfides, then the mobility of several of these metal sulfides is controlled by the solubility of the respective sulfides. In general, most of the metal sulfides exhibit low solubility in an aqueous medium. On the other hand, oxidizing conditions promote the oxidation of reduced forms of sulfur to sulfuric acid which, in turn, play a dominant role in the mobilization of acid-soluble metals. Furthermore, an acid pH drastically retards the capacity of the soil or clay liner of a landfill to attenuate the movement of heavy metals from the landfill.

Microorganisms also produce mobile water-soluble complexing agents such as organic acids, catechols, and hydroxamates, which may chelate certain metals. Certain heterotrophic bacteria, fungi, and actinomycetes are known to produce siderphores which are extremely effective and selective sequestering agents for Fe(III). The significance of microbially synthesized chelating agents in the mobilization of toxic metals in these disposal environments is poorly understood. For example, at the shallow-land, low-level radioactive waste disposal sites, the presence of highly solubilized organoradionuclide complexes due to synthetic and microbially produced complexing agents, and the highly anoxic character of the leachate, believed to cause many radionuclides to remain in low valence states, are possible contributing factors to mobilization of radionuclides that might otherwise be expected to be absorbed onto clay and soil particles.

A. Microbial Leaching of Metals from Waste

Microorganisms are known to solubilize various metals from ores and in soil by production of mineral acids, organic acids, and oxidizing agents. Much of the information on the microbial contribution to leaching of metals comes from studies of ore leaching processes. The responsible microorganisms, chemical and biochemical mechanisms involved in the microbial leaching or biomining of metals, have been the subject of intensive study during the past several years (Zajic, 1969; Murr et al., 1978; Kelly et al., 1979; Ehrlich, 1981; Brierley, 1982). Consequently, such microbiological processes are now being exploited on a commercial scale for extraction of copper and uranium from ores and to recover strategic metals.

Since the radioactive waste materials contain iron and sulfur either in the oxidized or reduced form, they are important in driving the overall chemical and biochemical reactions at the disposal environment in the presence of microorganisms. In many respects, the microorganisms and the basic principles involved in bioleaching would be operative in the waste. The environmental factors and the organic and inorganic composition of the waste would determine the type, rate, and extent of the reactions. In general, direct solubilization and leaching of metals are principally brought about by the activities of autotrophic bacteria. The role of heterotrophic bacteria in the solubilization and leaching by production of metabolites such as organic complexing agents has been reported, but their significance in metal mobilization from inorganic wastes has not been elucidated. These heterotrophic organisms may significantly contribute to the mobilization of metals from wastes containing organic materials.

Under appropriate conditions, nuclear and coal wastes including uranium mill tailings may undergo a series of chemical reactions that result in the solubilization of many minerals. In addition, this dissolution process may be brought about by direct or indirect bacterial action. Direct bacterial action entails an enzymatic attack by the bacteria on components of the mineral that are susceptible to oxidation. In the process of obtaining energy from the inorganic material, the bacteria cause electrons to be transferred from iron or sulfur to oxygen. In many cases, the more oxidized product is more soluble. On the other hand, indirect leaching can involve oxidation of soluble ferrous iron to ferric iron, a powerful oxidizing as well as solubilizing agent which reacts with other metals, transforming them into the soluble oxidized form in a sulfuric acid solution. In this reaction, ferrous iron is again produced and is rapidly reoxidized by the bacteria. In an acidic solution in the absence of bacteria, ferrous iron is stable and leaching mediated by ferric iron would be slow. Bacteria (*Thiobacillus ferrooxidans*) can accelerate such an oxidation reaction by a factor of more than a million (Brierley, 1982). In nature, it is difficult to differentiate between direct and indirect leaching by bacteria quantitatively due to the presence of some iron in most minerals.

The role of autotrophic bacteria, such as iron and sulfur oxidizers, in the solubilization of uranium and possibly thorium in mill tailings and coal wastes could be significant. For example, the biogeochemistry of uranium has been studied in light of the recovery of uranium from ores. The uranite is oxidized by ferric ions to UO_2^{2+}, which is soluble in water, while the iron-oxidizing microbes provide a continuous supply of ferric ions. Recently it has been demonstrated that *T. ferrooxidans* can directly oxidize tetravalent uranium to its hexavalent form. As an additional aspect of this work, mixed-substrate studies have indicated a competitive type of inhibition by uranium on the iron oxidation (Dispirito and Tuovinen, 1982). Acids produced by microorganisms such as H_2SO_4 by sulfur-oxidizing bacteria, and also organic acids, play important roles in metal mobilization. Strayer (1982) found that reduced sulfur forms (0.3–4%) present in the ash/slag of the coal gasification waste are readily oxidized by microorganisms.

Naturally occurring thorium is weakly radioactive, yet its chemical properties are similar to $^{239}Pu(IV)$. Some thorium isotopes and daughter nuclides are strongly radioactive and may be found in nuclear wastes. The highly charged positive ion Th^{+4} has a strong tendency to hydrolyze and form complexes with anions, which may be present in solution. The role of microorganisms in thorium solubilization is not known.

B. Microbial Chelation and Solubilization of Metals

Chelation and solubilization of metals are brought about by the activities of microorganisms in nature. Several mechanisms for microbial solubilization of insoluble metals have been proposed, including organic acid production (Duff et al., 1963), formation of chelating agents (Snow, 1970), and metabolism of the metal-associated anion (Silver and Torma, 1974).

1. *Microbial Production of Organic Acids*

Microbially generated dicarboxylic acids, polyhydroxy acids, 2-ketogluconic acid, and phenolic compounds such as protocatechuic acid, and salicylic acid are effective chelating agents of heavy metals and are known to accelerate the movement of metals in soils (Duff et al., 1963; Shindo and Kuwatsuka, 1977). Increased solubilization of heavy metal sulfides by heterotrophic bacteria under anaerobic incubations of soil or sludge (Kee and Bloomfield, 1962; Bloomfield and Pruden, 1975), and aerobically in culture media as a result of an undescribed solubilizing agent, have been reported (Cole, 1979). Bolter et al. (1975) found that organic acids from decaying leaf litter increased the solubility of heavy metals deposited in soil from smelters. Complexation of cadmium by organic components of sanitary landfill leachates was attributed to low- and high-molecular-weight compounds representing simple carboxylic acids and compounds containing hydroxyl groups (Knox and Jones, 1979).

2. *Microbial Production of Complexing Agents*

Chelating agents are produced by microorganisms that require iron or other essential metals for growth. Much is known about the chemistry, the biochemistry, the type of microorganisms, and the rate of production of the complexing agents which chelate iron and transport iron into the cell. When microorganisms are grown in an iron-deficient medium, they elaborate specific iron chelators into the medium. Iron exists predominantly as insoluble complexes of Fe(III) in aerobic environments. These chelating agents enhance the dissolution of the metals with which they complex and thus increase their mobility and bioavailability. The role of iron chelators, such as siderochrome and siderophore, has been established on the basis of promotion of microbial growth by specific iron-chelating compounds produced by the microorganism (Rodgers and Neilands, 1973; Cox and Graham, 1979). Examples of bacterial siderochromes include enterobactin and enterochelin, produced by several enteric bacteria, mycobactins, which are produced by *Mycobacterium* species, and schizokinen, which is produced by *Pseudomonas* species. The functional groups of these molecules involved in iron binding are the phenolic groups in enterobactin and pyochelin; the hydroxamate groups in schizokinen; the hydroxamate groups in pseudobactin; and both the phenolic and the hydroxamate groups in mycobactins.

Several of the antibiotics produced by microorganisms possess functional groups and stereochemistries suitable for chelation. Since a variety of microbial types are active in the waste and waste leachate, it is possible that these microorganisms may elaborate specific or nonspecific compounds capable of complexation with the radionuclides. It has been shown that some plant disease may be suppressed by bacteria that aggressively colonize plant root surfaces and increase plant growth and yield related to bacterial siderophore production in soil (Schroth and Hancock, 1982). As chemical and biochemical similarities have been observed between Pu(IV) and Fe(III) and between Th(IV) and Pu(IV), the iron-sequestering agents could play an important role in the complexation of Pu and other metals and thus increase their bioavailability. Recent studies show that dissolution of plutonium dioxide was enhanced in the presence of Desferol, a polyhydroxamate chelate produced by microorganisms (Barnhart et al., 1980). This clearly indicates the potential of plutonium and other radionuclides present in the waste for complexation by microbially produced chelating agents and thus an increased bioavailability and mobilization of radionuclides due to microbial activity. Wildung and Garland (1980) found that microorganisms grown in the presence of plutonium produced complexing agents of higher molecular weight than that of DTPA. Many of the cultures tested were capable of transporting plutonium into the cell, and the role of complexing agents with such a transport has been suggested (Beckert and Au, 1976; Wildung and Garland, 1980). Consequently, studies are under way to isolate specific sequestering agents produced by microorganisms superior to EDTA or DTPA for clinical use as well as for cleanup of radioactive spills.

V. CONCLUSION

Low-level radioactive wastes generated from nuclear and nonnuclear fuel cycle will continue to be buried in shallow-land waste disposal sites. Several of the burial sites have been closed because of the problems that developed as a result of poor site characteristics, types of waste buried, and a number of other environmental factors. Some of the problems encountered at these disposal sites can be traced to the activities of microorganisms. These include microbial degradation of waste forms resulting in trench cover subsidence, production of radioactive gases such as $^{14}CO_2$, $^{14}CH_4$, CH_3T, and $^{14}CH_3T$, production of microbial metabolites capable of complexation, solubilization, and bioaccumulation of radionuclides. Many of these microbial transformations relate to both aerobic and anaerobic heterotrophic microbial processes based on the organic materials in the wastes and these play an important role in the radionuclide mobilization. Consequently, improvements in waste disposal technology are being developed to minimize these problems. These include waste segregation, waste pretreatment, incineration, and solidification. In addition, microorganisms are known to catalyze the oxidation-reduction reactions of metals that affect the valence or ionic state of the metals and enhance their mobility in the environment by retarding the soil-binding characteristics. Microorganisms also precipitate and decrease the solubility, or enhance the soil-binding characteristics of the metals and thus attentuate metal movement from the disposal site. As an especially important point, little is known about the role of autotrophic microbial transformations of radionuclides. Such microbial processes may be significant in light of improved disposal procedures, which may result in reductions in the organic content of the waste disposed of at shallow-land burial sites.

Although many of these problems do not directly influence public safety and health at the present time based on available information, they have provided better insights into how to or how not to select, design, and operate low-level disposal sites based on microbial considerations. Although a regional network of state-controlled, commercial, low-level waste disposal sites is being considered, there still remains the unanswered question about the long-term performance and decommissioning of existing active disposal sites and of controlling or better managing possible microbial contributions to environmental concerns which may result from low-level radioactive waste disposal.

ACKNOWLEDGMENTS

This research was performed under the auspices of the United States Department of Energy (OER/OHER) under Contract No. DE-AC02-76CH00016.

REFERENCES

Adams, M. W. W., Mortensen, L. E., and Chen, J. S. (1981). Hydrogenase. *Biochim. Biophys. Acta 594*: 103–176.

Aickin, R. M., and Dean, C. R. (1978). Lead accumulation by microorganisms. *Microbios Letters 5*: 129–133.

Allison, M. J. (1978). Production of branched-chain volatile fatty acids by certain anaerobic bacteria. *Appl. Environ. Microbiol. 35*: 872–877.

Anand, S. R., and Krasna, A. I. (1965). Catalysis of the N_2-HTO exchange by hydrogenase. A new assay for hydrogenase. *Biochemistry 4*: 2747–2754.

Arunachalam, G., Oblisami, G., and Rangaswami, G. (1974). Effect of gamma radiation on certain microbial and chemical properties of two soil types. *Madras Agric. J. 61*: 992–996.

Avnimepech, Y., and Raveh, A. (1982). Decomposition of chelates leached from waste disposal sites. *J. Environ. Qual. 11*: 69–72.

Barnhart, B. J., Campbell, E. W., Martinez, E., Caldwell, D. E., and Hallett, R. (1980). Potential microbial impact on transuranic wastes under conditions expected in the waste isolation pilot plant (WIPP). Los Alamos National Laboratory, LA-8297-PR.

Beckert, W. F., and Au, F. H. F. (1976). Plutonium uptake by a soil fungus and transport to its spores. In *Transuranium Nuclides in the Environment*, IAEA-SM-199/72, pp. 337–345.

Belley, R. T., Lauff, C. C., and Goodhue, C. T. (1975). Degradation of EDTA by microbial populations from aerated lagoon. *Appl. Microbiol. 29*: 787–794.

Bloomfield, C., and Pruden, G. (1975). The effects of aerobic and anaerobic incubation on the extratabilities of heavy metals in digested sewage sludge. *Environ. Pollut. 8*: 217–232.

Bolter, E., Butz, T., and Arseneau, J. F. (1975). Mobilization of heavy metals by organic acids in the soils of a lead mining and smelting district. In *Trace Substances in Environmental Health*, D. D. Hemphill (Ed.). Univ. of Missouri, Columbia, Missouri, Vol. IX, 107–112.

Brierley, C. L. (1982). Microbiological mining. *Sci. Am. 242*: 44–53.

Brown, M. J., and Lester, J. N. (1979). Metal removal in activated sludge: The role of bacterial extracellular polymers. *Water Res. 13*: 817–837.

Bunch, R. L., and Ettinger, M. B. (1967). Biodegradability of potential organic substitutes for phosphates. Proc. Ind. Waste Conf. 22nd, Purdue Engineering Exten. Serv. No. *129*: 393–396.

Chakrabarty, A. M. (1976). Plasmids in *Pseudomonas*. *Ann. Rev. Genet. 10*: 7–30.

Chapin, J. A., and Cooper, C. S. (1979). Feasibility of using plants to decontaminate radioactive soils. TREE-1366, EG and G, Idaho, Inc.

Charley, R. C., and Bull, A. T. (1979). Bioaccumulation of silver by a multispecies community of bacteria. *Arch. Microbiol. 123*: 239–244.

Claesson, A. (1971). Anaerobic bacterial degradation of nitrilotriacetate NTA. *Vatten 27*: 410--411.

Clancy, J. J., Gray, D. F., and Oztunali, O. I. (1981). *Data Base for Radioactive Waste Management*. Dames and Moore, Inc., New York, NUREG/CR-1759, Vol. 1.

Cleveland, J. M., and Rees, T. F. (1976). Investigation of solubilization of plutonium and americium in soil by natural humic compounds. *Environ. Sci. Technol. 10*: 802--806.

Cleveland, J. M., and Rees, T. F. (1981). Characterizations of plutonium in Maxey Flats radioactive trench leachates. *Science 212*: 1506–1509.

Cole, M. A. (1979). Solubilization of heavy metal sulfides by heterotrophic soil bacteria. *Soil Sci. 127*: 313–317.

Cox, C. D., and Graham, R. (1979). Isolation of an iron-binding compound from *Pseudomonas aeruginosa*. *J. Bacteriol. 127*: 357–364.

Daniels, L., Fulton, G., Spencer, R. W., and Orme-Johnson, H. W. (1980). Origin of hydrogen in methane produced by *Methanobacterium thermoautotrophium*. *J. Bacteriol. 141*: 694–698.

de Marsily, G., Ledoux, E., Barbreau, A., and Margat, J. (1977). Nuclear waste disposal: Can the geologist guarantee isolation? *Science 197*: 519–527.

Department of Energy (DOE) (1979). Report to the President by the Interagency Review Group on Nuclear Waste Management. March 1979. TID-29442, DOE, Washington, D.C.

Dispirito, A. A., and Tuovinen, O. H. (1982). Uranium oxidation by *Thiobacillus ferrooxidans*: Kinetic evaluation, CO_2 fixation, and inhibitor studies. Abstr. American Microbiol. Soc. Ann. Meeting, p. 88, CCI, 117.

Duff, R. P., Webley, D. M., and Scott, R. O. (1963). Solubilization of minerals and related materials by 2-ketogluconic acid-producing bacteria. *Soil Sci. 95*: 105–114.

Eagon, R. G., and Carson, K. J. (1965). Lysis of cell walls and intact cells of *Pseudomonas aeruginosa* by ethylenediaminetetraacetic acid and by lysozyme. *Can. J. Microbiol. 11*: 193–201.

Ehrlich, H. C. (1981). *Geomicrobiology*. Marcel Dekker, Inc., New York.

Enfors, S. O., and Molin, N. (1973). Biodegradation of nitrilotriacetate (NTA) by bacteria. I. Isolation of bacteria to grow anaerobically with NTA as a sole carbon source. *Water Res. 7*: 881–888.

Eriksen, W. H., and Emborg, C. (1978). Increase of radiation resistance of a soil microflora exposed to long-term gamma irradiation. *Appl. Environ. Microbiol. 36*: 618–619.

Fallon, R. D. (1982). Influence of pH, temperature and moisture on gaseous tritium uptake in surface soils. *Appl. Environ. Microbiol. 44*: 171–178.

Fliermans, C. B., Hayes, D. W., and Johnson, N. D. (1978). Radon-222 in biologically produced gas from a reactor cooling pond. *Health Phys. 34*: 701–704.

Francis, A. J. (1982). Microbial transformation of low-level radioactive waste. In *Environmental Migration of Long-Lived Radionuclides*, IAEA-SM-257/72, Vienna, pp. 415–429.

Francis, A. J., and Quinby, H. (1981). Microbial degradation of spent reactor resins—a preliminary study. Brookhaven National Laboratory, Upton, New York, BNL 30897.

Francis, A. J., Dobbs, S., and Nine, B. J. (1980a). Microbial activity of trench leachates from shallow-land, low-level radioactive disposal sites. *Appl. Environ. Microbiol. 40*: 108–113.

Francis, A. J., Iden, C. R., Nine, B. J., and Chang, C. K. (1980b). Characterization of organics in leachates from low-level radioactive waste disposal sites. *Nucl. Technol. 50*: 158–163.

Francis, A. J., Dobbs, S., and Doering, R. F. (1980c). Biogenesis of tritiated and carbon-14 methane from low-level radioactive waste. *Nuclear Chem. Waste Manag. 1*: 153–159.

Francis, A. J. (1984). Anaerobic Microbial Transformations of Radioactive Wastes in Subsurface Environments. (Presented at the Workshop on the Role of Microorganisms on the Behavior of Radionuclides in Aquatic and Terrestrial Systems and their Transfer to Man.) Brussels, Belgium. BNL-34968.

Francis, A. J., Lin, M., Premuzic, E. T., Quinby, H. L., and Schubert, J., Chelation of Thorium and Uranium by Pseudomonas aeruginosa. (In preparation.)

Giesy, J. P., and Paine, D. (1977). Effects of naturally occurring aquatic organic fractions on ^{241}Am uptake by *Scenedesmus obliquus* (Chlorophyceae) and *Aeromonas hydrophila* (Pseudomonadaecae). *Appl. Environ. Microbiol. 33*: 89–96.

Grossbard, E., and Stranks, D. R. (1959). Translocation of cobalt-60 and cesium-137 by fungi in agar and soil cultures. *Nature 184*: 310–314.

Humphrey, B., and Vincent, J. M. (1962). Calcium in cell walls of *Rhizobium trifolii*. *J. Gen. Microbiol. 29*: 557–561.

Husain, L., Matuszek, J. M., Hutchinson, J., and Wahlen, M. (1979). Chemical and radiochemical character of a low-level radioactive waste burial site. In *Management of Low-Level Radioactive Waste*, M. W. Carter et al. (Ed.). Pergamon Press, New York, Vol. 2, pp. 883–900.

Johnson, J. E., Svalberg, S., and Paine, D. (1974). The study of plutonium in aquatic systems of the Rocky Flats environs. Technical Report, 1971–1974, Colorado State University.

Kee, N. S., and Bloomfield, C. (1962). The effect of flooding and aeration on the mobility of certain trace elements in soil. *Plant Soil 16*: 108–135.

Keith, C. L., Bridges, R. L., Fina, L. R., Iverson, K. L., and Cloran, J. A. (1978). The anaerobic decomposition of benzoic acid during methane fermentation. IV. Dearomatization of the ring and volatile fatty acids formed on ring rupture. *Arch. Microbiol. 118*: 173–176.

Kelley, D. P., Norris, P. R., and Brierley, C. L. (1979). Microbiological methods for the extraction and recovery of metals. *Soc. Gen. Microbiol. Symp. 29*: 263–308.

Kirby, L. J., Toste, A. P., and Wilkerson, C. L. (1982). Chemical species of migrating radionuclides at a shallow land low-level radioactive waste burial site. In "Environmental Migration of Long-Lived Radionuclides," IAEA-SM-275/55.

Kirk, P. W. W., Lester, J. N., and Perry, R. (1982). The behavior of nitrilotriacetic acid during the anaerobic digestion of sewage sludge. *Water Res. 16*: 973–980.

Kiselov, P. N., Kashkin, K. P., Boltaks, B., and Vitovskaya, G. A. (1961). Acquisition of radio resistance by microbe cells inhabiting media with increased levels of natural radiation. *Microbiology* (Engl. Trans.) *30*: 194–198.

Klibanov, A. M., and Huber, J. (1981). Application of immobilized hydrogenase for the detritiation of water. *Biotechnol. Bioeng. 23*: 1537–1551.

Knox, K., and Jones, P. H. (1979). Complexation characteristics of sanitary landfill leachates. *Water Res. 13*: 839–846.

Lowe, H. N., Jr., Lacy, W. J., Surkiewicz, B. F., and Jaeger, R. F. (1956). Destruction of microorganisms in water, sewage, and sewage sludge by ionizing radiations. *J. Amer. Water Works Assn. 48*: 1363–1372.

Lu, A. H., and Matuszek, J. M. (1978). Solute and gas transport from radioactive waste. *Trans. Am. Nucl. Soc. 30*: 94–95.

Mah, R. A., Ward, D. M., Baresi, L., and Glass, T. L. (1977). Biogenesis of methane. *Ann. Rev. Microbiol. 31*: 309–341.

Matsuyama, A., Okazawa, Y., Arai, H., and Mijune, M. (1973). Isolation of radio resistant vegetative bacteria from the high-background radioactive area. *Sci. Pap. Inst. Phys. Chem. Res.* (Jap.) *67*: 34–40.

Matuszek, J. M. (1980). Biochemical and chemical processes leading to radionuclide transport from low-level waste burial sites. *Trans. Am. Nucl. Soc. 34*: 155–156.

McFarlane, J. C., Rogers, R. D., and Bradley, D. V. Jr. (1979). Tritium oxidation in surface soils. A survey of soils near five fuel reprocessing plants. *Environ. Sci. Tech. 13*: 607–608.

McFarlane, J. C., Rogers, R. D., and Bradley, D. V. (1978). Environmental tritium oxidation in surface soil. *Environ. Sci. Tech. 12*: 590–593.

McLaren, A. D., and Petersen, G. H. (1967). *Soil Biochemistry*. Marcel Dekker, Inc., New York.

Means, J. L., Crerar, D. A., and Duguid, J. O. (1978). Migration of radioactive waste: radionuclide mobilization by complexing agents. *Science 200*: 1477–1481.

Means, J. L., Kucak, T., and Crerar, D. A. (1980). Relative degradation rates of NTA, EDTA, and DTPA and environmental implications. *Environmental Pollution (Series B) 1*: 45–60.

Menon, S. H., Wagner, N. Jr., and Tsan, M. -F. (1978). Studies on gallium accumulation in inflammatory lesions: I. Uptake by *Staphylococcus aureus*: concise communication. *J. Nucl. Med. 19*: 44–47.
Miles, A. A., and Khimji, P. L. (1975). Enterobacterial chelators of iron: their occurrence, detection, and relation to pathogenicity. *J. Med. Microbiol. 8*: 477–490.
Molecke, M. A. (1979). Gas generation from transuranic waste degradation: data summary and interpretations. Sandia Laboratories Report SAND79-1245, December 1979.
Murphy, C. E., Boni, A. L., and Tucker, S. P. (1976). The conversion of gaseous molecular tritium to tritiated water in biological systems. E. I. Du Pont De Nemours and Company, Savannah River Laboratory, Aiken, South Carolina, DP-1422, p. 15.
Murr, L. E., Torma, A. E., and Brierley, J. A. (Eds.) (1978). *Metallurgical Applications of Bacterial Leaching and Related Phenomenon.* Academic Press, London, New York.
Nuclear Regulatory Commission (NRC). (1982). Licensing Requirements for Land Disposal of Radioactive Waste. Federal Register Vol. 47, No. 248. December 27, 1982. pp. 57466–57484.
Oztunali, O. I., Re, G. C., Moskowitz, P. M., Picazo, E. D., and Pitt, C. J. (1981). *Data Base for Radioactive Waste Management.* Dames and Moore, Inc., New York. NUREG/CR-1759, Vol. 3.
Repaske, R. (1958). Lysis of gram-negative organisms and the role of versene. *Biochim. Biophys. Acta. 30*: 255–262.
Ridenour, G. M., and Armbruster, E. H. (1956). Effects of high-level gamma radiation on disinfection of water and sewage. *J. Am. Water Works Assn. 48*: 671–676.
Rodgers, G. C., and Neilands, J. B. (1973). Microbial iron transport compounds. In *Handbook of Microbiology*, Laskin, A. I., and Liechevalier, H. A. (Eds.). Chemical Rubber Co., Cleveland, Ohio, Vol. 2, pp. 823–830.
Saeger, V. W., Hicks, O., Kaley, R. G., Michael, P. R., Mieure, J. P., and Tucker, E. S. (1979). Environmental fate of selected phosphate esters. *Environ. Sci. Technol. 13*: 840–844.
Schnitzer, M., and Kahn, S. U. (1972). *Humic Substances in the Environment.* Marcel Dekker, Inc., New York.
Schroth, M. N., and Hancock, J. G. (1982). Disease-suppressive soil and root-colonizing bacteria. *Science 216*: 1376–1381.
Shindo, H., and Kuwatsuka, S. (1977). Behavior of phenolic substances in the decaying process of plants. V. Elution of heavy metals with phenolic acids from soil. *Soil. Sci. Plant. Nutr. 23*: 185--193.
Shumate, S. E., II, Strandberg, G. W., and Parrott, J. F., Jr. (1978). Biological removal of metal ions from aqueous process streams. Biotechnology and Bioengineering Symp. No. 8, 13–20. John Wiley and Sons, Inc., New York.
Silver, M., and Torma, A. E. (1974). Oxidation of metal sulfides by *Thiobacillus ferrooxidans* grown on different substrates. *Can. J. Microbiol. 20*: 141--147.

Snow, G. A. (1970). Mycobactins: Iron-chelating growth factors from Mycobacteria. *Bacteriol. Rev.* *34*: 99–125.

Spalding, B. P. (1981). Chemical treatments of soil to decrease radiotrontium leachability. *J. Environ. Qual.* *10*: 42–46.

Spencer, R. W., Daniels, L., Fulton, F., and Orme-Johnson, L. H. (1981). Product isotope effects on *in vivo* methanogenesis by *Methanobacterium thermoautotrophicum*. *Biochemistry* *19*: 3676–3683.

Spoelstra, S. F. (1978). Degradation of tyrosine in anaerobically stored piggery wastes and in pig feces. *Appl. Environ. Microbiol.* *36*: 631–638.

Strandberg, G. W., Shumate, S. E., II, and Parrott, J. R., Jr. (1981a). Microbial cells as biosorbents for heavy metals: Accumulation of uranium by *Saccharomyces cerevisiae* and *Pseudomonas aeruginosa*. *Appl. Environ. Microbiol.* *41*: 237–245.

Strandberg, G. W., Shumate, S. E., II, Parrott, J. R., Jr., and North, S. E. (1981b). Microbial accumulation of uranium, radium, and cesium. In *Environmental Speciation and Monitoring Needs for Trace Metal-Containing Substances from Energy-Related Processes*, F. E. Brinckman and R. H. Fish (Eds.). NBS Special Publication 618, November 1981, pp. 274–283.

Strayer, R. (1982). Microbial sulfur oxidation in ash/slag from coal gasification. Abstr. American Microbiol. Soc. Ann. Meeting, p. 230, Q128.

Takamatsu, T., and Yoshida, T. (1978). Determination of stability constants of metal-humic acid complexes by potentiometric titration and ion-selective electrodes. *Soil Sci.* *125*: 377–386.

Tiedje, J. M. (1975). Microbial degradation of ethylenediaminetetraacetate in soils and sediments. *Appl. Microbiol.* *30*: 327–329.

Tiedje, J. M. (1977). Influence of environmental parameters on EDTA biodegradation in soils and sediments. *J. Environ. Qual.* *6*: 21–26.

Tiedje, J. M., and Mason, B. B. (1974). Biodegradation of nitrilotriacetate (NTA) in soils. *Soil Sci. Soc. Am. Proc.* *38*: 278–283.

Tornabene, T. G., and Edwards, H. W. (1973). Microbial uptake of lead. *Science* *176*: 1334–1335.

Touhill, C. J., Martin, E. C., Fujihara, M. P., Olesen, D. E., Stein, J. E., and McDonnell, G. (1969). *J. Water Poll. Contr. Fed.* *41*: R44-R60.

Tsezos, M., and Volesky, B. (1981). Biosorption of uranium and thorium. *Biotechnol. Bioeng.* *23*: 583–604.

Ward, R. L., Yeager, J. G., and Ashley, C. S. (1981). Response of bacteria in wastewater sludge to moisture loss by evaporation and effect of moisture content on bacterial inactivation by ionizing radiation. *Appl. Environ. Microbiol.* *41*: 1123–1127.

Wild, R. E., Oztunali, O. I., Clancy, J. J., Pitt, C. J., and Picazo, E. D. (1981). *Data Base for Radioactive Waste Management*. Dames and Moore, Inc., New York, NUREG/CR-1759, Vol. 2.

Wildung, R. E., and Garland, T. R. (1982). Effects of plutonium on soil microorganisms. *Appl. Environ. Microbiol. 43*: 418–423.

Wildung, R. E., and Garland, T. R. (1980). The relationship of microbial processes to the fate and behavior of transuranic elements in soils, plants, and animals. In *Transuranic Elements in the Environment*, W. C. Hanson (Ed.). DOE/TIC-22800. Technical Information Center/U.S. Dept. of Energy, pp. 300–335.

Yano, K., Cho, B., Yoshinaka, T., and Yamaguchi, H. (1975). Surveys on radiation resistant asporogenic bacteria in natural environments. In *Radiation for a Clean Environment*. IAEA-SM-194/202, Vienna, pp. 85–98.

Yoshinaka, T., Yano, K., and Yamaguchi, H. (1973). Isolation of a highly radio resistant bacterium, *Arthrobacter radiotolerans* nov. sp. *Agric. Biol. Chem. 37*: 2269–2275.

Zajic, J. E. (1969). *Microbial Biogeochemistry*. Academic Press, New York.

Zurino, H., and Martin, J. P. (1977). Metal-binding organic molecules in soil: 2. Characterization of the maximum binding ability of the macromolecules. *Soil Sci. 123*: 188–202.

10

General Interactions of Microbes in Reclamation Processes

DONALD A. KLEIN / *Colorado State University, Fort Collins, Colorado*

ROBERT L. TATE III* / *University of Florida, Belle Glade, Florida*

The role of microorganisms and the implications of their activity on the potential success of a variety of reclamation situations have been discussed in the preceeding chapters. These examples have vividly depicted the varying degrees of understanding of microbial function in disturbed ecosystems; i.e., some processes are apparently well documented whereas with others, the hypotheses may best be described as being derived at least as much from inference and speculation as from sound field data. Because of the limitations discussed in the preface, all or even a representative sample of reclamation needs could not be discussed in this volume. Yet, with the objective of providing an overview of the microbial role in reclamation processes, some synthesis of the common aspects of the chapters is needed. Hence, this chapter will demonstrate the overall principles derived from the proceeding discussions and evaluate current and future needs for reclamation research.

The microbiologist's contribution to the evaluation of reclamation procedures includes the (1) evaluation of physical and chemical environments to predict, based on microbial characteristics, the potential for development prior to initiation of a reclamation program, (2) monitoring of reclamation progress, (3) measurement of the potential for toxin production and destruction, (4) prediction of the microbial modification of the environmental impact of various materials, and (5) determination of the need for pretreatment of waste materials prior to disposal. Some of these microbial interactions relate simply to the rate of ecosystem de-

**Current affiliation*: Cook College, Rutgers, The State University of New Jersey, New Brunswick, New Jersey

velopment following disturbance. Consideration of such parameters prior to implementation of on site work is not obligatory. Others, such as existence or production of toxic substances, in contrast, may preclude successful reclamation. Hence, failure to evaluate these parameters prior to development could have long-term negative effects.

The microbial role in the development of the physical and chemical conditions conducive to plant and animal community development includes a variety of soil forming processes, as well the biogeochemical cycles necessary for provision of a constant plant nutrient supply (Tate, Chap. 1). Soil structure developmental rates control ecosystem stability through interactions with soil moisture availability, aeration, as well as provision of a substratum adequate for plant root development. Although soil structure development has been shown to require from decades to centuries in some disturbed soils (Jenny, 1980), with proper management procedures, this time interval can be reduced to as few as 3–5 years. Visser (Chap. 7) noted in her studies of mineland reclamation in arid western environments that in these ecosystems where there had not been release and/or potentiation of toxic or inhibitory materials, native soil-forming processes and plant reestablishment can be accelerated through amendment with fertilizer and/or organic amendments. Management decisions relating to such amendments must include consideration of the effect of the organic amendments on long-term ecosystem stability. For example, with the inhibitory effect of fixed nitrogen on symbiotic nitrogen fixation, establishment of this important aspect of the nitrogen cycle in the disturbed soil may be delayed or even precluded by excessive fertilization (Tate, Chap. 1).

Evaluation of the progress of reclamation has typically involved assessment of the diversity and stability of the aboveground plant populations. Because of the nature of plant growth, such measurements of progress necessarily will involve months to years before the future of the site can be predicted. Such time delays may preclude corrective actions involving toxic or limiting conditions existent in the soil prior to reclamation. A more sensitive, short-term evaluation procedure is needed. Two such measures have been discussed in *Soil Reclamation Processes: Microbiological Analyses and Applications*. These are soil algal community structural analyses (Shubert and Starks, Chap. 3) and soil enzymatic activity analyses (Klein et al., Chap. 5). Both procedures were examined in a limited number of reclamation systems, but more generalized application appears favorable.

Shubert and Starks (Chap. 3) evaluated the use of soil algal community structural analysis to follow reestablishment of a plant-soil system on mine spoils. The algal community is an ecosystem biological component that is generally overlooked in terms of microbiological approaches to assessment of reclamation progress. The capability of monitoring both community structure by cultural procedures and algal biomass in terms of soil chlorophyll levels, when combined with a factor analysis approach, provides the potential to identify the limiting factors for belowground process development at particular reclamation sites.

Soil enzymatic activities are also sensitive to the presence of toxic substances and, hence, could provide a measure of reclamation progress. Due to the extreme disturbance of these soils, extracellular (abiontic) enzymatic activity is frequently essentially absent or severely reduced. Redevelopment of these enzymatic activities provides a measure of microbial ecosystem development (Klein et al., Chap. 5). These activities are particularly attractive for use because of the ease of measurement of enzymatic activity and enzyme sensitivity to substances potentially toxic to microbial growth. The effects of heavy metal contamination on soil enzymatic activity are well documented (for example, Doelman and Haanstra, 1979; Mathur et al., 1979; Lighthart et al., 1983). Klein et al. (Chap. 5) extended these observations to the evaluation of management procedures for retorted oil shale to allow plant growth and microbial functioning. Based on their studies, Klein et al. conclude that toxic components from underlying oil shales may cause decreased biogeochemical cycling in overlaying surface soils, and that fertilization, uses of mulches, and other types of management make it possible to establish plant communities, even if soil enzymatic activities are depressed. This observation stems from the fact that abiontic enzymatic activities under steady state conditions exceed the levels needed by the microbial community. Thus, the value of soil enzymatic activity measurements appears to be as a predictor of the existence of chemical stresses which may cause measurable shifts in plant communities. The long-term consequences of these plant population shifts, in term of plant function and shale component mobilization, remain to be documented.

Chemical stresses retarding or even precluding biotic development in disturbed ecosystems can result directly from the resource recovery process or from microbial transformation of the waste following deposition at the final disposal site. Examples of microbially synthesized environmental toxins were provided by Mills (Chap. 2) and Francis (Chap. 9). Acid mine waste damage results from biological oxidation of sulfides to sulfates. Successful reclamation is dependent upon control of this biological process. The release of biologically produced sulfuric acid, a process that is stimulated under oxygenated conditions, can be controlled, at least in principle, by sealing mine areas or by stimulating native biological sulfate reduction processes (Mills, Chap. 2). Sulfate reduction in aquatic sediments is a key process for total ecosystem recovery. Our understanding of the reductive process is such that mathematical models predictive of the recovery time required to transform sulfate to sulfide with particular aquatic systems can be developed. As the biochemistry of soil processes is further documented, predictive models will become a major tool for predicting reclamation success. Sulfate reduction is dependent on organic matter availability in anaerobic sediments where sulfate, after biological reduction to sulfides, will tend to be isolated from the aerobic aquatic community.

Wildung and Garland (Chap. 4) provided an example of recolonization of a terrestrial ecosystem containing toxic components. This is a clear example of development of microbial populations that tend to detoxify the soil system. In this case, the microbial community reduced toxicity through reduction or methylation of heavy metals and mineralization of toxic organic compounds found in the unamended retorted oil shale materials studied. The interactions between microbial processes and shale components in terms of carbon, nitrogen, and sulfur cycling, as well as trace components, such as arsenic, were considered. Methylation and complexation, which can lead to trace element mobilization, were discussed together with recarbonization in terms of oil shale weathering. Although the studies were of limited duration, it was apparent that microbial populations could be reestablished in these toxic materials.

Francis (Chap. 9) provides an example of the complications resulting from the failure to consider microbial modification of waste materials in situ. Thus, the microbial transformations of the waste following disposal resulted in an enhancement of the environmental impact. Francis' report relates specifically to radioactive waste disposal; yet the principles, and problems, are applicable to evaluation of the environmental impact of a variety of waste disposal processes. These include augmented metal mobility as well as problems with biochemical oxygen demand (BOD) and toxic organic compounds in leachates. Francis (Chap. 9) observed augmented metal (radionuclide) mobility resulting from microbial modification of site pH, redox potential, chelator production, and organic acid synthesis. In addition, volatile organic compound synthesis increased the mobility of carbon-14 and tritium. The overall conclusion is that ultimate control of toxicant mobility within disposal sites rests with development of procedures founded upon an understanding of the microbial reactions occurring within a disposal environment.

Informed waste management must necessarily involve consideration of pretreatment of waste materials prior to disposal and/or reclamation. Examples of microbial involvement were provided by Brierley and Lanza (Chap. 8) and Francis (Chap. 9).

Water recovery is of prime importance in phosphate clay flocculation processes, both from the view of resource recovery as well as that of acceleration of slag reclamation (Brierley and Lanza, Chap. 8). Microbial polymers, if they can be synthesized, can enhance clay flocculation. The critical constraints that limit polymer production appear to be nutrient availability and moisture levels. Resource recovery should be considered, including the use of "environmentally programmed" microorganisms, together with currently lost or under utilized nutrient sources, such as sewage.

In contrast to the positive aspects of microbial activity utilization in phosphate slag processes, failure to consider microbial limitations following organic waste disposal can result in enhancement of the environmental impact of the disposal process (Francis, Chap. 9). Radioactive wastes contain a variety of chelators that are not biodegradable under the anaerobic conditions prevalent in the disposal trenches. Breaching of the waste canisters containing the chelators allows movement of these compounds along with any metal solubilized into the soil water. Therefore, previously insoluble metal complexes may become solubilized; thereby, posing a threat to ground and surface waters. In this case, pretreatment or segregation of chelator containing waste materials is desirable (Colombo et al., 1982; Francis, Chap. 9).

A common element in each of these microbial functions is the demonstrated need to understand and manage the physical environment of the reclamation sites. This includes microbial energy sources, abiotic conditions, and trace element availability. Each of these controllable variables must be maintained at levels to allow or preclude microbial activity as the reclamation objectives dictate. With superficial analysis, total exclusion of biological activity appears to be desirable in the management of organic wastes, such as radioactive waste materials, but when long-term objectives are considered, it becomes readily apparent that total exclusion of the microbial community, although possible, is an economic impracticality. Thus, management of the microbial community and the resultant processes becomes most desirable.

Microbial community management primarily involves adjustment of the physical and chemical conditions to the range favorable to the desired microbial processes. But this manipulation may also include microbial inoculation to assure the presence of microorganisms capable of catalyzing the required reactions. To date, such inoculation has only been feasible in limited situations, such as that required for the development of symbiotic nitrogen fixing or mycorrhizal associations or in the use of especially selected microorganisms to improve phosphatic clay flocculation.

Danielson (Chap. 6) demonstrated the importance of mycorrhizal associations in alleviating the effects of chemical stresses on plant community development and the potential for exploiting mycorrhizal associations as part of an integrated reclamation program. Major questions addressed by Danielson involve the possibility that mycorrhizal infection may increase plant tolerance to extreme environmental stresses and the maintenance of mycorrhizal inocula in stored soils. The potential exists for the use of mycorrhizal fungi selected to impart a greater stress tolerance on plants established on particular waste materials. Currently, implementation of these procedures is limited by the problem of evaluating strain differences among mycorrhizal fungi and the emphasis that has been given to examination of successful ectomycorrhizal symbionts from warm climates. Additional research emphasis on mycorrhizal associations from colder climates or from ecosystems subject to unique stresses is needed. It appears that mycorrhizal inoculation will be most beneficial in environments where inoculum is not present, or where distinct chemical stresses (metals, etc.) must be managed.

Any consideration of more intensive use of microorganisms in reclamation will be subject to financial, time, and resource constraints. Prior to the last 15–20 years, most reclamation was conducted with empirical procedures where microbes and their contributions to ecosystem function were essentially disregarded. With less exotic reclamation demands, i.e., those involving normal soil genesis processes, it was possible to work successfully with this type of reclamation management. However, with the need to reclaim areas where chemical releases can occur, the need to stimulate or possibly to preclude significant microbiological activity may be critical in terms of reclamation success. Especially with the extended times over which microbial processes can occur in reconstructed systems, and the role of microorganisms in stimulating or inhibiting the dispersion of potentially environmentally troublesome chemicals, microbiological management should receive additional consideration as a part of environmentally (and cost) effective reclamation. Under these conditions, knowledge of microbiological processes and their contribution to ecosystem development is critical in predicting long-term effects of reclamation decisions. The discussions in *Soil Reclamation Processes: Microbiological Analyses and Applications* point in this direction and suggest the potential contributions that a greater consideration of microorganisms and their activities should play in future reclamation activities.

ACKNOWLEDGMENTS

This research was supported by the Department of Energy under Contract DE-AS02-76EV04018, awarded to the Department of Range Science, Colorado State University. New Jersey Agricultural Experiment Station Publication No. F-15187-2-83 was supported by state funds.

REFERENCES

Colombo, R., Tate III, R. L., and Weiss, A. J. (1982). Assessment of microbial processes on radionuclide mobility in shallow land burial. BNL 51574. Brookhaven National Laboratory, Upton, New York.

Doelman, P., and Haanstra, L. (1979). Effect of lead on soil respiration and dehydrogenase activity. *Soil Biol. Biochem. 11*: 475–479.

Jenny, H. (1980). *The Soil Resource*. Springer-Verlag, New York. 377 pp.

Lighthart, B., Baham, J., and Volk, V. V. (1983). Microbial respiration and chemical speciation in metal-amended soils. *J. Environ. Qual. 12*: 543–548.

Mathur, S. P., Hamilton, H. A., and Levesque, M. P. (1979). The mitigating effect of residual fertilizer copper on the decomposition of an organic soil in situ. *Soil Sci. Soc. Am. J. 43*: 200–203.

Index

A

D

E

F

W

X

Y

Z